Natural Philosophy

[Kepler's] *Harmonia Mundi* was for him a new kind of labour. Before, he had voyaged into the unknown, and the books he brought back were fragmentary and enigmatic charts apparently disconnected with each other. Now he understood that they were not maps of the islands of an Indies, but of different stretches of the shore of one great world. The *Harmonia* was their synthesis.

John Banville, *Kepler,* 179.

Natural Philosophy

On Retrieving a Lost Disciplinary Imaginary

ALISTER E. McGRATH

OXFORD UNIVERSITY PRESS

OXFORD
UNIVERSITY PRESS

Great Clarendon Street, Oxford, OX2 6DP,
United Kingdom

Oxford University Press is a department of the University of Oxford.
It furthers the University's objective of excellence in research, scholarship,
and education by publishing worldwide. Oxford is a registered trade mark of
Oxford University Press in the UK and in certain other countries

First Edition published in 2023
Impression: 1

Published in the United States of America by Oxford University Press
198 Madison Avenue, New York, NY 10016, United States of America

British Library Cataloguing in Publication Data
Data available

Library of Congress Control Number: 2022936280

ISBN 978-0-19-286573-1

DOI: 10.1093/oso/9780192865731.001.0001

Printed and bound by
CPI Group (UK) Ltd, Croydon, CR0 4YY

Acknowledgements

This book owes much to many. Charles A. Coulson, Jeremy R. Knowles, and R. J. P. Williams nurtured my love of the natural sciences at Oxford University back in the 1970s, while at the same time pointing me to a deeper vision of reality than that which the sciences could offer on their own. I am grateful to many colleagues at Oxford University in more recent years, especially at the Ian Ramsey Centre for Science and Religion, where many aspects of natural philosophy are regularly discussed. I am particularly grateful to two anonymous readers of the original typescript, who made some highly perceptive and helpful suggestions for its improvement, and to Joanna Collicutt, who helped me see how I could weave these ideas together coherently.

Rebecca Elson's poem 'Dark Matter', in the collection *A Responsibility to Awe*, is reprinted by kind permission of Carcanet Press, Manchester, UK.

The extract from John Banville's *Kepler*, published by Pan Macmillan, is reproduced with permission of the Licensor through PLSclear.

Contents

Introduction

On Retrieving Natural Philosophy

This book explores a concept which once stood at the forefront of progressive thought, reaching back to ancient Greece and arguably reaching its zenith in the best-known scientific writing of Isaac Newton, *Philosophiae Naturalis Principia Mathematica* ('The Mathematical Principles of Natural Philosophy', 1687), published in Latin to ensure it was read throughout the European 'Republic of Letters'.[1] This is widely seen as a milestone in what some still call 'The Scientific Revolution',[2] although this is better described as the gradual 'emergence of a scientific culture'.[3] This work established Newton's reputation as the leading natural philosopher of his day, demonstrating both the fundamental unity of celestial and terrestrial mechanics, and the capacity of the human mind to discern and represent this unity.

Many would now think of Newton as a *scientist*, often overlooking the recent origins of this term;[4] yet for Newton and his age, he was a *natural philosopher*, a leading representative of both a category of reflection and an understanding of philosophy which is both scientific, as we now understand this idea, and yet at the same time more than scientific.[5] For its leading representatives, such as Kepler and Boyle, natural philosophy was a conceptual

[1] For Newton's European following and impact, see Boran and Feingold, eds., *Reading Newton in Early Modern Europe*.

[2] Although this phrase was widely and approvingly used by intellectual historians in the late twentieth century, it is now widely accepted that we cannot speak of the seventeenth century as having developed or adopted a unitary set of approaches to the natural world directly analogous to what twenty-first century scholars would recognize as 'natural science'. For the problems with the older position, see Osler, 'The Canonical Imperative'; Harrison, 'Was there a Scientific Revolution?'

[3] For the reasons for this, see Gaukroger, *The Emergence of a Scientific Culture*, 1–8. For further reflections on the inadequacies of thinking in terms of a single 'scientific revolution', see Cohen, *How Modern Science Came into the World*, xv–xxxiii.

[4] The term 'scientist' began to be used in the 1830s, and achieved wide acceptance during the later nineteenth century: Ross, 'Scientist: The Story of a Word'.

[5] Note Justin Smith's lament for the loss of intimacy between science and philosophy, which he associates with early modern natural philosophy: Smith, *The Philosopher*, 25: 'One of the most significant developments in philosophy since the eighteenth century... is that it has gradually lost its institutional connection with science, and so also its self-understanding as not just being interested in science or "pro-science," but as including science.'

space, a disciplinary imaginary that brought together a way of 'beholding' or understanding the natural world, a method for exploring it, and an *ethos* or way of existing within it.[6]

Today, natural philosophy is seen as an historical anomaly, a way of envisaging and inhabiting our world which has faded into oblivion through the institutionalization of individual scientific disciplines.[7] Ann Blair notes that historians of science often use this phrase 'as an umbrella term to designate the study of nature before it could easily be identified with what we call "science" today'.[8] Yet she also makes the important point that early modern natural philosophy is to be seen as an 'actor's category', demanding that it be understood from the perspective of those who originally used this term.[9] What, we must ask, did they understand by this term, and how does it relate to what is today described as 'natural science'?

Simply stated, this book is concerned with exploring this lost conceptual space of 'natural philosophy',[10] leading into reflection on the wider contemporary significance of this disciplinary imaginary and the possibility of its partial retrieval. The origins of this book lie in a series of eighteen public lectures I gave in the City of London during the years 2015–18, while I served as Professor of Divinity at Gresham College, London, a position established in 1597. Since Gresham College played a significant role in the emergence of English natural philosophy in the seventeenth century, it seemed appropriate to explore a range of issues relating to theology and the philosophy of science which were integral to this imaginatively and intellectually spacious vision of the natural world. This book represents a substantial reworking and conceptual expansion of those lectures, allowing me to explore the history of natural philosophy, reflect on how it might be reframed in our own time, and consider the benefits it offers.

So why might this conceptual space, which brings together learning *about* nature and learning *from* nature, retain imaginative appeal and reflective coherence today? What difficulties does it face? Is such a retrieval of this disciplinary imaginary, no matter how critical, a futile exercise in metaphysical

[6] As we shall see, these took the form of personal syntheses of 'conceptual spaces', weaving together differing empirical and experimental methodologies, epistemological and metaphysical assumptions, and theological assumptions about the relation of God and the natural order.

[7] Lüthy, 'What to Do with Seventeenth-Century Natural Philosophy?', 166–7.

[8] Blair, 'Natural Philosophy', 365.

[9] Collins, 'Actors' and Analysts' Categories in the Social Analysis of Science'.

[10] I have used the term 'conceptual space', as it enfolds many of the issues I hope to address in this work. Other terms are used in the literature to refer to what today constitute cross-disciplinary undertakings, such as 'interdisciplines' (Graff, *Undisciplining Knowledge*, 11–15) and 'interdisciplinarities' (Klein, *Crossing Boundaries*).

manipulation, or might it offer an empirical lens that transmutes our understanding of the world and ourselves? Like Charles Taylor's 'exercises in historical retrieval',[11] this work will use historical analysis as a means of understanding the intellectual and social role of natural philosophy in the past, with a view to exploring whether its intellectual vision might enrich present discussions of human interaction and engagement with the natural world.

Inevitably, this historical conversation will be incomplete, in that there is simply not enough space to do justice to all the historical, gendered, and geographical voices, episodes, and concerns that might be expected to find their way into a full account of the history of natural philosophy.[12] What I offer is more limited—a critical reading of this narrative, which attempts to identify core themes and fundamental approaches that might find their way into a retrieved version of this way of engaging the natural world, laying the foundation for a recovery of such an approach in the present.

Natural Philosophy: A Lost Conceptual Space

Natural philosophy emerged as a significant intellectual domain in western Europe in the late modern period, before the fragmentation of disciplines led to the emergence of distinct communities of discourse, each with its own vocabularies and methods of investigation.[13] It was seen to designate a coherent intellectual and cultural domain, which wove together what would now be considered to be quite distinct disciplines—such as the natural sciences, mathematics, music, philosophy, theology, and the cultivation of personal and social virtues. It was a way of life, not merely an intellectual endeavour. Though natural philosophy would now be thought of as an 'interdisciplinary' field, this anachronistic judgement would make little sense in the early

[11] I borrow this phrase from Dyck, 'Sovereign Selves', 138.

[12] The most comprehensive survey to date is that of Stephen Gaukroger, whose four-volume history of the notion from 1210–1935 takes up some 2,000 pages. This account is supplemented by his detailed account of important episodes and figures in this history, such as Bacon and Descartes, and his recent reflection on the problems facing philosophy as a discipline. See Gaukroger, *The Emergence of a Scientific Culture*; Gaukroger, *The Collapse of Mechanism and the Rise of Sensibility*; Gaukroger, *The Natural and the Human*; Gaukroger, *Civilization and the Culture of Science*. On Descartes and Bacon, see Gaukroger, *Descartes' System of Natural Philosophy*; Gaukroger, *Francis Bacon and the Transformation of Early-Modern Philosophy*. On the relevance of the history of natural philosophy, see Gaukroger, *The Failures of Philosophy*.

[13] For a good account of this development, see Becher and Trowler, *Academic Tribes and Territories*, 14–19, 41–50.

modern period, which had yet to develop the disciplinary structures that are now regarded as normative in academic life.[14]

The development of specific intellectual disciplines, which arguably dates from the nineteenth century, has brought significant benefits—most notably, a focused set of research tasks, a manageable professional literature, the creation of specific communities of discourse, and the articulation of clear research methods, and intellectual norms. Yet this narrowing of intellectual focus inevitably means the fragmentation of what was formerly seen as a coherent intellectual territory into a group of regions, each with its own emerging professional languages, research methods, and epistemic norms, with no sense of being part of a greater whole.[15] Where earlier writers had affirmed the importance of the fundamental unity of nature, while acknowledging its 'great, and finely ordered, multiplicity',[16] disciplinary and professional specialization in the modern period has led to focus shifting towards each of the multiple aspects of nature, with a consequent loss of focus on its unity. It is helpful to distinguish between different fields or domains of human knowledge; this does not, however, entail their isolation, or prevent dialogue between them.

Leading natural philosophers of the early modern period wove together insights deriving from multiple sources to create their own distinct individual visions of this conceptual space, while seeing each of these as part of a greater whole which could be apprehended, if only in part. It was in pursuing this vision that the philosopher could achieve intellectual satisfaction, personal happiness, and learn to cultivate virtue.[17] So can this lost vision be retrieved and recast in terms of contemporary disciplinary structures, given the radical changes in terms of how we think about the natural world and the manners of its investigation in the intervening centuries? Where some interdisciplinary projects aim to 'fill any unoccupied [intellectual] spaces',[18] my proposal is to retrieve or recreate a larger conceptual and imaginative space, enabling deeper and richer conversations about the natural world.

[14] Hösle, 'How Did the Western Culture Subdivide Its Various Forms of Knowledge and Justify Them?'; Krohn, 'Interdisciplinary Cases and Disciplinary Knowledge'; McGrath, *Territories of Human Reason*, 19–89.

[15] For the issues that this raises, see Graff, *Undisciplining Knowledge*. Jacobs's critical assessment of 'disciplinary silos' is particularly relevant here: Jacobs, *In Defense of Disciplines*, 13–26.

[16] Lewis, *The Discarded Image*, 11. Although Lewis here describes what he observed in medieval writings, this concern for a theoretical unity underlying an empirical multiplicity can be seen in early modern natural philosophy.

[17] Compare this with the early Wittgenstein: 'In order to live happily I must be in agreement with the world. And that is what "being happy" means.' Wittgenstein, *Notebooks, 1914–1916*, 75.

[18] Frank, 'Interdisciplinarity', 141.

In undertaking this task, I have not used the 'building blocks approach' that some scholars recommend for dealing with complex cultural phenomena or conceptual spaces, such as religion.[19] Rather than trying to identify the individual components of a 'natural philosophy', and attempting to reassemble these, I intend to consider natural philosophy as an historical actuality, trying to understand how this was understood and enacted, and the benefits this was understood to bring. Yet before developing this point, a serious objection to this enterprise needs to be considered.

Retrieval: An Objection Considered

The objection is this: why bother resurrecting an obsolete form of thought, in the light of an inexorable scientific progress which discredits and renders unnecessary any engagement with the past?[20] Natural philosophy, it might be objected, has now been displaced as an outmoded form of discourse on the natural world by the natural sciences. Attempting its retrieval amounts to a pointless exercise in intellectual nostalgia, an unrealistic and possibly illegitimate attempt to recapture a vision of reality which was specific to its age, and has now, like the notions of 'caloric' and 'phlogiston', been discredited. I concede immediately that the modern natural sciences have refined and developed certain aspects of natural philosophy—such as the close observation of nature (present from ancient times) and the use of the experimental method (which became of increasing importance in early modern natural philosophy).[21] Yet it has also marginalized and suppressed certain concerns that were integral to early modern natural philosophy—such as behaving properly towards and within the natural order, and cultivating habits of attentiveness towards its beauty.

As has often been pointed out, the concept of 'science' has changed over time, and is too often defined polemically in response to perceived threats rather than constructively in terms of its specific working methods.[22] Philosophy, science, and religion—to mention three significant areas of

[19] See, for example, Taves, *Religious Experience Reconsidered*, 161–5. A similar strategy lies behind the 'fractionation' of complex phenomena into their components: Whitehouse, 'Cognitive Evolution and Religion'.

[20] Some would see this as belonging to what Todd Moody termed a 'necrology' of failed arguments and positions: Moody, 'Progress in Philosophy', 45.

[21] Gaukroger, *The Collapse of Mechanism and the Rise of Sensibility*, 150–225.

[22] Pigliucci, 'The Demarcation Problem'. For the factors leading to emergence of such a 'rhetoric of demarcation', see Taylor, *Defining Science*, 101–74.

cultural and intellectual engagement—have clearly undergone change over time,[23] making it problematic to essentialize any of them. Yet some studies of scientific advance are essentially ahistorical, offering little more than a selective account of the history of science which fails to take its historical location seriously. A good example lies in modern misreadings of German *Naturphilosophie*, which is regularly dismissed as a biological embarrassment.[24] Happily, more recent scholarship has shown how Schelling's articulation of *Naturphilosophie* set out a broad scientific research agenda which had a significant impact on early nineteenth-century German chemistry, physics, and physiology.[25]

Yet perhaps the best example of this flawed approach is found in Steven Weinberg's anachronistic 'Whiggish' account of the history of science, which presents premodern attempts to do science as a series of fumblings and blunders, eventually remedied by the genius of Newton.[26] Weinberg's method is to judge both philosophers and natural philosophers of the past by the standards of modern physics, as if the term 'science' had a permanent, essential meaning throughout history. This account of the history of science is that of a modern scientist,[27] someone who has succumbed to the definitional hegemony of the present, believing that it is obvious what science *ought* to be, and has no hesitation in judging the past somewhat severely for failing to anticipate the present.[28]

For Weinberg, science is primarily about a quest for order within nature, and has little, if anything, to say to those wishing to think about goodness, justice, or love. Science, Weinberg remarks, has had to 'outgrow' such pursuits and beliefs.[29] Yet this represents a professional narrowing of focus, which now requires reflective scientists to look beyond their own disciplines for their

[23] There is a large literature exploring (and explaining) the historical fluidity of the terms 'science', 'philosophy', and 'religion'. Useful entry points are Brown, 'On Why Philosophers Redefine Their Subject'; Harrison, 'The Pragmatics of Defining Religion in a Multi-Cultural World'; Harrison, *The Territories of Science and Religion*; Gaukroger, *The Failures of Philosophy*.

[24] For example, see Mayr, *The Growth of Biological Thought*, 132.

[25] On the science, see especially Zammito, *The Gestation of German Biology*; Gambarotto, *Vital Forces, Teleology and Organization*. On Schelling's *Naturphilosophie*, see Whistler, *Schelling's Theory of Symbolic Language*.

[26] Weinberg, *To Explain the World*, 215–54. For a critical assessment of Weinberg's reductionist agenda, see Wójcicki, 'Physics, Theoretical Knowledge and Weinberg's Grand Reductionism'.

[27] Weinberg notes that his assessment of certain issues and discussions reflect 'the perspective of a contemporary working scientist': Weinberg, *To Explain the World*, 146.

[28] Happily, there are some occasions when Weinberg makes some fair points—for example, his criticism of Aristotle's 'elaborate reasoning based on assumed first principles' which turn out to be based on 'only the most casual observation of nature, with no effort to test them' (Weinberg, *To Explain the World*, 27).

[29] Weinberg, *To Explain the World*, 45.

ideas about goodness, justice, and beauty. Weinberg's critical remarks raise an important question: is there a more capacious way of thinking which can enfold the natural sciences, while at the same time encouraging (and perhaps enabling?) scientists to develop a wider vision of reality than those resulting from the natural sciences on their own? Why should the empirical study of nature be separated from reflection about beauty, love, justice, or goodness? Early modern natural philosophy regarded these and other questions as integral to a proper human engagement with the natural world. Surely these concerns need to be acknowledged and engaged, especially in an age which is increasingly aware of the human capacity to degrade and destabilize the natural order?

Certainly, there have been gains in our understanding of many aspects of our world. But might we have *lost* something important along the way? Reading Weinberg's account of relentless scientific progress calls to mind Charles Taylor's remark about our fallible attempts to make sense of our world: we begin to realize that, despite progress in some areas, we have a sense that 'we are missing something, cut off from something, that we are living behind a screen.'[30] Roger Scruton pointed out that the realization that 'good things are more easily destroyed than created'[31] leads many to both want to hold on to wisdom that seems to be endangered, and to try to retrieve what should never have been allowed to be forgotten or suppressed in the first place.

Nicholas Maxwell has made a powerful plea for the reconnection of science and philosophy through a revitalized 'natural philosophy', rightly pointing out their close mutual relationship in the early modern period, and the intellectual benefits that would result from a recalibration of their working relationship.[32] In particular, Maxwell highlights the need to solve two problems, now seen as distinct and divergent, yet which were once seen as interconnected aspects of a 'natural philosophy'—namely, learning about the universe on the one hand, and how to create a good and wise world on the other.[33] While I here develop an approach and an agenda that diverge from that set out by Maxwell, we share a common concern to address some mistakes that have led to an impoverishment of our engagement with the natural world, which we both believe can be done through the retrieval of a 'natural philosophy'.

[30] Taylor, *A Secular Age*, 302. [31] Scruton, *Conservatism*, 127.
[32] Maxwell, *In Praise of Natural Philosophy*, 115–217.
[33] Maxwell, *In Praise of Natural Philosophy*, 218.

Against those who now locate an exclusive right to discursive justification across all domains of knowledge in the methods of the natural sciences, I propose a counter-narrative: that exploring early modern approaches to natural philosophy can help in the ongoing process of extending the vision and scope of the philosophical enterprise, and making plausible and productive reconnections across the disciplinary boundaries that subsequently emerged. To explore this point further, we need to consider the nature of progress in philosophy.

Philosophical Progress and the Retrieval of the Past

There is growing interest within the discipline of philosophy in the merits of studying its history, both as an intrinsically interesting academic exercise, but also as a means of stimulating and resourcing philosophical reflection in the present.[34] The study of past philosophical enterprises is to be seen as attempting to understand how philosophy was understood and conducted *under a specific set of historical conditions*, in part to stimulate modern forms of the enterprise. As Karl Marx pointed out, 'people make their own history, but they do not make it as they please under circumstances of their own choosing, but under circumstances that already exist, given and handed down to them.'[35] Marx's point is that, whether we like it or not, we do philosophy within a 'given' limiting historical context, precisely because we are *culturally and historically embedded*. Our historicity limits our options, while at the same time stimulating and informing our reflections.

Ian Hunter thus remarks that past philosophies, considered as objects of intellectual history, are not 'serial historical expressions of a universal human desire for knowledge and understanding, but regimes for inducing such a desire in those whom circumstance or chance have selected to cultivate a philosophical persona.'[36] Similarly, Stephen Gaukroger has offered an historical narrative of the development of philosophy which documents changes in the way in which philosophy has been understood over time, without using the value-laden terminology of 'progress' to describe such changes in its

[34] E.g., MacIntyre, 'The Relationship of Philosophy to Its Past'; Hatfield, 'The History of Philosophy as Philosophy'; Hunter, 'The History of Philosophy and the Persona of the Philosopher'; Antognazza, 'The Benefit to Philosophy of the Study of Its History'.

[35] Marx, *Der achtzehnte Brumaire des Louis Bonaparte*, 1: 'Die Menschen machen ihre eigene Geschichte, aber sie machen sie nicht aus freien Stücken, nicht unter selbstgewählten, sondern unter unmittelbar vorgefundenen, gegebenen und überlieferten Umständen.'

[36] Hunter, 'The History of Philosophy and the Persona of the Philosopher', 587.

'culturally specific modes of engaging with the world'.[37] Even analytic philosophy itself has undergone a 'historical turn', with some of its leading exponents moving away from its traditional ahistoricism towards an acknowledgement of the importance of the study of its own history.[38]

The growing recognition of the historical and cultural situatedness of the philosophical *persona* has been helpful in a number of ways.[39] It has heightened awareness of seemingly 'obvious' ideas having been shaped, to some extent, by historical and cultural factors. It has also led to a growing appreciation that the ideas of past philosophers should be reconstructed 'on their own terms' (that is to say, considering the importance of the historical context within which these ideas developed, if they are to be properly understood).[40]

In reflecting on where philosophy might go in the future, it is clearly helpful to grasp its different self-understandings over time, not least in that this undermines the view that there is some 'correct' definition or 'authentic' task for philosophy. This point appears to be recognized by Mary Midgley, whose final book proposed that our philosophies are best seen as interim responses to a particular historically situated concern.

> Philosophizing, in fact, is not a matter of solving one fixed set of puzzles. Instead, it involves finding the many particular ways of thinking that will be the most helpful as we try to explore this constantly changing world. Because the world – including human life – does constantly change, philosophical thoughts are never final. Their aim is always to help us through the present difficulty.[41]

Midgley's approach suggests that the historical study of how philosophy has engaged its 'present difficulties' in the past might help resource those same discussions in the present.[42]

[37] Gaukroger, *The Failures of Philosophy*, 1–12; quotation at p. 8. Note also Hacker, 'Philosophy', 130: 'If we examine the history of modern philosophy, it appears to be a subject in search of a subject matter.'

[38] See two important collections of essays: Reck, ed., *The Historical Turn in Analytic Philosophy*; Sorell and Rogers, eds., *Analytic Philosophy and History of Philosophy*.

[39] The term 'persona' is increasingly used to refer to an 'exemplary identity wrought by intellectual, moral, and even corporeal disciplines, one that represented an *office* (sometimes a noninstitutionalized one) in specific cultural spaces': Corneanu, *Regimens of the Mind*, 7.

[40] Hatfield, 'The History of Philosophy as Philosophy'; Lærke, 'The Anthropological Analogy and the Constitution of Historical Perspectivism'.

[41] Midgley, *What Is Philosophy For?*, 6.

[42] The views of Pierre Hadot should be noted here, especially his argument that the most authentic way of doing philosophy is to be a philosopher and a historian: see Force, 'The Teeth of Time'.

In a recent analysis, Maria Rosa Antognazza has pointed to another philosophical motivation for studying the history of philosophy—the capacity of the past to challenge present-day orthodoxies, pointing to alternative approaches which might result from the creative appropriation of the wisdom of the past. Studying the history of philosophy thus

> keeps us alert to the fact that latest is not always best, and that a genuinely new perspective often means embracing and developing an old insight. The upshot is that the study of the history of philosophy has an innovative and subversive potential, and that philosophy has a great deal to gain from a long, broad, and deep conversation with its history.[43]

At first sight, this might appear to represent a demand for regression, rather than progress, in philosophy. Surely this amounts to a nostalgic immersion in the past, which could easily become a distraction from the challenges and needs of the present? Yet Antognazza's suggestions can easily be accommodated within recent discussions of the complex and contested theme of 'progress in philosophy'. Philosophy can be said to 'progress' in a number of manners. Some would point to the emergence of new research traditions within philosophy—such as empirical philosophy (drawing on empirical science), or linguistic philosophy (drawing on the analysis of language)—as representing progress. Others, however, would point to the increasing conceptual refinement and precision of more recent discussions of classic themes, including the 'big questions' of 'How do we know about the external world? What are the fundamental principles of morality? Is there a god?'[44]

Yet despite this increasing precision in some areas, there remains an unease at the apparent lack of progress within philosophy, particularly the striking lack of convergence in philosophy when compared with the hard sciences, such as physics. One problem often cited in explaining the 'appearance of persistent and intractable disagreement' among philosophers is the difficulty in finding undeniable premises for philosophical arguments.[45]

Philosophical progress may involve the interrogation of widely accepted assumptions or narratives used to frame discussions of leading philosophical themes—such as the relation of 'knowledge' and 'belief'. Until about 2000, the

[43] Antognazza, 'The Benefit to Philosophy of the Study of Its History', 161.

[44] Chalmers, 'Why Isn't There More Progress in Philosophy?', 5. See also Brake, 'Making Philosophical Progress'; Stoljar, *Philosophical Progress*, 1–19; Frances, 'Extensive Philosophical Agreement and Progress'.

[45] Chalmers, 'Why Isn't There More Progress in Philosophy?', 16–22.

standard account of this development was framed in terms of the notion of 'knowledge as justified true belief' having dominated western reflection from Plato to Locke.[46] Given this contextualization of the discussion, Edmund Gettier's short paper of 1963 raised fundamental concerns for this tradition, which he assumed went back to Plato, in arguing that justified true belief was not sufficient for knowledge.[47] The scholarly consensus of that period initially endorsed Gettier's reading of the tradition,[48] leading some to suggest that the western tradition of knowledge had been thrown into crisis.[49] Others, however, subsequently offered a re-reading of Plato which raised questions about traditional accounts of Plato's views on the relation of knowledge and belief,[50] and hence of Plato's position within the overall narrative arc of discussion. Gettier's intervention could therefore be seen as having facilitated 'progress' by catalysing a re-evaluation of a traditional account, leading to a more reliable statement of its core themes.

The form of philosophical advance set out in the present work involves a critical and creative retrieval of older ways of thinking, which are correlated with contemporary concerns. As Antognazza points out, the history of philosophy has an 'innovative and subversive potential', in that it offers alternative approaches to 'problems of enduring philosophical relevance'.[51] Revisiting the past is an important way of injecting a new critical energy into contemporary discussions. A good example of such a critical retrieval of older ways of conceiving and enacting philosophy, and bringing them to bear on today's questions, can be seen what Edward Feser has called the 'Aristotelian revival'.[52] Aristotle's theory of practical reason has been taken up by philosophers such as David Wiggins, John McDowell, and Martha Nussbaum, as well as ethicists such as Alasdair MacIntyre, in that it is seen as making more sense of pervasive features of our ethical lives than some more recent alternatives.[53] There are also clear signs of interest in Aristotle's natural philosophy: for example,

[46] For a critical account of this tradition, see Ayers and Antognazza, *Knowledge and Belief from Plato to Locke.*

[47] Gettier, 'Is Justified True Belief Knowledge?'

[48] Such as Fine, 'Knowledge and Belief in Plato's Republic 5–7'; Fine, 'Knowledge and True Belief in the *Meno*'.

[49] Pollock, *Contemporary Theories of Knowledge*, 180. For a more guarded account of its importance, see Williamson, *Knowledge and Its Limits*, 4; Enskat, 'Ist Wissen der Paradoxe epistemische Fall von Wahrheit ohne Wissen?'

[50] Gerson, 'Platonic Knowledge and the Standard Analysis'; Sedley, *The Midwife of Platonism*; Ayers and Antognazza, *Knowledge and Belief from Plato to Locke.*

[51] Antognazza, 'The Benefit to Philosophy of the Study of Its History', 180.

[52] Feser, 'An Aristotelian Revival?' See also the case studies included in Miller, ed., *The Reception of Aristotle's Ethics.*

[53] For well-known and influential examples of this retrieval of Aristotle, see Nussbaum, *The Fragility of Goodness*; MacIntyre, *After Virtue.*

many contemporary philosophers continue to find Aristotle's hylomorphism fruitful for contemporary philosophical discussions (such as the mind–body problem).[54]

The Method to be Adopted

The present study was originally conceived as a celebration of the four hundredth anniversary of the publication of Johann Kepler's *Harmonices Mundi* ('The Harmonies of the World', 1619), perhaps one of the most important works of early modern natural philosophy, which embodies many of the core themes of this approach. What I propose is a critical conversation with representative past examples of natural philosophy, such as Kepler's, attempting to identify their core concerns, the methods they adopted, the manner in which their agendas were shaped by their historical and cultural specificity, and what might be appropriately and profitably reclaimed. This work has been made much easier through the surge of scholarly engagement with the concept of natural philosophy in the last two decades, engaged extensively in this study, which has clarified and corrected earlier studies of the distinct nature and focus of such movements.

The first major section of this work thus takes the form of a selective critical conversation with the tradition, which can be traced back to Aristotle. It represents a reading of the tradition which aims for scholarly accuracy on the one hand, and contemporary application on the other. After opening reflections on the tradition in Aristotle and the Middle Ages, the narrative turns to focus on a group of writers—above all, Kepler, Bacon, Boyle, and Newton—who spearheaded the conceptual development of natural philosophy during the height of its influence in the seventeenth and early eighteenth centuries, leading into reflection on its declining influence, and displacement by alternative approaches. The object here is not so much to establish the fine details of the historical trajectory of the movement, or enter into complex arguments about points of interpretation. While this has been the subject of considerable scholarly interest in recent years, my concern lies elsewhere. It is rather to ask what may be learned from our current understanding of this narrative, and

[54] Debate often centres on how much Aristotle's ideas need modification in the light of contemporary discussions: see the divergent positions of Jaworski, *Structure and the Metaphysics of Mind*; Marmodoro, 'Aristotle's Hylomorphism without Reconditioning'; Rea, 'Hylomorphism Reconditioned'.

how it can inform the quite different historical and cultural location of today's discussions.

This leads into the second part of the work, which offers a critical and original assessment of what a contemporary reconstructed natural philosophy might look like, and how it might engage a number of contemporary philosophical concerns and agendas, particularly the need for a recalibration of the relationship of philosophy with the natural sciences. Where the first part is largely analytical, the second is more constructive. It offers a nuanced and historicized reflection on how the intellectual virtues of earlier forms of natural philosophy might be recaptured, if only in part, not by some crude historical transplantation of the notion, but rather through ensuring at least some of those virtues can be recovered and integrated. It is quite possible that older understandings of natural philosophy may prove to lie beyond retrieval; yet perhaps we might be able instead to retrieve something of the *persona* of a natural philosopher.[55] The key difficulty we shall consider in this section is how a correlation of insights about the world that the early modern period took to be natural and intuitive can find a fresh rationale and application in the twenty-first century.

We therefore begin our critical conversation with the past by turning to consider the emergence of a 'natural philosophy' in the works of Aristotle.

[55] Gaukroger, *The Failures of Philosophy*, 132–8.

PART I

A CRITICAL CONVERSATION WITH THE TRADITION

1
The Origins of Natural Philosophy
Aristotle

Forms of thinking that deserve to be called 'philosophical' can be found in ancient Chinese and Indian traditions,[1] and include reflection on the natural world, and its significance for human beings.[2] The discipline now known as 'philosophy', however, is generally agreed to have emerged in Greece during the fourth century BCE, with the appearance of new forms of culturally located reasoning with the potential to transcend those specific historical contexts and constraints. While some locate the emergence of Greek philosophy in terms of the triumph of *logos* over *muthos*, others protest at the historical and intellectual simplifications attending this suggestion.[3] Plato, for example, was certainly critical of older myths, but was perfectly capable of developing his own to advance his philosophical agendas.[4]

Others point to the importance of transcending local cultural and communal norms, moving away from solidarity towards objectivity through the recognition of something beyond a local frame of reference as the basis for truth. As Richard Rorty summarized Plato's approach, 'the way to transcend skepticism is to envisage a common goal of humanity – a goal set by human nature rather than by Greek culture.'[5] To reason *peri phuseōs* ('according to nature') was to render oneself intellectually accountable to something beyond the local community and its norms.[6]

[1] See, for example, Lloyd, *Ancient Worlds, Modern Reflections*; Seaford, *The Origins of Philosophy in Ancient Greece and Ancient India*.

[2] For example, note the *Ru* tradition (Ruism, often still referred to as 'Confucianism'): Berthrong, 'Confucian Views of Nature'.

[3] For a classic statement of this position, see Nestle, *Vom Mythos zum Logos*; for corrections of its overstatements, see Morgan, *Myth and Philosophy from the Pre-Socratics to Plato*; Struck, 'The Invention of Mythic Truth in Antiquity'; Fowler, 'Mythos and Logos'.

[4] Morgan, *Myth and Philosophy from the Pre-Socratics to Plato*, 242–89.

[5] Rorty, 'Solidarity or Objectivity?', 21. Rorty himself rejected the notion of some 'transcultural rationality', and affirmed instead the quest for rational solidarity within a community: see Hill, 'Solidarity, Objectivity, and the Human Form of Life'.

[6] For the various meanings of this phrase, particularly the associated form *historia peri phuseōs*, see Naddaf, *Greek Concept of Nature*, 11–35.

Our story begins with Aristotle's scientific observations on the Aegean island of Lesbos, where he acquired an extraordinary knowledge of fishes and their behaviour around the Gulf of Kalloni, which was documented and analysed in his *Historia Animalium.*[7] Aristotle argued that the complex observed patterns of the migratory movements of living creatures could be explained on the basis of a few general principles: the food supply, temperature changes, and issues related to breeding.[8]

Aristotle's natural philosophy rests on the assumed capacity of the human mind to discern an existing, rather than imposed, order within the natural world, and the ability of human culture to offer analogies by which this may be explained and understood.[9] This process of exploration was catalysed by a sense of wonder, which expands our intellectual and imaginative horizons.

> It is through wonder that people now begin and originally began to philosophize; wondering in the first place at obvious perplexities, and then by gradual progression raised questions about greater questions, such as the phenomena of the moon and of the sun, about the stars and about the origin of the universe.[10]

Aristotle's fascination with nature created a 'drive to capture and recapture rhetorically that feeling of wonder (*thaumazein*) that is the origin and promise of all inquiry'.[11]

Aristotle was not the first thinker to advocate or undertake an investigation of the natural world. At the time, there already existed an established tradition of inquiry into nature, which is clearly delineated in by Plato in the *Timaeus*,[12] even if it is imperfectly and incompletely developed (for example, Plato seems surprisingly uninterested in the non-human biological world). Aristotle's agenda is broader, aiming to identify what is seen in the natural order, and how this may be accounted for. Aristotle does not use the term *phusis* to correspond to 'nature', as this term is now widely used, but to rather to designate the 'nature [of things]'. The term *kosmos* began to be used from

[7] For an excellent account of Aristotle's period on Lesbos, see Leroi, *The Lagoon.*

[8] *Historia Animalium* 596b 20.

[9] For what follows, see especially Falcon, *Aristotle and the Science of Nature*; Lennox, '*Bios* and Explanatory Unity in Aristotle's Biology'; Leunissen, *Explanation and Teleology in Aristotle's Science of Nature.*

[10] Aristotle, *Metaphysics*, 982 b12–16. See further Miller, *In the Throe of Wonder*, 11–52.

[11] Poulakos and Crick, 'There Is Beauty Here, Too', 298.

[12] See the analysis in Johansen, *Plato's Natural Philosophy*, 69–176.

the sixth century to mean something like a 'unified world order',[13] loosely corresponding to what is today designated 'nature'.

Theōria: The Nature of and Benefits of Explanation

As Aristotle conceives it, natural philosophy is an autonomous and self-contained discipline. In the *Meteorology*, natural philosophy is understood as a systematic inquiry into the different elements of the natural world, in order to discover the connections between its different parts, thus leading to an *understanding* of nature, which transcends a mere *knowledge* of nature.[14] This idea is often linked with the notion of *theōria*—a mental beholding of the natural order. As this notion is important for the retrieval of natural philosophy, we shall consider it in some detail.

There is a growing recognition that a 'novel and subversive claim' can be seen as lying at the heart of the intellectual enterprises of both Plato and Aristotle: the belief that the 'supreme form of wisdom is *theōria*, the rational "vision" of metaphysical truths'.[15] The emergence of *theōria* as a means of articulating the task of philosophy and the status of its outcomes is an important staging-post in the narrative of the emergence of natural philosophy. It is important to note that this classical understanding of *theōria* was somewhat more sophisticated than some modern interpreters suggest—think, for example, of Martin Heidegger's reduction of this notion to metaphysical optics.[16] We begin by noting its cultural location, and then move on to consider its role in Plato, before turning to its implementation in Aristotle's project of 'theorizing' the natural world.

While it is clearly problematic to define, or even describe, the nature of fourth-century Greek philosophy in terms of a transition from *muthos* to *logos*, there are good reasons for suggesting that a core task for the new cultural phenomenon of 'philosophy' was to defend the status of its claims to knowledge in the face of its rivals originating from religious or cultically privileged sources. The use of the term *theōria* had cultural overtones at this time, being linked with facilitating connections between individual Greek

[13] Marrow, 'Κόσμος in John', 90–3; Slatkin, 'Measuring Authority, Authoritative Measures', 28.

[14] See especially Falcon, *Aristotle and the Science of Nature*, 85–122. See also Moravcsik, 'What Makes Reality Intelligible?'; Winslow, *Aristotle and Rational Discovery*.

[15] Nightingale, *Spectacles of Truth in Classical Greek Philosophy*, 3.

[16] Heidegger interpreted *theōria* as 'observation' (*Betrachtung*), which aims to fix, limit, or entrap its object. For an extended criticism of this approach, see Foltz, *The Noetics of Nature*.

city-states and the broader Hellenic world through networks linking individuals and cults in different locations. A city—such as Athens—would participate in other Greek festivals and sanctuaries by sending delegates to witness the events in question.[17] The idea of *theōria* as a 'sacred beholding' was particularly important in sustaining a shared religious practice across the Greek world—for example, in relation to the pronouncements of oracles.[18]

In the fourth century, Plato used the term *theōria* to characterize the role of the philosopher, speaking of a 'mind habituated to thoughts of grandeur and the contemplation (*theōria*) of all time and all existence'.[19] The *bios theōretikos* of a philosopher was not based on religious rituals or privileged oracles, but on the wise use of reason to discern and contemplate a grander vision of the world. The transfer of this notion of 'contemplation' from the religious or cultic domain to that of philosophy can be seen as intimating its oracular or revelatory status—and hence had a right to be considered as a public philosophy, not simply as a private persuasion or pursuit.[20] 'Plato identifies the philosopher as a new kind of *theōros*, an intellectual ambassador who makes a journey to a divine realm to see the spectacle of truth and then brings a report of his findings back home.'[21]

This view of the philosopher is set out in the allegory of the Cave, perhaps the best-known passage of Plato's *Republic*, where the theme of a beholding of a grander reality that needs to be declared and enacted plays a critical role.[22] For our purposes, the central narrative of the allegory is a group of people, entrapped within a dark cave, knowing only a world of shadows cast on its walls, who are unaware of a grander world beyond the cave. The philosopher is someone who has escaped from this intellectual bondage, and has seen the sunlit world beyond the cave—and thus, like a *theōros*, has a responsibility to bear witness to that world to those remaining within the cave, despite being

[17] Rutherford, *State Pilgrims and Sacred Observers in Ancient Greece*; Nightingale, *Spectacles of Truth in Classical Greek Philosophy*, 40–71.

[18] Rutherford, *State Pilgrims and Sacred Observers in Ancient Greece*, 93–109. For an important earlier account of the philosophical significance of this cultural practice, see Rausch, *Theoria*.

[19] Plato, *The Republic*, 6; 486a. For comment, see Nightingale, *Spectacles of Truth in Classical Greek Philosophy*, 72–93. There are hints of this idea in the pre-Socratic tradition: Bénatouïl and Bonazzis, '*Θεωρια* and *Βιοσ Θεωρητικοσ* from the Presocratics to the End of Antiquity', 3–4. Nightingale notes that two of Plato's metaphysical dialogues (the *Phaedo* and the *Republic*) open with scenes involving civic *theōria*: Nightingale, *Spectacles of Truth in Classical Greek Philosophy*, 74, n.8.

[20] Bénatouïl and Bonazzis, '*Θεωρια* and *Βιοσ Θεωρητικοσ* from the Presocratics to the End of Antiquity', 5–7. However, some considered that such philosophical contemplation led to some form of privatized existence: see the critical evaluation of this trend in Brown, 'Contemplative Withdrawal in the Hellenistic Age'.

[21] Nightingale, *Spectacles of Truth in Classical Greek Philosophy*, 82.

[22] *Republic*, VII 514a2–517a7. For the problems of interpreting this analogy, see Karasmanis, 'Plato's Republic: The Line and the Cave'; Johansen, 'Timaeus in the Cave'.

overwhelmed by the enormity of this acquired vision of reality, and its obvious dissonance with the known and familiar world of the cave. For Plato, the philosopher is not primarily someone who has acquired a body of insight through teaching; rather, the philosopher has caught a vision of reality, or experienced a transformative beholding of the world.

Yet Plato has surprisingly little interest in the specifics of this world beyond the cave. The philosopher beholds this world, noting its 'plants and animals and the light of the Sun';[23] yet Plato's concern does not lie in reflecting on the natural world, but rather grasping and inhabiting the grander *theōria* that the escape from the cave makes possible. This becomes particularly clear in a brief discussion about looking at the stars. Glaucon believes this involves looking upwards; Socrates retorts that it means looking downwards, arguing that we should 'behold (*theōrein*) with the mind, not with the eyes'.[24] While Plato's account of *theōria* deserves further elaboration,[25] for our purposes there are two points that merit close attention. First, philosophy entails a new way of beholding reality, including the world of nature; and second, Plato does not appear to see this *theōria* as entailing or even encouraging a closer engagement with the world of nature. Aristotle concurs on the importance of *theōria*, but is willing to engage the natural world from its perspective, and to recognize the role of nature in developing theoretical perspectives. The process of engaging with the natural world, which is integral to Aristotle's natural philosophy, particularly in dealing with questions such as the observation of change (*kinēsis*) or apparent goal-directed behaviour within nature, involves abstraction from the realm of nature to a realm of essences, which are clearly understood to be part of the natural world rather than belonging to some separate domain.

But what is the point of gaining such an understanding? What intellectual, moral, or practical benefits does it bring? How does it contribute to the pursuit of happiness or wellbeing (*eudaimonia*), which Aristotle commends as a human ideal? This is an important question, given the persistent anxiety within Aristotle scholarship that the pursuit of such theoretical knowledge might turn out to be pointless, or at least to be incompatible with Aristotle's statements concerning a good life.[26] Unlike some later theoreticians, Aristotle does not hold that *theōria* holds the key to mastering the natural world,

[23] *Republic*, VII 532b. [24] *Republic*, VII 529b.

[25] See, for example, Nightingale, *Spectacles of Truth in Classical Greek Philosophy*, 94–138.

[26] For some important reflections, see Nightingale, *Spectacles of Truth in Classical Greek Philosophy*, 187–251; Engelmann, 'Scientific Demonstration in Aristotle, "Theoria", and Reductionism'; Roochnik, 'What Is *Theoria*?'

allowing it to be transformed for the benefit of human beings. Contemplation may indeed be the ultimate goal of human rational action and reflection; yet it does not appear to serve any useful or productive purpose.[27]

This problem can be addressed in several ways. Perhaps the most satisfying would be to demonstrate that a closer reading of Aristotle's works shows that *theōria* has a role that has hitherto been overlooked. Matthew Walker, for example, develops an interesting argument to this effect, picking up on Aristotle's maxim that 'nature does nothing in vain'. If this is indeed the case, then there must be some reason for, and benefit from, the human desire to understand our world.[28]

An alternative approach is to locate *theōria* within a wider interpretative framework, arguing that the reflective beholding of the natural world serves a deeper *telos* that lies beyond the scope of Aristotle's system. Perhaps the most significant development of this approach lies in early Christian theology, which held that the 'beholding' of the natural world was religiously significant, in that it was an indirect beholding of God, thus facilitating both an understanding of reality and appropriate forms of worship.[29] Any perceived inadequacy in Aristotle's approach was thus remedied by a Christian expansion of his understanding of the theoretical beholding of nature to embrace a theoretical beholding of God's work of creation, which exhibited the wisdom and order of its creator.[30] This, of course, raises the important question of the interface between a natural *philosophy* and a natural *theology*, to which we shall return at several points in this work.[31]

Observational Respectfulness and Natural Philosophy

One of the most interesting aspects of Aristotle's natural philosophy is his attentiveness to observing natural entities—such as plants, animals, and stars.

[27] For a good summary of these concerns within recent Aristotle scholarship, see Walker, *Aristotle on the Uses of Contemplation*, 1–12; Nightingale, 'On Wondering and Wandering'.

[28] Walker identifies some potential areas of theoretic utility, such as a self-awareness of the human animal's status between gods and beasts: Walker, *Aristotle on the Uses of Contemplation*, 183–205.

[29] A good example of this lies in the theologically informed 'intellectual contemplation of nature' in the writings of Maximus the Confessor. For this approach, see Harrington, 'Creation and Natural Contemplation in Maximus the Confessor's *Ambiguum* 10:19'; Steel, 'Maximus Confessor on Theory and Praxis'.

[30] Foltz, *The Noetics of Nature*, 158–74.

[31] For this relationship in the ancient world, see Enders, *Natürliche Theologie im Denken der Griechen*.

Perhaps the best-known statement of this observational respectfulness is found in the opening lines of Book VII of his *Ethics*:[32]

> We must, as in all other cases, set the observed facts (*phainomena*) before us and, after first discussing the difficulties (*diaporēsantas*), go on to prove, if possible, the truth of all the common opinions about these affections of the mind, or, failing this, of the greater number and the most authoritative; for if we both refute the objections and leave the common opinions (*endoxa*) undisturbed, we shall have proved the case sufficiently.

The Greek phrase *sōzein ta phainomena*, best translated as 'saving the appearances (or phenomena)', is often used to designate this respect for observation, and is variously and vaguely attributed to both Plato and Aristotle.[33] In fact, the Greek phrase *sōzein ta phainomena*, popularized by Pierre Duhem in the first decade of the twentieth century, is more characteristic of later commentators rather than of Aristotle himself. It is difficult to translate the phrase into English, in that the Greek term *phainomena* resists easy renderings, such as 'appearances', 'observations', or 'observed facts'.[34] This phrase now tends to be associated with an instrumentalist understanding of scientific theory, which holds that a theory capable of accommodating existing observations is not necessarily true, in the sense of somehow 'corresponding' to reality.[35]

While this principle of respecting phenomena is not *explicitly* stated in Aristotle's writings using this phrase, the idea itself is clearly present. Both science and scientific explanation—as Aristotle understands these—are to be construed *realistically*, in that science must mirror reality, and therefore theory always must cohere with observation and empirical investigation.[36] It is, however, innocent of the instrumentalist associations that later came to be associated with the phrase, even if it can be argued to be consistent with them. Aristotle is concerned with the identification of the *archē* or *archai*—the 'first principles' that lie behind what is observed in natural phenomena and living

[32] Aristotle, *Nicomachean Ethics* (1145 b2–7); as translated by W. D. Ross, *The Nicomachean Ethics*, 118. See the extended discussion in Nussbaum, *Fragility of Goodness*, 240–63.

[33] The phrase began to secure scholarly traction following the publication of the historian of science Pierre Duhem's monograph *Sozein ta phainomena: Essai sur la notion de théorie physique de Platon à Galilée* (1908). Duhem's interpretation of his Greek sources is suspect at multiple points: see especially Lloyd, 'Saving the Appearances'. There are also some important observations in Goldstein, 'Saving the Phenomena'; Repellini, 'Platone e la salvezza dei fenomeni'.

[34] Note Nussbaum's comments on this problem: *Fragility of Goodness*, 240–3.

[35] For a good account of Duhem's position and its intrinsic difficulties, see Dion, 'Pierre Duhem and the Inconsistency between Instrumentalism and Natural Classification'.

[36] A point developed and documented in detail in Hankinson, *Cause and Explanation in Ancient Greek Thought*, 160–200.

beings. He clearly considers these to have a real existence independent of these phenomena, which underlies their explanatory capacity in relation to those phenomena.[37] Aristotle's persistent emphasis on the importance of observation of nature may occasionally lead him to some false conclusions (such as the highly influential theory of spontaneous generation);[38] nevertheless, his empirical instincts must be acknowledged.

Yet Aristotle is not entirely consistent here. As Eckard Kessler points out, there is an apparent (and important) methodological pluralism within Aristotle's writings on natural philosophy concerning the identity of natural entities: a largely non-empirical metaphysical approach (found in the *Physics*), and the more empirical approach found in *On Generation and Corruption* and the *Meteorology*.[39] This led to the emergence of tensions in the later Renaissance between metaphysical and empirical approaches to natural knowledge. The emergence of a 'strong naturalistic tendency as early as the beginning of the sixteenth century' does not need to be seen as a rejection of Aristotle, but can rather be seen as a methodological refocusing within the Aristotelian corpus.[40]

In the end, the empirical method won out, mainly because it proved capable of dealing with questions that simply could not be resolved by an appeal to first principles. Ptolemy's geocentric view of the solar system, though arguably based on the best observational evidence of his age,[41] had to give way to the heliocentric systems of Kepler and Galileo, because of the observational evidence that could not adequately be accommodated by the older theory.[42] Yet the empirical method would undergo significant development in the early modern period, partly through the rise of the experimental method, and partly through the use of instruments such as microscopes and telescopes, which extended the capacity for both observation and measurement, and in doing so, undermined some of the more questionable aspects of Aristotle's natural philosophy.

[37] Irwin, *Aristotle's First Principles*, 4; Gotthelf, *Teleology, First Principles, and Scientific Method in Aristotle's Biology*, 153–214. On the dangers of conflating Aristotle's approach with those developed in modern metaphysical realisms, see Winslow, *Aristotle and Rational Discovery*, 7–8.

[38] As noted by Hankinson, *Cause and Explanation in Ancient Greek Thought*, 171–3.

[39] Kessler, 'Metaphysics or Empirical Science?' See further Speer, 'Zwischen Naturbeobachtung und Metaphysik'.

[40] Kessler, 'Metaphysics or Empirical Science?', 80. During the fourteenth century, writers such as William of Ockham and John Buridan adopted a more explicitly empirical approach to natural philosophy, arguably on account of their changed understanding of the notion of *substantia*: see Lagerlund, 'The Changing Face of Aristotelian Empiricism in the Fourteenth Century'.

[41] Swerdlow, 'The Empirical Foundations of Ptolemy's Planetary Theory'; Goldstein, 'Saving the Phenomena'.

[42] For Copernicus's reasoning at this point, and his relation to the Aristotelian tradition, see Goddu, *Copernicus and the Aristotelian Tradition*, 207–2.

Aristotle's own specific ways of making sense of the natural world reflect the limiting conditions of his historical and cultural location. It is not difficult to identify what appear to be errors of judgement on Aristotle's part,[43] some of which became petrified as he came to be regarded as an 'authority' in Renaissance science. For example, in *On the Parts of Animals*, Aristotle remarked that a philosopher could take pleasure in contemplating 'things constituted by nature' *kata theōrian*, including both the 'ungenerated, imperishable, and eternal' phenomena of the heavens and the generated and changeable realities of the world.[44] Yet this absolute distinction between the changing 'sub-lunar' world and the perfect, regular, and unchanging 'celestial' realm caused his later interpreters some difficulties. Tycho Brahe's observation of the appearance of a bright new star in the constellation of Cassiopeia in November 1572, and Christopher Scheiner's observation of sunspots in 1611,[45] both suggested that the celestial realm was subject to imperfection and change.

This discussion, however, lay in the future. Our concern now focuses on what Aristotle appears to have understood by the 'explanation' of nature.

Explaining Nature: Aristotle on Induction

So what does Aristotle understand by explaining nature? We have already noted the importance of identifying 'first principles (*archai*)' underlying the phenomena of the natural world. Aristotle's natural inclination is to turn to teleology as an explanatory framework.

> Teleological explanations are a central feature of Aristotle's investigation of nature and reflect the importance he attributes to final causality in the coming to be and presence of regular natural phenomena. In Aristotle's view of the world, everything that exists or comes to be "by nature" comes to be or changes, unless prevented, for a purpose and towards an end, and is present for the sake of that purpose or end.[46]

[43] For some obvious examples, see Viano, 'Mixis and Diagnôsis'; Johnson, *Aristotle on Teleology*.

[44] Aristotle, *On the Parts of Animals*, IV.10; 645a8–10.

[45] For the issues, see Methuen, 'This Comet or New Star'; Reeves, 'From Dante's Moonspots to Galileo's Sunspots'; Engvold and Zirker, 'The Parallel Worlds of Christoph Scheiner and Galileo Galilei'.

[46] Leunissen, *Explanation and Teleology in Aristotle's Science of Nature*, 2. For a detailed account, see also Gotthelf, *Teleology, First Principles, and Scientific Method in Aristotle's Biology*.

As is well known, Aristotle identifies four causes which may be discerned as underlying phenomena in the world or human culture—such as the creation of a bronze statue:[47] a material cause; a formal cause; an efficient cause; and a final cause. In the cause of a bronze statue, the efficient cause (or principle that produces the statue) is the art (*technē*) of bronze-casting the statue.[48]

This raises a difficulty that has long been recognized in Aristotle's approach: the assumed analogy between the causal processes that generate *natural* and *cultural* entities.[49] It is quite easy to point to the causal processes that underlie the production of a cultural artefact, such as a bronze statue. Yet the unevidenced assumption that such causal processes also underlie the generation of natural entities, such as plants and animals, is deeply problematic. Aristotle in effect speaks of each internal causality within nature as an 'artist (*demiourgos*)'.[50] Furthermore, there are points in Aristotle's scientific reflections, particularly in trying to understand the structures of plants and animals, when certain non-empirical grounding assumptions are brought into play, including the belief that 'nature does nothing in vain'.[51]

Although there are clear parallels between Aristotle's understanding of 'science (*epistēmē*)' and that of Plato,[52] there are significant points of divergence, generally reflecting Aristotle's adaptation of Plato's approach to enable an engagement with the phenomena of the natural world. Perhaps the most important is Aristotle's use of the process of induction (*epagōgē*) in engaging with these phenomena.[53] For Aristotle, scientific knowledge is based on an empirical apprehension of particulars, which requires an inductive method of investigation.

At several points, Aristotle defines *epagōgē* in terms of 'proceeding from particulars up to a universal',[54] which he often contrasts with *sullogismos* ('deduction'), in which the reasoning process begins from established universals.

[47] This account is set out in *Physics* II 3 and *Metaphysics* V 2; see further Leunissen, *Explanation and Teleology in Aristotle's Science of Nature*, 10–16.

[48] *Physics*, 195a6–8. The other causes are as follows: material (the bronze from which the statue is made); formal (the shape the sculptor imposes on the statue); and final (the sculptor's purpose in making the statue).

[49] On which see Broadie, 'Nature and Craft in Aristotelian Teleology'.

[50] For some reflections on how such difficulties might be mitigated, see Leunissen, *Explanation and Teleology in Aristotle's Science of Nature*, 16–18.

[51] Lennox, 'Nature Does Nothing in Vain'; Leunissen, *Explanation and Teleology in Aristotle's Science of Nature*, 63–70; 121–35.

[52] Bolton, 'Science and Scientific Inquiry in Aristotle'.

[53] For documentation, see Winslow, *Aristotle and Rational Discovery*, 19–32.

[54] For example, *Topics* I, 105a10–14. For the suggestion that Aristotle treats *epagōgē* as a form of inference which has a specifiable logical structure, see Hintikka, 'Aristotelian Induction'; Kakkuri-Knuuttila and Knuuttila, 'Induction and Conceptual Analysis in Aristotle'. Hintikka particularly notes Aristotle's linking of *epagōgē* and concept formation.

This, of course, raises questions about imperfect inductive generalizations,[55] and whether such generalizations can really be considered as *epistēmē* rather than simply *doxa*. One important (and influential) outcome of the application of this form of reasoning within Aristotle's writings concerns the eternity of the world. This belief is clearly not an empirical observation in itself; it is a view that is based on Aristotle's understanding of the relation of matter and forms, neither of which are empirical notions.[56] As the subsequent discussion of this issue made clear, Aristotle's arguments for the eternity of the world proved vulnerable.[57] The 'eternity of the world' was *doxa*, rather than *epistēmē*.

The debate about the correct interpretation of Aristotle's notion of *epagōgē* and its relevance for a natural philosophy continues. I shall note two significant engagements with Aristotle's account by leading philosophers of science, noting how both found Aristotle profitable yet underdeveloped on this point. In both cases, a strong case can be made for asserting that these are legitimate readings of Aristotle's concept of *epagōgē* and its associated methods for studying the natural world, which are defensible alternatives to the 'standard' reading of Aristotle at this point.[58]

The Victorian philosopher William Whewell (1794–1866) held that Aristotle anticipated his own view that induction involves not only the accumulation and generalization of facts, but also the discovery of new concepts through a process of active interpretation of the natural world.[59] For Whewell, induction was a process of reflection that added something essential to what was otherwise simply a process of enumeration. 'There is a New Element added to the combination [of instances] by the very act of thought by which they were combined.'[60] Whewell's main criticism of Aristotle was that he

[55] Note the argument of Upton, 'Infinity and Perfect Induction in Aristotle'. There is a debate about whether Aristotle is entirely consistent in his understanding of *epagōgē*. Some passages (e.g., *Prior Analytics*, 68b15) seem to suggest that Aristotle there understands *epagōgē* as *sullogismos ex epagōgēs*. For comment on this apparently anomalous use of the term, see McCaskey, 'Freeing Aristotelian Epagōgē from *Prior Analytics* II 23'.

[56] Aristotle, *Physics*, VIII, 1, 251a8–b10. For a good discussion of the issues, see McGinnis, 'The Eternity of the World', 272–4.

[57] Dales, *Medieval Discussions of the Eternity of the World*, 50–198.

[58] I take this 'standard' reading to be that represented in W. D. Ross's influential and frequently republished introduction, first published in 1923: Ross, *Aristotle*, particularly 38–41.

[59] Whewell's views on Aristotle's method of induction are set out in his 1850 paper 'Criticism of Aristotle's Account of Induction'. For comment and assessment, see Niiniluoto, 'Hintikka and Whewell on Aristotelian Induction', 52–4. For the debate between Whewell and Mill on the nature of induction, which reflects Whewell's assessment of Aristotle, see Snyder, *Reforming Philosophy*, 95–155.

[60] Whewell, *The Philosophy of the Inductive Sciences*, vol. 2, 48. For a more detailed account of Whewell's approach, see Snyder, *Reforming Philosophy*, 33–94; Morrison, *Unifying Scientific Theories*, 52–4.

failed to give adequate attention to the need to 'introduce' new ideas or concepts that were 'distinct from sensation'[61]—such as some kind of 'organizing principle', which was integral to the process of scientific explanation.

The American philosopher Charles S. Peirce (1839–1914) is noted for his development of 'abductive' modes of thought in making sense of the natural world. While much attention has been given to the specifics of this 'creative act of making up explanatory hypotheses',[62] which is now widely referenced within the philosophy of science, not enough attention has been given to its origins in Aristotle.[63] Peirce himself held that the origins of his abductive method lay in his reading of *Prior Analytics* II.25, which he considered to be defective as it stood, yet generative in terms of the possibilities that it suggested. Although Peirce's reconstruction of Aristotle's argument is open to criticism at points,[64] it is clearly a legitimate interpretation and extension of his approach.

For example, it could be argued that the *Posterior Analytics* suggests that Aristotle is feeling his way towards offering what is essentially a psychological description of the mental process of grasping of a universal form in particular cases through 'intuitive induction' or intuitive reason.[65] Even the 'standard' reading of Aristotle concedes the imaginative or intuitive elements in his account of induction: it is 'a process not of reasoning, but of direct insight, mediated psychologically by a review of particular instances'.[66] In an early lecture, Peirce described abduction as

> that process in which the mind goes over all the facts the case, absorbs them, digests them, sleeps over them, assimilates them, dreams of them, and finally is prompted to deliver them in a form, which, if it adds something to them, does so only because the addition serves to render intelligible what without it, is unintelligible.[67]

[61] Niiniluoto, 'Hintikka and Whewell on Aristotelian Induction', 53.

[62] Davis, *Peirce's Epistemology*, 22. For accounts of Peirce's abductive accounts of scientific discovery, see McKaughan, 'From Ugly Duckling to Swan'; Rodrigues, 'The Method of Scientific Discovery in Peirce's Philosophy'; Schurz, 'Patterns of Abduction'.

[63] The importance of Aristotle is highlighted in Anderson, 'The Evolution of Peirce's Concept of Abduction'.

[64] See especially Flórez, 'Peirce's Theory of the Origin of Abduction in Aristotle', which both queries Peirce's own reading of *Prior Analytics* II.25, while pointing to other Aristotelian texts that are supportive of his abductive approach.

[65] Niiniluoto, 'Hintikka and Whewell on Aristotelian Induction', 50.

[66] Ross, *Aristotle*, 41.

[67] MS 857: 4–5; cited McKaughan, 'From Ugly Duckling to Swan', 466. Peirce states that be believes that his idea of 'abduction' corresponds to what Aristotle intended to denote by the corresponding Greek term *epagoge*.

This brief discussion of Aristotle's natural philosophy sets the scene for its later development. The full story of how Aristotle's natural philosophy was favourably received within the Islamic world,[68] before coming to be rediscovered and reappropriated in western Europe, is fascinating, and still not fully understood. How, for example, did Aristotle's views on astronomy, supposedly *empirical* in their foundations, come to be received as *authoritative* in European universities, thus causing such difficulties for empirical innovators such as Galileo? Although these are important historical questions, they are not all relevant for our purposes.

The most appropriate point at which to reconnect with the western tradition of natural philosophy is in the thirteenth century, when this discipline emerged as a distinct and potentially autonomous field of inquiry, which would gradually loosen its connections with Aristotle and his commentators, in favour of a more explicitly empirical approach.

[68] See the important collection of studies in Alwishah and Hayes, eds., *Aristotle and the Arabic Tradition*.

2

The Consolidation of Natural Philosophy

The Middle Ages

From the thirteenth to the sixteenth century, Aristotelian natural philosophy dominated western attempts to make sense of the world of nature.[1] This is not to say that western scholars were inattentive to the natural world in earlier centuries. The eighth-century theologian and historian Bede, based in the kingdom of Northumbria in north-eastern England, represents an important tradition of natural philosophy, though based on biblical exegesis than Aristotle.[2] Yet Aristotle was seen to offer natural philosophy an informing intellectual framework, thus stimulating major translation projects throughout the twelfth and thirteenth centuries, in which Aristotle's works and those of his commentators were translated from Greek and Arabic into Latin.[3] While this tradition of commentary emerged in the ancient world,[4] it was retrieved and expanded in the Middle Ages and Renaissance, allowing Aristotle's intellectual vision to be continuously evaluated, corrected, and extended. Edward Grant's assessment of medieval Aristotelian interpretation is helpful in understanding its appeal and persistence: 'It was always a domain of both traditional and innovative concepts and interpretations and was, therefore, inevitably elastic and absorbent. Hence its most interesting feature was a capaciousness that knew few limits.'[5]

While some historians appear to consider that the appearance of this literature itself was the cause of the emergence of natural philosophy or even the natural sciences (though this latter term was not used frequently in the early Middle Ages),[6] it seems clear that this interest in Aristotle and his interpreters

[1] For excellent accounts of this development, see Grant, *A History of Natural Philosophy*, 130–78; Gaukroger, *The Emergence of a Scientific Culture*, especially 47–153.

[2] For Bede's natural philosophy, see Wallis, '*Si Naturam Quaeras*'. For a good study of Bede's *de natura rerum*, see Ahern, *Bede and the Cosmos*.

[3] For details of such works, see Lindberg, 'The Transmission of Greek and Arabic Learning to the West'.

[4] Tuominen, 'Philosophy of the Ancient Commentators on Aristotle'.

[5] Grant, 'Ways to Interpret the terms "Aristotelian" and "Aristotelianism" in Medieval and Renaissance Natural Philosophy', 352.

[6] For example, see Crombie, *Augustine to Galileo*, 19–43.

was either the result of, or an essential accompaniment to, something more fundamental—a new interest in the philosophical and religious importance of the natural world.[7] As a result of this enterprise, the concept of *philosophia naturalis* was widely discussed within the western church and the universities. While the Latin term *natura* could be used to translate Aristotle's *phusis*, it came to have wider associations by the end of the twelfth century, when the concept of 'nature' had developed a range of meanings, including some close approximations to what modern readers understand by the term.[8]

For M. D. Chenu, the twelfth century witnessed a growing interest in the phenomenal world. Nature was now seen both as an object of study in itself, but also as a means of stimulating religious devotion through seeing nature as God's creation.[9] Chenu suggested that the twelfth century laid the foundations for this new interest in engaging the natural world, displacing a symbolic understanding of nature with a new concern to understand the structure, constitution, and underlying principles which seemed to underlie physical reality.[10] Where traditional understandings of the natural world proceeded from a theological perspective, Chenu argued that such motivations were supplemented by a growing interest in acquiring an understanding of nature using methods that were not grounded upon or governed by theological principles.

Subsequent research has confirmed that a renewed interest in the study of nature emerged in the mid-twelfth century,[11] and that scholars of this period drew on classical sources in developing a natural philosophy. This development is clearly related to the emergence of distinctively Christian, Islamic, and Jewish forms of natural philosophy during this period. Each found ways of drawing on the emerging tradition of natural philosophy, while accommodating this to the religious traditions that they represented. We shall return to this point later in this chapter (pp. 36–8). Yet it is important to appreciate that similar approaches emerged in Asia around this time, as forms of natural philosophy emerged in China, many of which placed an emphasis upon

[7] A point stressed by Speer, *Die entdeckte Natur*; Speer, 'Zwischen Naturbeobachtung und Metaphysik'.

[8] See further Sprandel, 'Vorwissenschaftliches Naturverstehen und Entstehung von Naturwissenschaften'; Wegmann, *Naturwahrnehmung im Mittelalter im Spiegel der lateinischen Historiographie des 12. und 13. Jahrhunderts*, 35–128. For a useful survey of the range of meanings now associated with the English word 'nature', see Lewis, *Studies in Words*, 24–74.

[9] For the importance of this point, see Ritchey, 'Rethinking the Twelfth-Century Discovery of Nature'.

[10] Chenu, 'La découverte de la nature'. Cf. Haskins, *The Renaissance of the Twelfth Century*, 303–40; Speer, 'Secundum Physicam'.

[11] Poirel, *Livre de la nature et débat trinitaire au XII*[e] siècle; Mews, 'The World as Text'.

tian-ren-he-yi (the unity or harmony of heaven and humanity),[12] thus giving rise to forms of thought that parallel comparable developments in western Europe during the Middle Ages and Renaissance. Although Chinese natural philosophers such as the twelfth-century writer Chu Tsi did not deal directly with some of the more abstract discussions that characterize modern science, there is a clear trajectory from his close attentiveness to nature to such an approach.[13]

This developing European interest in natural philosophy was well adapted to some significant trends within western religious and academic culture around this time, most notably the political separation of the ecclesiastical and secular realms resulting from the Investiture Controversy of the eleventh century, and the increasing structural separation—and hence intellectual isolation—of faculties of philosophy and theology in medieval universities.[14] The growing interest in the concept of 'natural law' in the academic culture of the early medieval period also points to the interest in legal notions which were not grounded theologically.[15] Although this concept was partly grounded in the notion of right reason, it also reflected a metaphysical belief in the ordering of the natural world, which was seen as lying beyond human intervention and manipulation.

For some of its interpreters, the task of natural law is to 'interrelate systematically practical reason with a philosophy of nature'.[16] A natural law theory therefore requires 'a commitment to law as in some way "natural," and nature as in some way normative'.[17] For Aquinas, the concept of 'natural law' ultimately rests on a metaphysical realism of structured natural kinds and an epistemological realism by which this realist structure can be known by human observers of the natural world.[18] This theme attracted much attention during the Renaissance, and was widely seen as a legitimate and generative topic of exploration.[19]

[12] Chen and Bu, 'Anthropocosmic Vision, Time, and Nature'.

[13] Kim, *The Natural Philosophy of Chu Tsi, 1130–1200*, 308–13.

[14] For reflections on the philosophical significance of these developments, see Gaukroger, *The Failures of Philosophy*, 111–26.

[15] On which see Saccenti, 'The *Ministerium Naturae*'.

[16] Hittinger, *A Critique of the New Natural Law Theory*, 7–8. For the points at issue, see Crowe, 'Metaphysical Foundations of Natural Law Theories'.

[17] Hittinger, *A Critique of the New Natural Law Theory*, 8.

[18] For a full discussion of Aquinas's metaethics, see Stump and Kretzmann, 'Being and Goodness'.

[19] For example, consider its widespread engagement and exploration in the literature of the English Renaissance, particularly William Shakespeare: White, *Natural Law in English Renaissance Literature*, 134–215.

The Universities and Early Medieval Natural Philosophy

The foundation of universities in western Europe is widely seen as a turning point in intellectual history, marking the relocation of the pursuit of knowledge from the piety of the monastic schools of theology to a new academic space, theoretically independent of monarchical and ecclesiastical control.[20] By far the most important such intellectual centre was the University of Paris. By the beginning of the thirteenth century, it is thought that there were around 4,000 students in residence at Paris, making up a tenth of the city's population. The University of Paris was a complex corporate entity, a federation of three main collegiate bodies: nations, faculties, and colleges. Tensions arose over many issues within the Parisian Faculty of Arts, particularly in the first half of the thirteenth century, when religious orders—such as the Dominicans and Franciscans—became involved in university teaching, and were seen as a threat to the autonomy of Faculty members (such as secular priests) who were not linked with these orders.

Yet one of the most significant debates at Paris during the thirteenth century arose from a turf war between the disciplines of philosophy and theology and their associated Faculties, each of which considered itself to be intellectually distinct and autonomous.[21] Aristotle's natural philosophy came to demarcate a form of knowledge production that was potentially a threat to theology's claim to intellectual supremacy, grounded in its privileged access to divine revelation.[22] This tension was experienced within Islam in previous centuries, as growing interest in Aristotelian works led to strained relations with Islamic theology, especially on questions relating to creation and the eternity of the world.[23] It was clear that Aristotle affirmed certain natural philosophical doctrines—such as the eternity of the universe—which contradicted, or at least were at least in tension with, certain traditional commitments of Jewish, Christian, and Islamic theology.[24]

[20] For a good study of the tensions this created, see Bernstein, 'Magisterium and License'. For the Parisian cathedral school of theology at this time, see Andrée, 'Peter Comestor's Lectures on the Glossa "Ordinaria" on the Gospel of John'.

[21] It is, however, important to note that certain writers and texts were common to the study and teaching of both Faculties of Arts and Faculties of Theology: see Luscombe, 'Crossing Philosophical Boundaries *c.* 1150–*c.* 1250'.

[22] For a full account of the significance of these developments for the fortunes of Aristotle's natural philosophy, see Gaukroger, *The Emergence of a Scientific Culture*, 59–86.

[23] For example, consider Al-Fārābī's defence of the eternity of the world in the first half of the tenth century: Vallat, 'Al-Fārābī's Arguments for the Eternity of the World'.

[24] Giletti, 'The Journey of an Idea'; Dales, *Discussions of the Eternity of the World*, 86–198. The case of medieval Spain is especially important, given the presence of both Islam and Christianity in this

While Thomas Aquinas rightly pointed out that Aristotle's belief in the eternity of the world rested on questionable lines of reasoning, he nevertheless held that Aristotelian ideas were helpful in the articulation of Christian theology.[25] However, Aquinas's argument for the critical appropriation of Aristotle was met with a counter-demand for a general rejection of his ideas. In 1277, the bishop of Paris, Etienne Tempier, issued a series of condemnations of Aristotle's ideas, in an ultimately unsuccessful attempt to curb their influence. Yet Tempier's attempts to suppress natural philosophy failed to satisfy either theologians or philosophers at Paris. A better strategy was clearly required.

With the benefit of hindsight, the contribution of the Dominican Parisian theologian Albertus Magnus may be seen to have paved the way for a solution to this problem, which resulted in the rise of Aristotelian philosophy in the great universities of England, France, Germany, and Italy. An acknowledged expert on Aristotle, he drew a distinction between the realms of the natural and supernatural, arguing that philosophy offered an appropriate means of studying the former, and theology the latter. The breadth of his intellectual vision was such that he was able to hold together natural philosophy and theology, while delineating their respective realms of authority and inquiry.[26] He was doubtless helped in this matter by Gregory IX's bull *Parens scientiarum* (1231), which is widely credited with bringing about a decisive change in university culture, especially in encouraging the study of Aristotle's *Libri naturales*.[27]

For Albertus, no special divine illumination is required in order to acquire a reliable knowledge of the natural world using intrinsic human powers of reasoning. The study of natural things (*naturalia*) can be conducted using a 'natural light'; the study of revealed truth, however, requires divine grace and illumination. It is instructive to consider Albertus's changing views on human nature, which show an increasing tendency to move away from a traditional Augustinian approach to human identity, and draw extensively on Aristotelian works (particularly his treatise *On the Soul*). Perhaps more importantly, he

region, and the growth of an Islamic tradition of interpretation of Aristotle: see Giletti, 'Aristotle in Medieval Spain'.

[25] For Aquinas's criticism of Aristotle on eternity, see Elders, 'St Thomas Aquinas's Commentary on Aristotle's "Physics"'. For his critical appropriation of Aristotelian themes—for example, in his theology of grace and sacraments—see Emery and Levering, eds., *Aristotle in Aquinas's Theology*.

[26] For an excellent overview, see Ashley, 'St. Albert and the Nature of Natural Science'. See further Miteva, 'Intellect, Natural Philosophy, Finality'; Collins, 'Albertus, Magnus or Magus?'; de Asúa, 'War and Peace'.

[27] Young, *Scholarly Community at the Early University of Paris*, 64–101.

tends to weave citations from theological sources into wider conversations,[28] rather than treating them as privileged.

In his assessment of the development of western philosophy, Gaukroger suggests that Albertus Magnus's significance for the development of natural philosophy thus lay partly in a careful delineation of magisterial boundaries: 'natural philosophy was concerned with natural truths, and theology with supernatural ones.'[29] Natural philosophy was thus something that could be pursued independently of theology, in effect constituting a distinct *scientia*. This was not simply a matter of disciplinary specificity; it had implications for the location of this discipline within a university's Faculty structures, and for the scholarly aptitudes and dispositions that accompanied such studies.[30] To simplify what Albertus Magnus has to say on this matter, divine grace and illumination are not required for matters of philosophy, but are necessary for theology, which deals with the interpretation and apprehension of divine reality.[31]

A similar view was taken by Thomas Aquinas during his period at Paris. For Aquinas, who can be seen as the successor to Albertus Magnus, philosophy works in its own distinct manner; yet its conclusions may be seen as congruent with those of theology. This theme of convergence of outcomes, but not identity of methods and norms, is seen particularly in Aquinas's *Summa contra Gentiles*, in which he deploys a natural philosophy, independent on Christian theology, in order to enable him to engage Muslims and pagans who would not accept the authority of the Christian Bible, or any arguments resting upon it.[32] A natural philosophy thus offered what could be seen as a shared common natural language, enabling conversations across both disciplinary and religious divides.

[28] On these points, see Blankenhorn, 'How the Early Albertus Magnus Transformed Augustinian Interiority', especially 355–6.

[29] Gaukroger, *The Failures of Philosophy*, 120–1.

[30] For the emergence of the Faculty of Theology at the University of Paris, see Young, *Scholarly Community at the Early University of Paris*, 20–63. On the Parisian Faculty of Arts, see Verger, 'La Faculté des Arts'. Note the comments of Gaukroger, *The Failures of Philosophy*, 118: 'The new Arts Faculties of the universities were staffed by a new kind of clerical *magister*, who now not only immersed himself in secular learning, particularly the liberal arts, but saw this very much as part of his identity.'

[31] The situation is somewhat more complex than this, as Albertus allows that there are truths concerning the divine being that are demonstrable and knowable without revelation. For a good account of this, see Führer, 'Albertus Magnus' Theory of Divine Illumination', 152–4.

[32] For detailed comment, see Kretzmann, *The Metaphysics of Creation*; Kretzmann, *The Metaphysics of Theism*.

Natural Philosophies: Christian, Islamic, and Jewish

The rise of the University as a discursive community, consisting of Faculties with different understandings of the nature of knowledge production, in effect created an environment in which natural philosophy and theology offered different means of making sense of the natural world and human nature, as well as engaging with questions relating to God. Thomas Aquinas illustrates this point well, in his discussion of the idea of 'the empyrean heaven'. This, he argued 'cannot be investigated by reason... but is held by authority'.[33]

'Natural philosophy' here is perhaps best understood as forms of reasoning which—following Albertus Magnus—rely on 'natural' capacities and resources, not requiring any special divine 'illumination'. This development could thus be understood as a form of intellectual empowerment, shifting some important areas of discourse beyond the sole authority of the church or a Faculty of Theology. There are clear parallels with Anselm of Canterbury's eleventh-century so-called 'Ontological Argument' for the existence of God (although as Anselm frames it, this is neither 'ontological' nor an 'argument'), which offers a gateway to some modest theological reflections using the natural light of human reason, devoid of any specifically theological presuppositions.[34] While some might see this as a threat to the authority of theology, most would see it as offering an expansion of intellectual possibilities, based on the traditional notion of *ancillae theologiae*—'assistants to theology', who are able to offer conceptual or methodological grist to theology's mills, enabling refinement and development of its ideas.[35] It could therefore be argued that a natural philosophy is a stimulus or gadfly to theology, perhaps encouraging a welcome interdisciplinarity.

One important intellectual and sociological factor stands out in this respect—the creation of a conceptual space for reflection on the grand questions of God, meaning, and nature which was not controlled or shaped by the church or any other religious body. We have already noted the importance of Christian writers such as Albertus Magnus and Thomas Aquinas in developing a viable relationship between Christian theology and natural philosophy (pp. 33–4). Yet such a critical religious appropriation of Aristotle was not a

[33] For this and other examples, see Grant, 'How Theology, Imagination, and the Spirit of Inquiry Shaped Natural Philosophy in the Late Middle Ages'.

[34] On this, see Leftow, 'The Ontological Argument'; Schumacher, 'The Lost Legacy of Anselm's Argument'.

[35] The idea of *philosophia* as the *ancilla theologiae* can be traced back to the patristic period, and is often (though questionably) attributed to Peter Damien: see Solère, 'Avant-Propos'; van den Hoek, 'Mistress and Servant'.

specifically Christian undertaking; similar projects were being developed by Jewish and Islamic writers in western Europe around this time. While medieval discussion of natural philosophy was often linked to the interpretation of Aristotelian texts, this was nuanced in different ways by Christian, Islamic, and Jewish writers.

Although many religious writers of this period expressed concerns about Aristotle's views on the eternity of the world, they also appeared to have viewed his works as a religiously neutral resource capable of being quarried for the development of their own theological and philosophical agendas. This shared interest in Aristotle led to a philosophical ecumenism which was remarkable in the light of the political and religious tensions of this period. A good example of this related to the twelfth-century Andalusian Islamic philosopher Avicenna, a noted practitioner of *falsafa*, an Arabic philosophical tradition that considered itself as an 'immediate heir and continuation of a Neoplatonized Aristotelianism'.[36] Averroes's commentaries were even translated into Hebrew so that Jewish interpreters of Aristotle might benefit from them.[37]

Judah ben Samuel Halevi, the leading Hebrew poet of his generation in medieval Spain, developed a natural philosophy which distinguished between the God who was disclosed through the natural order, and the God of Abraham, Isaac, and Jacob, who was disclosed through history.[38] Halevi's most significant work was originally written in Arabic, and later translated into Hebrew. In this work, Halevi contrasts the God of Aristotle, whom he designates as 'Elohim' and is known by reason and through the fixed order of nature, with the God of Abraham, known as 'Adonai', who is known through history, as interpreted through prophetic illumination. 'One passionately yearns for Adonai with a passion that involves both "taste" and witness, while attachment to Elohim is by way of speculation.'[39]

Other topics that were debated within medieval Jewish natural philosophy included the question of whether there existed a plurality of universes, and

[36] McGinnis, *Avicenna*, 3.

[37] Harvey, 'The Hebrew Translation of Averroes' *Prooemium* to his Long Commentary on Aristotle's *Physics*'. See also Glasner, *Averroes' Physics*.

[38] Kogan, 'Judah Halevi and His Use of Philosophy in the *Kuzari*'.

[39] I here draw on the translation and commentary on this work found in Kriesel, *Judaism as Philosophy*, 3–4. As Kriesel argues, Maimonides can be interpreted as taking a similar position on the relation of Elohim and the God of Abraham, despite indications of a preference for a God of nature, known through reason: Kriesel, *Judaism as Philosophy*, 5–7. Kriesel also notes the influence of natural philosophy on the thought of the fourteenth-century Jewish writer Nissim of Marseille: Kriesel, *Judaism as Philosophy*, 176–8.

how Aristotle was to be interpreted on this matter.[40] Yet medieval Judaism was divided in its attitudes towards Aristotelianism. Jewish communities in the northern communities of Ashkenaz and Northern France were sceptical of its merits; those in the southern communities of Spain, Provence, and Italy were considerably more positive.[41]

Despite their theological differences, many Jewish, Muslim, and Christian natural philosophers of this era thus shared a belief in the importance of Aristotle as a foundational resource for a natural philosophy. Yet this shared respect for Aristotle was in decline by the end of the Middle Ages,[42] with important consequences for the future directions of natural philosophy.

The Decline of Aristotelian Natural Philosophy

One important factor which contributed to the decline of Aristotle's influence in the later Middle Ages was growing unease over the adequacy of his empirical observations—for example, the dogma of the 'uninhabitability of the torrid zone'.[43] Aristotle held that the habitable world was confined to the temperate zone between the Tropics and the northern frigid zone, which was clearly inconsistent with the growing body of knowledge arising from the 'voyages of discovery' of the late fifteenth and early sixteenth centuries, which made it clear that the 'habitable' world extended far south of the Equator.[44] Cumulative observation by multiple contemporary observers increasingly came to be seen as providing a sufficient evidentiary foundation for overturning the once-authoritative judgments of Aristotle. This helps explain the growing importance of commentators on Aristotle in the Renaissance, who often nuanced or corrected his judgements on matters of importance for natural philosophy, including moving this tradition in a more empirical direction.[45]

[40] Feldman, 'On Plural Universes'.

[41] For the particular respect for Aristotle among Spanish medieval Jewish philosophers, see Glasner, 'The Peculiar History of Aristotelianism among Spanish Jews'.

[42] For the multiple levels of this 'subversion' of Aristotle in the sixteenth century, see Martin, *Subverting Aristotle*; Del Soldato, *Early Modern Aristotle*.

[43] *Meteorology* 362b 6–9. See the criticism of this view by José de Acosta, in his *Historia Natural y Moral de Las Indias* (1590), based on his observations in the Caribbean islands, Bolivia, Ecuador, Mexico, and Guatemala. Other Aristotelian dogmas which were seen as unsustainable in the light of empirical observation by the late sixteenth century include the immutability of the heavens, called into question by the appearance of comets and *novae*: see, for example, Methuen, 'This Comet or New Star'.

[44] For the cartographic consequences of these doubts about Aristotle's views, see Sanderson, 'The Classification of Climates from Pythagoras to Koeppen'.

[45] Kessler, 'Metaphysics or Empirical Science?'; Lagerlund, 'The Changing Face of Aristotelian Empiricism in the Fourteenth Century'.

Yet others were concerned that Aristotelian natural philosophy failed to articulate a moral vision, whether social or personal, enabling deeper questions of purpose and value in life to be addressed. The fourteenth-century Italian humanist Petrarch was particularly critical of the moral failures of the natural philosophers of his day, as well as their failure to engage deeper questions. 'What use is it, I ask, to know the nature of beasts and birds and fish and snakes, and to ignore or neglect our human nature, the purpose of our birth, or whence we come whither we are bound?'[46] Petrarch's more fundamental concern was that philosophy fails to motivate us to act virtuously, which required inspiration rather than mere intellectual analysis.[47] Petrarch here reflects the Renaissance immersion in classical antiquity, which led him to see philosophy as encouraging aspiration to moral excellence and the achievement of happiness.

The moral shortcomings of natural philosophy might, of course, be remedied by an appeal to natural law. As noted earlier (p. 32), a good case can be made for suggesting that natural philosophy lays the foundation for an understanding of *lex naturalis*—a concept of goodness or human obligation that is grounded in the deeper structures of reality, rather than being a human construct or invention.[48] While this idea of a cosmic moral order with implications for human social ordering was of interest to university Faculties of Law, it was clearly also of public interest, and was often debated in Renaissance literature—for example, in Milton's *Paradise Lost*, where the notion of natural law plays a 'defiantly and definitely central' role.[49]

Yet there are important questions about the wider cultural impact of natural philosophy at this time in western European history, especially in relation to the *affective* aspects of the natural world. Natural philosophy wrestled with questions of understanding nature, such as what first principles might be identified as lying behind natural structures and processes. Yet there is a substantial literature which points to both popular and academic interest in engaging matters such as the beauty and mystery of nature, the possibility that nature disclosed certain arcane ideas or mystical secrets, or that it might

[46] Petrarca, *Invectives*, 239. On Petrarch's cultural significance, see Witt, *In the Footsteps of the Ancients*, 230–91.

[47] A similar point was made by Philip Sidney in the late sixteenth century in urging the superiority of 'poesy' over philosophy: for discussion, see Campana, 'On Not Defending Poetry'.

[48] For the crossover between the academic fields of law and theology on this notion, see Pennington, '*Lex Naturalis* and *Ius Naturale*'. Perhaps reflecting the influence of Gratian, this term came to be replaced with *ius naturale* from the thirteenth century.

[49] White, *Natural Law in English Renaissance Literature*, 216–42; quote at p. 217.

hold the key to personal transformation.[50] Natural philosophy seemed to offer a somewhat dry and cerebral account of a natural world, whose beauty and complexity seemed to cry out for an engagement at the imaginative and emotional level. This approach to natural philosophy persisted into the early modern age. In his *Ancilla philosophiae* (1599), the Oxford Aristotelian philosopher John Case set out the object of natural philosophy in traditional Renaissance style as 'the fullest knowledge of all things'. Not only does Case show no awareness of the importance of interpreting nature,[51] he has little interest in engaging with the affective aspects of the natural world.

To appreciate the importance of this point, we may consider a quite different approach to the natural order developed in the twelfth century at the Parisian Abbey of St Victor. Hugh of St Victor developed a theology of the natural world which allowed its affective aspects to be appreciated, and incorporated into a Christian doctrine of creation. The study of nature as God's creation led, not simply to a better *understanding* of the world, but to the enhancing of human *wisdom*.[52] This has remained a classic theme of human engagement with nature: not simply understanding how nature works, but appreciating its deeper meaning and how this might make people wiser.[53] Even in the early Middle Ages, we can see hints of a recognition of the incompleteness of purely rationalist assessments of the natural order, which seem to offer an inadequate account of the human experience of nature.

The limits of such rational accounts of the natural order is highlighted in Hildegard of Bingen's remarkable *Liber divinorum operum* ('The Book of Divine Works'), which explores an approach to natural philosophy that highlights the importance of affective and imaginative engagement with the world.[54] Hildegard explores the affective links between God, human beings, and the cosmos through the notion of *symphonia*—an auditory image which highlights the harmony of the universe, and the pleasure that it gives to those who contemplate it.[55] The contemplation of the natural order thus goes beyond intellectual reflection, in that it elicits pleasure, praise, and devotion,

[50] For a good account of the multiple levels of popular engagement with nature in the twelfth century, see Ritchey, 'Rethinking the Twelfth-Century Discovery of Nature'.

[51] I here draw on the criticism of Case set out in Serjeantson, 'Francis Bacon and the "Interpretation of Nature" in the Late Renaissance', 686.

[52] See the analysis in Angelici, *Semiotic Theory and Sacramentality in Hugh of Saint Victor*, 95–117.

[53] As we shall see, this is a persistent theme in Romanticism and the post-Romantic period. See, for example, Bubel, 'Nature and Wise Vision in the Poetry of Gerard Manley Hopkins'. Hopkins here interprets an ambiguous nature through the 'appropriation of the Judeo-Christian sapiential tradition', which offers a lens leading to clarity of vision.

[54] Rabassó, '*In Caelesti Gaudio*'. See also the older study of Schipperges, 'Kosmologische Aspekte der Lebensordnung und Lebensführung bei Hildegard von Bingen'.

[55] Leigh-Choate et al., 'Hearing the Heavenly Symphony'.

while giving the beholder a sense of place within the universe. The intensity of Hildegard's engagement with nature, and the imagery she uses in expressing this, raise some important issues about the role of gender in engaging an allegedly 'objective' natural world, which have featured prominently in some recent feminist theological discussions.[56]

Hugh of St Victor also played an important role in developing the idea of nature and the Christian Bible as 'God's Two Books', which could be read alongside each other in mutually illuminating ways.[57] Although this was an interesting development in its own right, its importance for an understanding of natural philosophy relates to the question of the *interpretation* of nature—a matter that was rarely discussed in the Middle Ages, but which became important in the sixteenth century. It is now generally agreed that the imagery of 'God's Two Books' led theologians to suggest that the well-established hermeneutical tools used to interpret the Bible could also be applied to the natural world,[58] thus opening the way to the new interpretative empirical approaches to nature that emerged in the late sixteenth century. A set of familiar hermeneutical tools were available for the interpretation of the text of the book of nature, once the moment was right for their deployment. Yet although the conceptual foundations for such an interpretative approach to nature had been laid by 1200, their application dates from much later.

The Utility of Natural Philosophy: Alchemy and Magic

Aristotle's view that *theōria* was of little practical value was not shared by all. While early medieval natural philosophy offered a restrictively rationalized account of the natural world, others turned to the disciplines of alchemy to expand the scope of a philosophy of nature. In 1994, Pamela Smith pointed out how Renaissance alchemy, and the hermetic philosophy that informed it, contributed significantly to the 'habits of mind and practice that formed early modern philosophy'.[59] The Renaissance interest in *chrysopoeia* (transmuting

[56] For a good summary of the issues, see Anderson, 'What's Wrong with the God's Eye Point of View'.

[57] Mews, 'The World as Text', 99–102. Mews corrects the older study of Hubert Herkommer at points of importance: Herkommer, 'Buch der Schrift und Buch der Natur'. Note also the points made by Hünemörder, 'Traditionelle Naturkunde, realistische Naturbeobachtung und theologische Naturdeutung in Enzyklopädien des hohen Mittelalters'.

[58] See Harrison, *The Bible, Protestantism, and the Rise of Natural Science*; Gaukroger, *The Emergence of a Scientific Culture*, 129–53. This point will be developed further later in this work: see pp. 60–2.

[59] Smith, 'Alchemy as a Language of Mediation at the Habsburg Court', 2.

base metals into gold) that was evident in, for example, the alchemical *Liber Lucis* of John of Rupescissa,[60] was not seen as something that was irrational or occult, but rather as a legitimate and entirely rational way of moving beyond an understanding of the natural world to controlling it, and repurposing natural processes to deliver desirable outcomes. *Theōria* offered an understanding of natural processes, which could lead to their redirection to enhance human well-being.

Many alchemical works of the early Middle Ages, such as Robert of Chester's *Liber de compositione alchemiae* (1144), understood alchemy as an attempt to understand the fundamental constitution of the natural order, making it possible both to transform nature and heal humanity. While most now consider alchemy to be an arcane non-science with strongly mystical tendencies, its early medieval practitioners saw it as a form of natural philosophy, originating in Arabia.[61] Alchemists and the practitioners of magic were sometimes described as 'the perfect philosophers', who had mastered the arts of natural philosophy, geometry, music, and astronomy necessary to a transformation of the natural world.[62] The 'philosopher's stone' was thus not seen simply as an agent of theoretical understanding, but of personal transformation—a point particularly evident in the English poet George Herbert's poem 'The Elixir'.[63]

Alchemy remained a significant element of western European engagement with nature until the end of the seventeenth century, when growing concern about the absence of reliable witnesses to supposed acts of 'transmutation' led to an erosion of their plausibility.[64] Nevertheless, alchemical motifs can easily be discerned in early modern natural philosophy, such as that of Robert Boyle, which we shall consider presently. It was not until the development and general acceptance of Dalton's 'atomic hypothesis' in the late eighteenth century that the notion of transmutation of metals was recognized as scientifically impossible.[65]

This leads us to consider the place of 'magic' in the Middle Ages. It is important to appreciate that writers of the Middle Ages did not make the easy distinction between 'magic' and 'science' that seems self-evidently true in the twenty-first century. William of Auvergne, who was Bishop of Paris from

[60] DeVun, *Prophecy, Alchemy, and the End of Time.*

[61] Carusi, 'Alchimia Islamica e religione'; Viano, ed., *L'alchimie et ses racines philosophiques.*

[62] See Zambelli, *White Magic, Black Magic in the European Renaissance*; Hughes, *The Rise of Alchemy in Fourteenth-Century England*; Klaassen, *The Transformations of Magic.*

[63] See the analysis in McGrath, 'The Famous Stone'.

[64] Principe, 'Transmuting Chymistry into Chemistry'.

[65] For a good account of this development, see Rocke, 'In Search of El Dorado'.

1228 until his death in 1249, saw magic as part of the *naturalis scientiae*.[66] Much magical literature of the Middle Ages was concerned with identifying the networks of natural forces in the universe and explaining how these can be manipulated to human advantage.[67] These works included 'lapidaries' (studies of the magical powers attributed to different stones) and treatises on the medicinal uses of plants. For many in the Middle Ages and Renaissance, this offered an attractive alternative to the intellectual abstractions of Aristotelian natural philosophy. For reasons that are not completely understood, the domain of magic came to have increasingly feminine associations in the later Middle Ages, possibly reflecting its subversion of traditional gendered associations of intellectual authority.[68]

None of these considerations, of course, can really be considered to call into question the intellectual legitimacy of natural philosophy, as this emerged as a distinct way of reasoning and as a community of discourse in the early medieval period. It is simply to note some concerns which became increasingly obvious over the following centuries—namely, that this discipline was seen to offer an incomplete account of nature, thus raising the question of whether this deficiency could be remedied from within the discipline, either by expansion of its conceptual scope, or by creating alliances with other modes of human reflection which might alleviate these difficulties. While the process that Max Weber termed 'Disenchantment' is generally associated with the modern period, its roots can arguably be traced back much further.

This book is not a history of the concept of natural philosophy, but is rather an interpretative reflection on its past forms and present possibilities. While there is much more that needs to be said about the fortunes of natural philosophy in the Middle Ages, the specific purposes of this work are best served by moving directly to consider what might be considered to be the 'Golden Age of Natural Philosophy', beginning with the remarkable contribution of the late Renaissance philosopher Johannes Kepler. It was during this period that an interpretative empirical approach to natural philosophy, particularly in relation to the use of an experimental method, became firmly established—with significant results for how natural philosophy was understood and practised.

[66] Kieckhefer, 'The Specific Rationality of Medieval Magic', 819.

[67] See Page, 'Medieval Magic'; Copenhaver, *Magic in Western Culture*; Klaassen, *The Transformations of Magic*.

[68] See, for example, Bailey, 'The Feminization of Magic'.

3

Skywatching

The Natural Philosophy of Copernicus, Kepler, and Galileo

The sixteenth century is widely seen as witnessing a reconfiguration of the notion of natural philosophy. In part, this was due to increasing concern about the evidential basis of Aristotelian natural philosophy, which appeared to many to have tempered Aristotle's empirical instincts, leading to a petrification of his philosophical system. Aristotelianism was often treated as a definitive interpretation of the natural world, rather than as a provisional account of the nature of the world which was to be augmented and corrected in the light of a continuing engagement with the natural order. A growing interest emerged in expanding knowledge empirically—for example, by observing nature in newly discovered parts of the globe, or by making increasingly accurate observations of the natural world. In this chapter, we shall reflect on the importance of such empirical observations of the heavens, particularly the movement of the planets against the background of fixed stars.

The heavens have always fascinated human beings,[1] not least because they create a mind-expanding sense of what Albert Einstein described as 'rapturous amazement'.[2] Yet alongside this affective impact of the night sky, many recognized its utility. In Babylonian times, the study of the heavens was seen as practically important (for example, in the prediction of solar and lunar eclipses),[3] while in ancient Egypt it served to predict the annual flooding of the Nile or allow the correct orientation of temples.[4] Others were intrigued by the more cognitive questions it raised. The movements of the five 'wandering stars' (Greek: *planētes*) against the background of seemingly changeless 'fixed

[1] Hübner, 'The Professional *Astrologos*'. The modern distinction between 'astrology' and 'astronomy' dates from the eighteenth century; in classical Greek writings, the terms *astronomia* and *astrologia* are both encountered, the former primarily in Pythagorean and Platonist contexts, and the latter in Aristotle.

[2] Einstein, *Ideas and Opinions*, 38.

[3] Rochberg, 'Reasoning, Representing, and Modeling in Babylonian Astronomy'.

[4] Belmonte, 'In Search of Cosmic Order'; DeYoung, 'Astronomy in Ancient Egypt'.

stars' were observed with particular interest. What were they? And how could their slightly irregular movements across the sky be explained?

The model of planetary motion developed by Claudius Ptolemy of Alexandria in the second century achieved widespread acceptance. In his *Amalgest*, Ptolemy argued that the observed patterns of motion of the planets could be explained, and their future movements predicted, in terms of a series of nested circular orbits around the earth. Although this model is often described as 'forced' or 'contrived' by its modern critics, it is very difficult to sustain these criticisms in the light of the measurements of planetary positions available to Ptolemy at this time.[5] Ptolemy's planetary theory is 'strictly empirical', setting out a rigorous mathematical derivation of numerical parameters based on observations, aiming to 'preserve the phenomena'.[6]

Ptolemy's account of planetary motions remains the best theoretical approximation to the observational data before Johannes Kepler's model of the early seventeenth century. Perhaps its greatest achievement was to demonstrate that a fixed and constant framework could be discerned behind what seemed like somewhat irregular planetary motions. It is important not to judge Ptolemy in the light of today's more precise observational knowledge, but rather to adopt a historical approach which tries to understand 'how and why Ptolemy and his successors formulated the problems the way they did', and how successful this formulation was in accounting for the observable phenomena of their time, 'within the restrictions of their ability to observe'.[7]

So did Ptolemy's mathematical planetary model—or others that might arise to displace it—aim to *explain* the real world, or was the role of such mathematical models essentially to *describe* and accurately *predict* planetary motions? Many classical and medieval writers considered that the substances of nature did not conform to the abstractive precision that was characteristic of mathematics,[8] and that the deductive structure of mathematical

[5] Goldstein, 'Saving the Phenomena'; Swerdlow, 'The Empirical Foundations of Ptolemy's Planetary Theory'. Note in particular Swerdlow's interpretation of Ptolemy's 'bisection of the eccentricity'.

[6] Swerdlow, 'The Empirical Foundations of Ptolemy's Planetary Theory', 249–50: 'Every hypothesis that Ptolemy uses is justified empirically, by appeal to observations that show that the hypothesis is correct, that it corresponds to or produces the characteristic apparent motions upon which it is based.'

[7] Saliba, 'Arabic versus Greek Astronomy', 331. A similar point is made by Westman, *The Copernican Question*, who urges avoiding retrojecting current scientific vocabularies and assumptions onto the past, and instead recognizing that categories and criteria of knowledge are historically situated, bound to specific times and locations.

[8] On this point, see Modrak, 'Aristotle on the Difference between Mathematics and Physics and First Philosophy'.

demonstration could not adequately capture the complex causal relationships among natural bodies.[9] Aristotle had raised concerns about the mixing of disciplines on the grounds that qualities that might be essential from the perspective of one discipline might be considered accidental from another disciplinary perspective.[10] While Aristotle was satisfied that mathematics could be mixed in an acceptable manner with astronomy, optics, and harmonics, he did not consider it could be mixed with physics.[11]

One of the most interesting debates on the interpretation of Ptolemy on this point took place in the city of Toledo in Moorish Spain during the twelfth century, when Islamic philosophers discussed the status of his model.[12] The Toledan school of astronomy offered an instrumentalist approach to Ptolemy, where others preferred a realist interpretation.[13] This debate also highlighted the role of natural philosophy as an *interpretation*, not simply an *aggregation*, of observations.[14] European intellectuals of the sixteenth and seventeenth centuries increasingly came to understand the study of both the Bible and nature as mutually illuminating hermeneutical exercises.[15]

In this chapter, we shall consider the interpretative natural philosophy which emerged from the works of the three leading astronomers of the early modern period: Copernicus, Kepler, and Galileo. We begin by considering Nicolaus Copernicus's proposal for a heliocentric understanding of planetary motion.

[9] Many Aristotelians of the early modern period thus denied that mathematical accounts of nature could provide substantive (i.e., causal) accounts of natural processes: see Mancosu, *Philosophy of Mathematics and Mathematical Practice in the Seventeenth Century*.

[10] *Posterior Analytics* 75a27–75b2. Cf. Modrak, 'Aristotle on the Difference between Mathematics and Physics and First Philosophy'.

[11] *Posterior Analytics* 76a23–25. In his criticism of the mathematical cosmology of Plato's *Timaeus*, Aristotle insists that it is physics rather than mathematics which is appropriate to investigate the nature and motions of sublunary bodies: see the detailed analysis in Cleary, *Aristotle and Mathematics*, 71–142.

[12] For this neglected centre of astronomical debate, see Forcada, 'Astrology in Al-Andalus during the 11th and 12th Centuries'. On the question of how realist and instrumentalist approaches to astronomy were understood within the Toledan school, see Forcada, 'Saphaeae and Hay'āt'.

[13] A similar debate later developed concerning how to interpret Copernicus, with Andreas Osiander offering an instrumentalist interpretation, possibly to fend off theological critics of the new approach: Roelants, 'The Physical Status of Astronomical Models before the 1570s'. For the importance of the distinction between instrumentalism and realism for astronomy in the early modern period, see Freudenthal, '"Instrumentalism" and "Realism" as Categories in the History of Astronomy'; Barker and Goldstein, 'Realism and Instrumentalism in Sixteenth Century Astronomy'.

[14] A point stressed in Howell, 'The Hermeneutics of Nature and Scripture in Early Modern Science and Theology'.

[15] A point stressed by Harrison, 'The "Book of Nature" and Early Modern Science'. See also Kelter, 'Reading the Book of God as the Book of Nature'.

Copernicus and Tycho Brahe

Ptolemy's account of the relation of the planets, earth, sun, and moon secured wide acceptance for nearly a thousand years. Yet not all were persuaded. A treatise of Martianus Capella (fl. *c*.475), which attracted little attention at the time of its composition, began to be studied seriously with the emergence of the great cathedral schools of learning in the ninth century.[16] Capella argued that Ptolemy was correct in asserting that Mars, Jupiter, and Saturn orbited the earth; the observational data suggested, however, that Mercury and Venus orbited the sun. Capella's work was known to the Polish astronomer Nicolaus Copernicus, and appears to have played a significant role in the position Copernicus set out in his *De Revolutionibus orbium coelestium* ('On the Revolutions of Heavenly Orbs', 1543). By then, scholars such as Johannes Regiomontanus had raised significant concerns relating to many core assumptions of medieval astronomy, creating an environment that was open to reconsideration of traditional approaches.[17]

Although Copernicus himself carried out some observational work, his main contribution to the development of natural philosophy lay in re-evaluating existing data from a different perspective.[18] Copernicus's earliest discussion of planetary theory, known as the *Commentariolus*, probably dates from 1514. In this work, Copernicus noted the views of Capella, and appears to have been stimulated to extend this earlier writer's approach.[19] Bernard Goldstein's careful reconstruction of Copernicus's evolving heliocentrism is based on three core assumptions, all of which are stated clearly in the *Commentariolus*:

1. The sun is stationary, and the earth moves around it (in other words, the earth is recategorized as a planet);

[16] For the best study, see Eastwood, 'Astronomical Images and Planetary Theory in Carolingian Studies of Martianus Capella'.

[17] Swerdlow, 'Regiomontanus on the Critical Problems of Astronomy'.

[18] The best study of Copernicus, which is attentive to the cultural context of the time, is Westman, *The Copernican Question*. Westman rightly points out that astronomy and astrology are best seen as constituting a compound category in the sixteenth century, which he terms 'the science of the stars'.

[19] I here follow the argument of Goldstein, 'Copernicus and the Origin of His Heliocentric System', 220–3. For the background to Copernicus's reflections, see Finocchiaro, *Defending Copernicus and Galileo*; Goddu, *Copernicus and the Aristotelian Tradition*; Omodeo, *Copernicus in the Cultural Debates of the Renaissance*. The evidence indicates that the *Commentariolus* was circulated to colleagues, but never published during Copernicus's lifetime.

2. The sun is the centre of motion for all six planets (i.e., the five original planets, plus the earth itself);[20]
3. The moon is not a planet; it is rather a 'fellow traveller' with the Earth around the sun.

Copernicus's mature statement of his theory, set out in his posthumously published work *De Revolutionibus*, envisaged six planets (one of which is the earth) rotating around the sun.[21] The earth, Copernicus asserted, moved in a circular orbit at constant speed, where the other planets moved in eccentrically placed orbits in non-uniform speed. The closer a planet was to the sun, the faster it moved. The earth was the sole exception to this rule; although it had an eccentric orbit, it was nevertheless held to move at a constant speed.

The 'logic of discovery' underlying Copernicus's heliocentric model of the solar system remains unclear.[22] What was the fundamental motivating factor which led him to make this proposal, given that it failed to offer more accurate observational predictions in comparison with the older Ptolemaic model?[23] To put the matter bluntly, the traditional geocentric Ptolemaic cosmology and the 'new-fangled heliocentric Copernican arrangement' predicted 'essentially the same planetary positions'.[24] Copernicus's new theory might have seemed to be more elegant and economical at the qualitative level; closer examination, however, revealed a deeper quantitative complexity that compromised this simplicity.[25]

In November 1572 the Danish astronomer Tycho Brahe observed that a new star had appeared in the constellation of Cassiopeia. Using a sextant and

[20] For those preferring a more accurate statement of Copernicus's views, the central point around which the planets orbit is not the sun itself, but the centre of the earth's orbit, which is slightly displaced from the position of the sun. See the discussion in Swerdlow and Neugebauer, *Mathematical Astronomy in Copernicus's De Revolutionibus*, vol. 1, 159–61.

[21] For a good popular introduction to this work and its significance, see Gingerich, *The Book Nobody Read.*

[22] For a good summary of the issues, see Goddu, *Copernicus and the Aristotelian Tradition*, 325–60. As Goddu rightly notes, the interpretation of Copernicus at this point has been obscured by the rhetorical superimposition of progressive narratives of scientific discovery on his research programme. See further Moss, 'Rhetoric and Science', 429–30. For the concept of a 'logic of discovery', see Maher, 'Prediction, Accommodation, and the Logic of Discovery'; Chauviré, 'Peirce, Popper, Abduction, and the Idea of Logic of Discovery'.

[23] There are significant differences on this point within the research literature. Some suggest that Copernicus saw his theory as offering simply an accommodation of the observational data; others that he used a traditional Aristotelian *regressus* argument. There are also problems in understanding the amalgam of classical and medieval thought that may have shaped Copernicus's views: see especially Knox, 'Ficino and Copernicus'. Ficino's views indicate the close intermingling of what would now be described as 'astronomy' and 'astrology' in the late Renaissance. See Akopyan, *Debating the Stars in the Italian Renaissance.*

[24] Gingerich, 'The Great Martian Catastrophe and How Kepler Fixed It', 52.

[25] Barbour, *The Discovery of Dynamics*, 227–46.

cross-staff as measuring instruments, he was able to determine that the star was located among the 'fixed stars', which Aristotle had declared to be perfect and unchanging.[26] It was an important observation, which raised concerns about the reliability of some core assumptions of Aristotle's natural philosophy. Yet for our purposes, the importance of Tycho's increasingly precise observational instrumentation relates to his determination of the daily positions of planets.[27] Tycho's observations were made using instruments such as his 'equatorial armillary', completed in 1585—perhaps the last such instrument constructed which did not make use of the optical technology later used in telescopes.[28]

Tycho accumulated a vast repository of accurate determinations of the positions of the planets, particularly Mars. These observations, widely agreed to be the most accurate and comprehensive in pre-telescopic astronomy, needed to be *interpreted* using a model of the planetary system. Noticing increasing disparities between his observations of planetary locations and those predicted by the Copernican model,[29] Tycho proposed an alternative model, according to which the five planets rotated around the sun, which in turn rotated around the earth.[30] While this hybrid model is often interpreted as a conservative attempt to hold on to the traditional Ptolemaic model of the solar system, Tycho's own account suggests that the model was a response to the failure of Copernicus's new system to offer a good fit with the observational data. Given that the competing hypotheses offered by Copernicus and Tycho seemed to be observationally equivalent, those trying to adjudicate between rival theoretical claims often seemed to depend on non-empirical considerations in coming to a decision, such as their apparent metaphysical implications, or their theological foundations or consequences.[31]

Up to this point, Tycho had enjoyed the patronage of the Danish monarchy. However, Tycho fell out of favour with Christian IV, and was forced to leave

[26] For the importance of this event for contemporary natural philosophy, see Methuen, 'This Comet or New Star'; Pumfrey, Stephen, '"Your Astronomers and Ours Differ Exceedingly"'.

[27] For the development of Brahe's observatory at Uraniborg, see Thoren, *The Lord of Uraniborg*, 144–219.

[28] The optical telescope seems to have been first used the Netherlands around 1608, and its astronomical importance partly appreciated. Galileo first used the telescope for systematic astronomical observations in 1609. See van Helden et al., eds., *The Origins of the Telescope*.

[29] Blair, 'Tycho Brahe's Critique of Copernicus and the Copernican System'. For reflections on the issues that led to Kepler's resolution of the difficulties, see Carman and Recio, 'Ptolemaic Planetary Models and Kepler's Laws'.

[30] For a good account of this approach, see Thoren, *The Lord of Uraniborg*, 236–64.

[31] As late as 1670, Robert Hooke was working out how to 'furnish the Learned with an *experimentum crucis* to determine between the Tychonick and the Copernican Hypotheses', based on the observation of stellar parallaxes: Nauenberg, 'Robert Hooke's Seminal Contribution to Orbital Dynamics', 18.

Denmark in 1597. The Habsburg emperor Rudolf II welcomed him to Prague, and helped him to set up an observatory at a manor house at nearby Benatky, situated on high ground, and offering a clear view of the horizon in every direction.[32] There, he was joined by a rising star in the world of astronomy, Johannes Kepler, whose *Mysterium cosmographicum* ('The Mystery of the Universe', 1596) attracted international attention, partly on account of its explicit acceptance of the Copernican theory, but also because of the elegance of its central theme of 'nested spheres' enfolding geometrical solids, such as the cube, tetrahedron, and dodecahedron.[33] Following Tycho's death in 1601, Kepler became Imperial Mathematician in Prague, and gained access to Tycho's observational data, accumulated during his time in Denmark. These observations would be the foundation of his *Rudolphine Tables*, published in 1627, which predicted the positions of all the planets. Yet they also proved to be of critical importance in his philosophical reflections on the structure of the solar system.

Kepler on the Harmonies of Nature

Despite the cultural and religious disruptions caused by the Protestant Reformation, there was continuing interest in the study of the natural world throughout the regions which adopted Protestantism.[34] By the 1560s, the two major constituencies within the emerging Protestant tradition, traditionally referred to as 'Lutheranism' and 'Calvinism', had developed their own distinct theological frameworks which were hospitable towards the natural philosophies of the period. Calvinism achieved such an accommodation through its distinction between the 'natural knowledge of God' and a 'revealed knowledge of God',[35] where Lutheranism relocated the field of natural philosophy within its theological dialectic between 'law' and 'gospel'.[36] Like Albertus Magnus in the thirteenth century, the Lutheran theologian Philip Melanchthon

[32] For this final period of Brahe's life, see Thoren, *The Lord of Uraniborg*, 416–70. For the intellectual atmosphere in Rudolf's court, see Marshall, *The Magic Circle of Rudolf II*.

[33] Thoren, *The Lord of Uraniborg*, 432–42. For a good account of the theological background to Kepler's *Mysterium cosmographicum*, see Barker and Goldstein, 'Theological Foundations of Kepler's Astronomy'; Martens, *Kepler's Philosophy and the New Astronomy*, 39–56. For Kepler's geometric approach to the 'harmony' of the solar system, see Stephenson, *The Music of the Heavens*, 128–40.

[34] Note the points made by Crowther-Heyck, 'Wonderful Secrets of Nature'.

[35] Léchot, 'Calvin et la connaissance naturelle de Dieu'.

[36] Kusukawa, 'The Natural Philosophy of Melanchthon and His Followers'. For a more detailed account of this contextualizing framework of 'law and gospel', see Kusukawa's earlier study, *The Transformation of Natural Philosophy*, 27–74. For Luther's early criticism of Aristotle's philosophical methods, see Dieter, *Der junge Luther und Aristoteles*, 257–75.

made a critically important distinction between the truths of revelation, and those which resulted from natural inquiry. This development created a conceptual space that was hospitable to natural philosophers such as Kepler.[37]

After Tycho's death, Kepler began to engage the problem of planetary motion, aiming to make sense of Tycho's accumulated observations of the position of the planet Mars. Whatever theory he derived would have to agree with the planetary positions that were actually observed, and so carefully tabulated by Tycho. Using a process of fitting by trial and error, which Newton later considered to be 'rife with compromises and unsupported opportunistic assumptions',[38] Kepler came to the conclusion that Mars orbited the sun in an ellipse.[39] While this hypothesis fitted the observational data, it remains unclear how Kepler arrived at this conclusion, in that this does not appear to have been a totally reliable inference from the observational evidence.[40] Kepler could not have established an elliptical orbit on the basis of these observations alone, in that 'the uncertainty in the data is too great to plot out the points geometrically and fit a curve through them.'[41] But Kepler was looking for more than empirical correspondence, thus 'preserving the phenomena'; he was looking for 'plausible physical explanations for his astronomical hypotheses'.[42]

So why did Mars orbit the sun elliptically? In the *Mysterium cosmographicum* Kepler suggested this was due to some efficient force emanating from the sun; in the *Astronomia nova* (1609), this is put down to some *virtus motrix*. This latter idea was based on Kepler's reading of William Gilbert's *De magnete* (1600), which led him to suggest that some form of magnetic force existed

[37] For the importance of Lutheran thinkers at the University of Tübingen in shaping Kepler's theological and philosophical ideas, see Methuen, *Kepler's Tübingen*. For the earlier Lutheran tradition and Copernicanism, see Westman, 'The Melanchthon Circle, Rheticus and the Wittenberg Interpretation of the Copernican Theory'. For Luther's attitudes towards natural philosophy and science in general, see Frank and Rhein, eds., *Melanchthon und die Naturwissenschaften seiner Zeit*; Fink-Jensen, 'Medicine, Natural Philosophy, and the Influence of Melanchthon in Reformation Denmark and Norway'. For the impact of Melanchthon's views on natural philosophy on the Lutheran universities of Königsberg and Helmstedt, see Pozzo, 'Wissenschaft und Reformation', especially 103–5.

[38] Gal and Hodoba Eric, 'Between Kepler and Newton', 244.

[39] For the details, see Wilson, 'Kepler's Derivation of the Elliptical Path'; Baigrie, 'The Justification of Kepler's Ellipse'; Goldstein and Hon, 'Kepler's Move from Orbs to Orbits'; Miller, '*O male factum*', 44–54. For William Whewell's interpretation of Kepler's discovery, see Lugg, 'History, Discovery and Induction'.

[40] As Baigrie points out, Newton was sceptical about the empirical pedigree of Kepler's ellipse, feeling that the application of his methodology ought to have led Kepler to conclude planetary orbits were oval: Baigrie, 'The Justification of Kepler's Ellipse', 636–7. For Kepler's earlier assessment of oval orbits, see Donahue, 'Kepler's Approach to the Oval of 1602 from the Mars Notebook'.

[41] Voelkel, 'Commentary on Ernan McMullin', 324.

[42] Miller, '*O male factum*', 45.

between the planets and the sun, which resulted in some kind of orbital equilibrium.[43]

Kepler also corrected a problem with Copernicus's model of planetary motion, which rested on the assumption, apparently carried over from Ptolemy, that the earth moved in a circular orbit at constant speed. Tycho's observations of the position of the planet Mars were so detailed and accurate that Kepler was able to show, using Tycho's records for observations of Mars that were precisely 687 days (the orbital period of Mars) apart, that the earth also moved in an elliptical orbit.

What are now known as 'Kepler's three laws of planetary motion' may be set out as follows, summarizing the basic patterns that Kepler had identified through close examination of the Tychonian observational records:[44]

1. The planets move in elliptical orbits with the sun located at one of the two focal points.
2. The radius vector from the sun to a given planet sweeps out equal areas in equal times.
3. The square of the period of any planet is proportional to the cube of the semi-major axis of its orbit.

The important point, so easily overlooked, is that these 'laws' are summaries of what Kepler observed, and was able to represent mathematically; they are not themselves *explanations*, but are actually observations that themselves require explanation. Why, for example, is the square of the period of any planet proportional to the cube of the semi-major axis of its orbit?[45]

Kepler himself was not satisfied with inductions from pure observation, believing that there must exist a physical explanation for these general principles, grounded in some deeper ordering within nature which might not be entirely accessible to human reasoning. He offered a plausible causal explanation of planetary motion based on a rectilinearly oriented magnetic force. Although Kepler was somewhat vague about the details, and was clearly

[43] On this, see Bialas, *Johannes Kepler, Astronom und Naturphilosoph*, 92–9. For the wider context of this idea, see Boner, *Kepler's Cosmological Synthesis*.

[44] Kepler himself did not describe these as 'laws'; this way of referring to them appears to have emerged during the late eighteenth century. For the historical development of the notion of 'laws of nature', which emerged later than Kepler, see Harrison, 'The Development of the Concept of Laws of Nature'.

[45] For discussion, see Katsikadelis, 'Derivation of Newton's Law of Motion from Kepler's Laws of Planetary Motion'.

uneasy about the concept of some kind of magnetic force acting at a distance,[46] he nevertheless believed that the observed behaviour of the planets was best explained in terms of forces at work within the physical structure of the universe.[47] In the end, of course, Newton would later offer a better explanation, based on the concepts of inertia and universal gravitation.

Kepler's contribution to the mathematization of natural philosophy is of enormous significance. His success in the mathematization of the solar system encouraged the growing trend to see mathematics as the language of the natural world. 'Although astronomy had always been deemed a mathematical science, few in the early sixteenth century would have envisioned a reduction of physics – that is, of nature as motion and change – to mathematics.'[48] Yet our concern here is not with documenting the details of Kepler's mathematical success, but rather with reflecting on its wider importance, especially in relation to a reframing of the enterprise of natural philosophy. Why does mathematics seem able to describe some fundamental structures of the universe? As we shall see, this question would be developed further by Galileo and others.

Kepler's achievement was to reconceive the world of the planets as part of the natural order. To appreciate this point, we need to step into the bygone imagined world of the Renaissance, bifurcated into two domains by *Luna*, the moon. In his classic study *The Discarded Image*, C. S. Lewis offers an explanation of the significance of this division:

> At Luna we cross in our descent the great frontier which I have so often had to mention; from aether to air, from heaven to nature, from the realm of gods (or angels) to that of daemons, from the realm of necessity to that of contingence, from the incorruptible to the corruptible. Unless this great divide is firmly fixed in our minds, every passage in Donne and Drayton or whom you will that mentions translunary or sublunary will lose its intended force.[49]

As Lewis rightly observes, the 'sublunary world' of Aristotelian natural philosophy constituted 'Nature in the strict sense'. With the erosion of the

[46] See further Granada, '"A quo moventur planetae?"'; Regier, 'Kepler's Theory of Force and His Medical Sources'.

[47] The importance of such a rectilinearly oriented magnetic force in Kepler's thinking is explored in Miller, '*O male factum*', 55–9.

[48] Mahoney, 'The Mathematical Realm of Nature', 702.

[49] Lewis, *The Discarded Image*, 108.

artificial Aristotelian bifurcation of what we now take to be 'nature' as a whole, the scope of natural philosophy was significantly extended and enriched.[50]

A consequence of this development needs to be highlighted. Some popular accounts of the rise of Copernicanism suggest that it was resisted because it was seen to challenge any privileged status for humanity or the earth as the centre of things.[51] This is a somewhat modernist view, which is not sufficiently attentive to the Aristotelian intellectual framework which was so influential in the sixteenth century and beyond. Given the Aristotelian dichotomy between the perfection of the heavens and the imperfection of the sublunar regions, to relocate the earth in the realm of the heavens significantly *enhanced* its status. John Wilkins, an early English advocate of Copernicanism, reported that one of the most common objections he had encountered to the Copernican system was that it elevated human beings above their proper station.[52]

Kepler's Correlation of Theology and Natural Philosophy

Two themes in Kepler's natural philosophy require further discussion, in the light of its place in our overall project of reflection and recovery of natural philosophy. Kepler is an important witness to what some might now consider to be an inappropriate transgression of disciplinary boundaries, but which appeared to Kepler himself as natural and legitimate moves within a wider understanding of knowledge. Every natural philosophy is shaped by a wider framework of understanding; in the case of Kepler, this was framed by a Lutheran understanding of the place of a knowledge of nature, and a Trinitarian understanding of the nature of God.[53] Kepler did not see these as competing, but rather wove them into his overall vision of the purpose and place of natural philosophy, enfolding domains of knowledge which might now seem disconnected, yet were to him aspects of a greater whole.

Kepler's mathematical metaphysics was based on the informing theological belief that God had created the natural world according to certain 'archetypes'

[50] For development of this important point, see Gaukroger, *The Emergence of a Scientific Culture*, 169–95, noting especially the points made at pp. 172–4.

[51] For discussion and criticism, see Danielson, 'Myth #6'.

[52] For this and other examples of positive changes in human self-esteem resulting from a heliocentric worldview, see Brooke, 'Wise Men Nowadays Think Otherwise', 202–4.

[53] For an important reflection on the place of Christian theology in Kepler's analysis, see Hon, 'Kepler's Revolutionary Astronomy'. More generally, see Kozhamthadam, 'The Religious Foundations of Kepler's Science'.

expressed through an essentially geometric order. For Kepler, the human mind, which bears the 'image of God',[54] is shaped in such a manner that sensitizes it to patterns within the created order, and their potential implications. In his *Harmonices Mundi* ('The Harmony of the World', 1619), Kepler argued that, since geometry had its origins in the mind of God, it was only to be expected that the created order would conform to its patterns, and that humanity was endowed with faculties intended to allow these archetypal patterns to be discerned.

> In that geometry is part of the divine mind from the origins of time, even from before the origins of time (for what is there in God that is not also from God?) has provided God with the patterns for the creation of the world, and has been transferred to humanity with the image of God.[55]

This leads to the second point of interest. The reader of Kepler's main works—especially *Harmonices Mundi*—is struck by its pervasive theme of harmony, in which the universe is perceived to possess an archetypal pattern which is 'mathematical in nature and aesthetic in character'.[56] The use of the musical concept of 'harmony' is critically important here, not simply as an *analogy* for cosmic order, but as a fundamental aspect of that ordering, and the capacity of humanity to recognize it. Kepler himself suggested that the movements of the heavens could be understood as an 'everlasting polyphony'—a harmony which, though not audible, was certainly intelligible.[57] The complex interconnections that Kepler recognizes between music, mathematics, aesthetics, and astronomy has significant roots in the classical tradition, being seen as having some deeper rooting in the structures of the cosmos.[58]

For Kepler, the notion of harmony did not merely assist natural philosophy in interpreting the world; it offered a pattern for its relationship which called

[54] For the theological importance of this notion, see Middleton, *The Liberating Image*, 15–90.

[55] Kepler, *Gesammelte Werke*, vol. 6, 233.

[56] Martens, *Kepler's Philosophy and the New Astronomy*, 39. It is interesting to relate this general point to Kepler's short work *De nive sexangula* ('The Six-Cornered Snowflake', 1611), which shows a fascination with the mathematical archetypes discerned within nature. See Corrales Rodrigáñez, 'The Use of Mathematics to Read the Book of Nature'.

[57] For reflections on this theme, see Pesic, *Music and the Making of Modern Science*, 73–88; Bialas, *Johannes Kepler, Astronom und Naturphilosoph*, 120–48; Gingras, 'Johannes Kepler's *Harmonices Mundi*'; Clark, 'The Voice and Early Modern Historiography'; Fend, 'Historical Alternatives to a Poststructuralist Reading of Johannes Kepler's Harmony of the World'. Pythagoras held that the music of the cosmos was audible, a view that was rejected by Aristotle. For the implications of the discovery of Einsteinian gravitational waves for this discussion, see Neves, 'Einstein contra Aristotle'.

[58] See, for example, Radice, 'Ordine musica bellezza in Agostino'; Harrison, *On Music, Sense, Affect, and Voice*. For the wider issues, see Chua, *Absolute Music and the Construction of Meaning*.

out to be put into action—in other words, to make the structures of the social world parallel the harmony of the natural world. The motif of 'harmony' can be shown to inform Kepler's scientific, theological, and political thinking at a time of social upheaval and communal tensions.[59] This point is of wider importance, as Kepler himself appreciated. As Daniel Chua has pointed out in an important study of the musician Vincenzo Galilei (the father of Galileo Galilei), musical harmony was seen in this cultural context as holding everything together within the 'great chain of being', so that music 'was not simply an object in the natural world, but the rational agent of enchantment itself'.[60] Kepler's reflections on the 'harmony of the world' reflect a perception of the interconnections of astronomy, music, and mathematics that reaches back to the Hellenic era,[61] and was consolidated through the inclusion of these fields in the medieval university *quadrivium*.[62] Kepler clearly saw an association—whether intuited or culturally acquired—between these multiple disciplines, and incorporated these into his natural philosophy.

Kepler's contributions to natural philosophy chiefly take the form of reflections on Tycho Brahe's astronomical observations, undertaken with the naked eye. Although Kepler made significant contributions to the understanding of optics,[63] he did not use telescopes to augment natural human vision, or reflect on the issues of interpretation that arose from this technological extension of the range of the human eye. In turning to consider the contributions of Galileo Galilei, we are confronted with a question that has become of increasing importance in modern philosophy of science: the question of the observability of nature.[64]

[59] For a full account of this theme in Kepler's thought, see Rothman, *The Pursuit of Harmony*, especially 257–82. For the minority view that the language of cosmic harmony was a mere literary device favoured by poets, see Hollander, *The Untuning of the Sky*. Hollander here focuses on English poets, rather than works of natural philosophy. For Hollander's comments on Kepler, see *The Untuning of the Sky*, 38–40.

[60] Chua, 'Vincenzo Galilei, Modernity and the Division of Nature', 22. See further Chua's remarkable study of the ongoing philosophical significance of music: Chua, *Absolute Music and the Construction of Meaning*.

[61] For a thorough analysis, see Pesic, *Music and the Making of Modern Science*, especially 9–20.

[62] Heilmann, *Boethius' Musiktheorie und das Quadrivium*, 23–151.

[63] See, for example, Cardona, 'Kepler: Analogies in the Search for the Law of Refraction'; Chen-Morris, *Measuring Shadows*; Dupré, 'Kepler's Optics without Hypotheses'.

[64] The question of the observability of nature and the implications of the technological enhancement of human vision is particularly associated with Bas van Fraassen: see Leeds, 'Constructive Empiricism'.

Galileo: Natural Philosophy and the Extension of Human Vision

While Galileo made significant contributions to the development of the Copernican theory, and is regularly cited in discussions about the relation of science and religion,[65] his significance for the purposes of this present study lies in his use of the telescope to study the heavens. This optical instrument extended the range of human vision beyond its natural limits, allowing access to a greater horizon of entities and events.[66] The invention of the telescope is generally dated to 1608. It is known that several Dutch telescopes had been distributed in Italy between May and August 1609.

A year later, Galileo constructed his own telescope, and used it to study the heavens.[67] It was a small instrument by today's standards.[68] Galileo may not have been the first to use such an instrument to observe the heavens; he was, however, able to interpret his observations in ways that significantly advanced discussion of important issues in natural philosophy. Galileo was a good artist, and was able to draw the patterns of light and shade that he observed on the face of the moon,[69] as well as note the dense starfields of the Milky Way, and the changing positions of the four faint starlike objects he observed close to the planet Jupiter between December 1609 and January 1610.

Each of these three sets of observations deserves closer attention. The first set related to the moon. Galileo's telescope disclosed that the face of the moon was not perfectly smooth, thus calling into question the Aristotelian belief that the celestial bodies, such as the moon, were perfect. Although most accounts of this discovery report that Galileo 'observed mountains on the moon',[70] this simple statement needs to be nuanced significantly. What Galileo

[65] The best contextualizing study is Ponzio, 'Perera, Bellarmino, Galileo e il "Concordismo" tra Sacre Scritture e ricerca scientifica'. See also McMullin, 'Galileo's Theological Venture'; Blackwell, *Galileo, Bellarmine and the Bible.*

[66] The new horizons opened up by both telescope and microscope caused much discussion as to whether such enhanced vision was prone to distortion or might be an instrumental artefact. For Robert Hooke, these were 'artificial organs' that strengthened and improved upon natural human capacities for perception: Chapman, '"Micrographia" on the Moon'. Milton's use of the 'telescope' motif in *Paradise Lost* suggests it came to have unhelpful associations with Satan, made worse by 'morally and epistemologically ambiguous "spots"': see Brady, 'Galileo in Action'.

[67] For the background to this event, and reflections on the technological developments that led to it, see Reeves, *Galileo's Glassworks*; Bucciantini et al., *Galileo's Telescope*.

[68] The instrument's objective lens was 38 cm in diameter, and is now known to have had a capacity to magnify by a factor of 21 (Galileo suggested it was 30). For details of Galileo's first telescope, see Strano, 'Galileo's Telescope'.

[69] For a detailed illustrated account, see Bredekamp, *Galilei der Künstler*, 149–76.

[70] See, for example, North, *Mastering Astronomy*, 101.

actually *observed* were two dimensional patterns of light and shade; he *interpreted* these as lunar mountains and valleys. Galileo used a sophisticated analogical argument, embodying Renaissance theories of perspective,[71] which allowed him to conclude that these changing patterns of light and dark on the surface of the moon were evidence for the existence of mountains and valleys, analogous to those found on earth.[72]

The second set of observations related to the Milky Way and Ptolemaic nebulae—such as the fuzzy patch of light in the constellation of Cancer, now known as 'Messier 44'. In every case, objects that seemed 'nebulous' turned out to consist of individual stars, which could be resolved individually through the increased light-gathering power of Galileo's telescope. Galileo rightly concluded that certain objects that appeared to be misty patches of light were actually made up of individual stars; yet he seems to have wrongly believed that all such patches of nebulae could be resolved in this way. As is well known, the eighteenth-century British astronomer John Herschel changed his mind on the idea of 'nebulosity', initially believing that such patches of light were simply 'clusters of stars in disguise', and would be resolved into individual stars by sufficiently powerful telescopes, before realizing that some such objects could not be resolved in this way.[73]

The third set of observations concerned the planet Jupiter which appeared as a disc rather than as a point of light when seen through a telescope. What attracted Galileo's attention, however, were four stars arranged in a straight line close to the planet Jupiter, which changed their positions nightly. What Galileo *observed* were four starlike objects whose positions changed over time with respect to Jupiter; Galileo then *interpreted* these four objects as moons that were orbiting the planet.[74]

At that time, debates about the solar system focused on two rival models: Kepler's view that the earth and other planets orbited the sun, initially set out in his *Mysterium cosmographicum* (1596) and refined in *Harmonices Mundi* (1619), and Tycho Brahe's modified geocentric cosmology, set out in his 1588 work *De mundi aetherei recentioribus phaenomeni* ('Concerning Recent Phenomena in the Celestial World', 1588). Once he had concluded that these starlike objects were best understood as moons orbiting Jupiter, Galileo

[71] Edgerton, *The Mirror, the Window and the Telescope*, 151–68.

[72] For an excellent account of the analogical argument that led Galileo to conclude that he was actually observing mountains on the surface of the moon, and not simply shifting patterns of light and darkness, see Spranzi, 'Galileo and the Mountains of the Moon'. See also Hamou, *La mutation du visible*, 63–6; Shea, 'Looking at the Moon as Another Earth'.

[73] For a detailed discussion of Herschel's changing views, see Hoskin, *The Construction of the Heavens*, 34–72.

[74] Gingerich and Van Helden, 'How Galileo Constructed the Moons of Jupiter'.

realized that this interpretation had important implications for the evaluation of Kepler's model in relation to Tycho's. He therefore published his findings, with appropriate illustrations, in his pamphlet *Siderius Nuncius* ('Starry Messenger') in March 1609. Galileo named these four moons 'the Medicean Stars' in honour of his former student Cosimo of Medici; in the end, however, the four names proposed by Simon Marius in 1614—Europa, Io, Ganymede and Callisto—found wider acceptance.[75]

Galileo's discovery subverted a classic objection to Kepler's heliocentric model: the uniqueness of the earth in possessing a moon. In the Ptolemaic and Tychonian models, the moon was only one of a number of bodies to orbit the earth; according to Kepler, the moon was the *only* object to orbit the earth. This exceptionalism was seen by some critics as indicating that Kepler's system was implausible. Galileo's demonstration that Jupiter also possessed satellites indicated that the earth was not alone in this respect. Other observations confirming Kepler's model followed after the publication of *Siderius Nuncius*. In December 1610, Galileo observed the changing phases of the planet Venus, very similar to those of the moon. For Galileo, these phases only made sense within the context of a Copernican model of the solar system, in which Venus orbited the sun.

Galileo's application of the telescope to extend discussion of the themes of natural philosophy was paralleled several decades later by the use of the microscope to disclose a hitherto unobserved biological world. This led the English natural philosopher John Wilkins to remark that ''tis certain that our senses are extreamly disproportioned for comprehending the whole compasse and latitude of things.'[76] Other natural philosophers of this age, such as Margaret Cavendish, were, however, apprehensive of substituting artificial technology for natural human vision.[77] The point remains important today; the philosopher of science Bas van Fraassen continues to express concerns about transgression of the limits of human vision, noting its implications for a realist account of the natural world.[78]

[75] The possibility that Marius observed these moons before Galileo continues to intrigue scholars: see Pasachoff, 'Simon Marius's *Mundus Iovialis* and the Discovery of the Moons of Jupiter'.

[76] Wilkins, *Mathematicall Magick*, 115–16. For the significance of this work for natural philosophy, see van Dyck and Vermeir, 'Varieties of Wonder'.

[77] Keller, 'Producing Petty Gods'; Lawson, 'Bears in Eden'.

[78] See the analysis in Dicken and Lipton, 'What Can Bas Believe?'

Galileo and the Language of Mathematics

Yet Galileo is of importance to the development of natural philosophy in another way—the suggestion that mathematics is the language of nature.[79] This point is made clearly in his defence of natural philosophy over and against humanly constructed philosophies. In arguing that natural philosophy is grounded in the natural order, rather than being a human invention, Galileo turns to consider the language in which this natural philosophy is written.

> [Natural philosophy] is written in this grand book – I mean the universe – which stands continually open to our gaze, but it cannot be understood unless one first learns to comprehend the language and interpret the characters in which it is written. It is written in the language of mathematics, and its characters are triangles, circles, and other geometrical figures, without which it is humanly impossible to comprehend a single word of it.

Galileo here draws on the Renaissance metaphor of 'God's Two Books'—nature and the Christian Bible—which share the same author, but take different forms.[80] To read either requires knowledge of the languages in which it is written, and the appropriate hermeneutical tools to make sense of it.[81] The Bible was written in Hebrew and Greek; the natural order is written in the 'language of mathematics'.

Galileo's statement that the book of nature is written in mathematical language is open to several interpretations. It might, for example, be construed as an affirmation of the intellectual autonomy of natural philosophy,[82] or perhaps of the accountability of the natural philosopher to the objectivities under

[79] For a thorough discussion, see Palmerino, 'The Mathematical Characters of Galileo's Book of Nature'; Palmerino, 'Reading the Book of Nature'; Peterson, *Galileo's Muse.* For Galileo's use of the image of the 'language of nature' (to be distinguished from the 'language of Scripture'), see Stabile, 'Linguaggio della natura e linguaggio della Scrittura in Galilei'. Note also the discussion in Bagioli, 'Stress in the Book of Nature'. The older study of Stillman Drake is still valuable: Drake, 'Galileo's Language'.

[80] Tanzella-Nitti, 'The Two Books Prior to the Scientific Revolution'; Harrison, 'The "Book of Nature" and Early Modern Science'.

[81] For Galileo's familiarity with theories of biblical interpretation, including the 'accommodationist' approach of Paolo Antonio Foscarini, see Carroll, 'Galileo and the Interpretation of the Bible'; Bieri and Masciadri, *Der Streit um das Kopernikanische Weltsystem im 17. Jahrhundert*, with particular reference to Galileo's use of the notion of accommodation. For the development of the notion of God's self-accommodation to human weakness in divine self-disclosure, see Benin, *The Footprints of God*; Huijgen, *Divine Accommodation in John Calvin's Theology*, 47–105.

[82] Finocchiaro, *Defending Copernicus and Galileo*, 115–16.

consideration. Yet Galileo's approach is more naturally to be interpreted as a simple statement of mathematical realism, holding that 'mathematical entities are ontologically independent from us and that the physical world has a mathematical structure.'[83] Yet the implications are clear: to understand nature, one must master its language.

Natural philosophy is thus a human response to the 'book of nature' which is more trustworthy and secure than other philosophical or theological approaches.[84] The problem is that *lex divina*, being disclosed through a text, requires interpretation—and it is often difficult to secure closure on such interpretative disputes. But *lex natura* transcends such disputes; there is only one way of interpreting nature, so that the matter can be resolved.[85] Having already insisted that natural philosophy is something discovered or discerned, not invented, Galileo then makes an interpretative move which further privileges the natural philosopher. The 'Book of Scripture', Galileo suggests, is accessible to everyone on account of divine self-accommodation in revelation; yet this leads to interpretative pluralism and uncertainty.[86] The 'Book of Nature' is only accessible and intelligible to those who are mathematically competent, in that it offers no accommodation to those who are unwilling or unable to learn its mathematical language. If nature is to disclose its wisdom, there is a significant restrictive entry barrier.[87]

The importance of this point is best appreciated by considering Francis Bacon's critique of the specialist trade guilds and the universities, both of which he considered to have erected inappropriate barriers to the wider social dissemination of knowledge and wisdom. Bacon was particularly critical of 'the exclusivity both of the guilds, where practical information is esoteric by virtue of keeping knowledge or techniques within a trade or profession to which access is then restricted, and of the universities, where an esoteric and often convoluted language renders information inaccessible to all but those

[83] Palmerino, 'Reading the Book of Nature', 32.

[84] This point is stressed by Stabile, 'Linguaggio della natura e linguaggio della Scrittura in Galilei', 55–6.

[85] Palmerino, 'Reading the Book of Nature', 33–4.

[86] Carroll, 'Galileo and the Interpretation of the Bible'. Galileo was well aware of the divergent interpretations of key biblical passages offered by Protestant and Catholic biblical interpreters at this time, and their wider social implications. For the hermeneutical issues, see McGrath, *The Intellectual Origins of the European Reformation*, 119–80, especially 148–66; Klepper, 'Theories of Interpretation'; for the impact of rival theories of biblical authority, see the material in Holcomb, ed., *Christian Theologies of Scripture*.

[87] A similar point was later made in relation to both Newton's *Principia* and J. S. Bach's harmonics, in that both required a significant degree of competence on the part of their audiences: Wolff, 'Bach's Music and Newtonian Science'.

accepted into the university system.'[88] Galileo's vigorous defence of the necessity of mathematical literacy on the part of natural philosophers pointed towards natural philosophy becoming the preserve of an educated social elite.

The significance of this point is perhaps best appreciated by reading a popular work of natural theology, which was widely copied and circulated in the Iberian peninsula during the sixteenth century. In his *Liber naturae sive creaturarum* ('Book of Nature and the Creatures') the Catalan writer Raymond de Sebonde (*c*.1385–1436) spoke of the importance of the accessibility of knowledge of God through contemplation of the natural world for those who were unable to read, or did not have access to the Bible or theological works.[89] While Galileo did not deny that most could appreciate the beauty and wonder of the natural world, the task of understanding it properly was restricted to those with the necessary mathematical tools.

We now turn to consider the experimental natural philosophies that emerged in England during the earlier modern period, focusing on Francis Bacon, Robert Boyle, and Isaac Newton.

[88] Gaukroger, *Francis Bacon and the Transformation of Early-Modern Philosophy*, 9.

[89] Various spellings of this name are found in the literature, including the Catalan form 'Ramon Sibiuda'. See further Puig, *La filosofia de Ramon Sibiuda*; McGrath, *Re-Imagining Nature*, 13–16.

4
English Natural Philosophy
Bacon, Boyle, and Newton

The seventeenth century is widely regarded as marking a 'Golden Age' of natural philosophy in England. Its leading advocates and practitioners—particularly Francis Bacon, Robert Boyle, and Isaac Newton—propelled the field to new heights, generating considerable interest both in the study of nature itself, and how this could be represented theoretically. While this troubled century in English history witnessed growing religious tensions, as well as the rise of forms of atheism—some tentative and reluctant, others confident and assertive[1]—the dominant sentiment of this age was belief in a God who created an ordered universe which the human mind could grasp, however imperfectly. Natural philosophy came to be seen as enfolding both the discovery of the rational order of the natural world, and allowing this to act as a gateway to 'a community of values that was not dependent upon any particular religious confession.'[2] Yet such an understanding of the natural world was not seen as an end in itself, but as part of a process of intellectual, moral, and spiritual refinement of the natural philosopher. Like their counterparts in ancient Greece, many English natural philosophers saw a clear link between acquiring a theoretical understanding of the natural world, and improving the moral and spiritual quality of human life.[3]

Political stability began to return to England after the extended period of the Civil War and Puritan Commonwealth (1642–60). The restoration of the monarchy was seen by many as a sign of a new era of progress and intellectual development. One of Charles II's first acts was to grant a royal charter to the

[1] Ryrie, *Unbelievers*, 127–37.

[2] Haakonssen, 'Early Modern Natural Law Theories,' 76. Haakonssen here focuses on legal values; the same point was often made in relation to moral and social values at the time.

[3] There are parallels here with the 'spiritual alchemy' that emerged in England in the early modern period, which saw alchemy not primarily as a transmutation of the external world, but as a transformation of human perceptions, attitudes, and behaviour: see Schuler, 'Some Spiritual Alchemies of Seventeenth-Century England'; Merkur, 'The Study of Spiritual Alchemy'; Morrisson, *Modern Alchemy*, 135–83. This interest in a 'spiritual alchemy' is especially evident in George Herbert's poem 'The Elixir': see McGrath, 'The Famous Stone.'

Royal Society of London, which quickly established itself as the institutional embodiment of the new natural philosophy, and made frequent statements to the effect that it had adopted Baconianism as its working method.[4] A meeting on 28 November 1660 at Gresham College, London set out 'a design of founding a college for the promotion of physico-mathematical experimental learning'.[5] Although the new Society wanted to establish its own dedicated premises, accommodation difficulties resulting partly from the Great Fire of London (1666) led it to continue its association with Gresham College until the early eighteenth century.

So what was this 'physico-mathematical experimental learning' that the Royal Society adopted, and which was seen by its early apologists as defining its intellectual and cultural mission?[6] To find out, we shall consider the approach to natural philosophy set out by Francis Bacon, especially in his *Novum Organum* (1620).

Francis Bacon: Natural Philosophy and the Improvement of Humanity

To the best of our knowledge, the phrase 'experimental science' was first used in English in 1570 to translate the Latin term *scientia experimentalis*, particularly as this phrase was found in the writings of Nicolas of Cusa.[7] While many earlier studies of Bacon's natural philosophy detached his philosophical method from his wider cultural concerns,[8] there is a growing appreciation that Bacon's natural philosophy has both theological and political aspects. Bacon distinguishes his 'philosophy of nature' from that of the 'Grecians', which he considers to largely consist of debates about words, and that of the

[4] Gaukroger, *Francis Bacon and the Transformation of Early-Modern Philosophy*, 2–3. Note the considered judgement of Lynch, 'A Society of Baconians?', 173–4: 'Bacon's influence was the primary touchstone for the early Royal Society.'

[5] For these developments, see Hall, *Promoting Experimental Learning*, 24–65. Gresham College was founded in the City of London through the estate of Thomas Gresham in 1597, establishing professorships of Law, Physics, Music, Divinity, Geometry, Rhetoric, and Astronomy.

[6] See further Hunter, 'Robert Boyle and the Early Royal Society'; Anstey, 'Philosophy of Experiment in Early Modern England'; Anstey and Vanzo, 'The Origins of Early Modern Experimental Philosophy'.

[7] See Nagel, 'Scientia Experimentalis'. The English work in question is John Dee's 'Mathematicall Præface' to the writings of the geometer Euclid. For some cautionary remarks about this phrase, see Anstey and Vanzo, 'The Origins of Early Modern Experimental Philosophy', 507–8.

[8] A good example is found in William Whewell's excellent, though historically decontextualized, account of Bacon's empirical methods: Whewell, *Philosophy of the Inductive Sciences*, vol. 2, 226–51. For discussion, see Snyder, 'Renovating the "Novum Organum"'. For Whewell's own views on induction, see Dethier, 'William Whewell's Semantic Account of Induction'.

'Alchemists', which he considers to rest on 'imposture', 'auricular traditions and obscurity'.[9]

While it is difficult to capture the essence of Bacon's distinctive approach, Stephen Gaukroger helpfully identifies one of its central themes: a concern with 'the understanding of and reshaping of natural processes'.[10] To understand nature is to be able to redirect nature for the improvement of the human condition. The idea that an attentiveness to the natural world could be a 'propaedeutic to self-betterment' is encountered in many writers of the late fifteenth and sixteenth centuries.[11] While Bacon would not dissent from this judgement, his emphasis lay on 'social-betterment', achieved through an understanding of the natural world that led into the use of nature for the advancement of human ends.[12] This aspect of Bacon's natural philosophy has led some to interpret him as an apologist for the exercise of human power and dominion over nature.[13]

Bacon considered natural philosophy to be a means to an important end. Knowledge of nature was not something 'barren', but had the potential to produce outcomes that were of advantage to humanity. This goal was grounded on an experimental, natural-historical, and broadly inductive approach to the natural sciences, coupled with the institutionalization of scientific experimentation and knowledge production in a quest for objectivity. An important distinction should be drawn here between two forms of natural philosophy, 'speculative' and 'experimental', that are encountered at this time. Peter Anstey explains this distinction as follows:

> Speculative natural philosophy is the development of explanations of natural phenomena without prior recourse to systematic observation and experiment. By contrast, experimental natural philosophy involves the collection and ordering of observations and experimental reports with a view to the development of explanations of natural phenomena based on these observations and experiments.[14]

[9] Bacon, *Works*, vol. 8, 123–4. Bacon's attitude to alchemy is actually somewhat more complex than this dismissive reference might suggest: see Joly, 'Francis Bacon, the Reformer of Alchemy'.

[10] Gaukroger, *Francis Bacon and the Transformation of Early-Modern Philosophy*, 5.

[11] See the discussion in Lancaster, 'Natural Knowledge as a Propaedeutic to Self-Betterment', especially 185–7.

[12] For the relation of Bacon's natural philosophy and action—as opposed to contemplation—see Pérez-Ramos, *Francis Bacon's Idea of Science and the Maker's Knowledge Tradition*, 141–5 (Bacon uses the term 'work' rather than 'action').

[13] For example, Rodríguez-García, 'Scientia potestas est'.

[14] Anstey, 'Experimental versus Speculative Natural Philosophy', 215.

Like many English thinkers of his time who took an interest in what would now be understood as science, Bacon held that religion and natural philosophy 'should be kept separate', while at the same time holding that 'the two were complementary to one another'.[15] We thus find Bacon entwining religious and scientific themes at points, even though he clearly regards them as distinct. For example, in his *Novum Organum*, Bacon set his basic principle of working for human improvement against an informing (though not determining) theological context:

> There cannot but follow an improvement in man's estate, and an enlargement of his power over nature. By the Fall, man fell from both his state of innocence and from his dominion over creation. But even in this life both of those losses can be made good; the former by religion and faith, the latter by arts and sciences.[16]

Although Bacon pressed the case for an experimental natural philosophy, like many at the time, he drew on theological accounts of the human situation in offering his solution for its improvement.

There is a clear shift here from the earlier Renaissance idea of natural knowledge as edification, something that was good in itself, to the idea of communally gained natural knowledge as a means of securing and exercising power over nature.[17] For Bacon, this includes the possibility of extending the human lifespan.[18] Bacon uses the language of *instauratio* to speak of this great change in the place and role of humanity within the world.[19]

> The first [ambition] is of those who desire to extend their own power in their native country; which kind is vulgar and degenerate. The second is of those who labour to extend the power of their country and its dominion among men. This certainly has more dignity, though not less covetousness. But if a man endeavour to establish [*instaurare*] and extend the power and

[15] Zagorin, *Francis Bacon*, 44–51; quote at p. 49. A similar position is associated with Robert Boyle: Wojcik, *Robert Boyle and the Limits of Reason*, 121–2. As Funkenstein observes, for many at this time, the boundaries between 'science, philosophy, and theology' were somewhat porous: Funkenstein, *Theology and the Scientific Imagination from the Middle Ages to the Seventeenth Century*, 3.

[16] Bacon, *Works*, vol. 4, 247–8.

[17] A point stressed by Ogilvie, *The Science of Describing*, 110. For a careful account of Bacon's views on human 'dominion over nature', see Gaukroger, *Francis Bacon and the Transformation of Early-Modern Philosophy*, 166–220.

[18] For the importance of this theme of the 'restitution and renovation of things corruptible' in Bacon's natural philosophy, see Giglioni, 'The Hidden Life of Matter'.

[19] For an influential account of this notion, see Webster, *The Great Instauration*.

> dominion of the human race itself over the universe, his ambition (if ambition it may be called) is without doubt both a more wholesome thing and a more noble than the other two. Now the empire of man over things depends wholly on the arts and sciences.[20]

It will be clear that the goals of Bacon's natural philosophy diverge from those of earlier writers in this tradition. But what of the *methods* that Bacon uses in developing this philosophy? For many, Bacon represents a landmark statement of an experimental natural philosophy, which needs to be given careful consideration.

Bacon's Experimental Natural Philosophical Method

The third part of Bacon's *Novum Organum* is devoted to the question of natural philosophical method.[21] Bacon locates his discussion within the cultural context of increasing anxiety about the possibility of securing reliable knowledge. For Bacon, such knowledge is best acquired through 'a *Natural and Experimental History*', characterized by the investigation of nature, and the ordered presentation of what is observed. 'We are not to imagine or suppose, but to *discover*, what nature does or may be made to do.'[22] We cannot predetermine nature's structures or processes, but must discover them through reflective observation and experimentation. Nature must be 'ranged and presented to view in a suitable order', so that an inductive method can be used to interpret these observations.[23]

Although some have treated Bacon as setting out a 'naïve empiricism', which is inattentive to the role of generating hypotheses in advancing knowledge, this judgement seems unfair. Bacon clearly indicates that some degree of speculation or hypothesis-building is necessary in order to move beyond the aggregation of observational particularities into the development of natural philosophy in a fuller sense of the term.[24] For Bacon, it was possible to derive axioms 'from the senses and particulars, rising by a gradual and

[20] Bacon, *Works*, vol. 4, 114. For the possible relation of these 'ambitions' to the rise of colonialism, see Scalercio, 'Dominating Nature and Colonialism'.

[21] For a close analysis of this methodological section of the *Novum Organum*, see Gaukroger, *Francis Bacon and the Transformation of Early-Modern Philosophy*, 132–65.

[22] Bacon, *Works*, vol. 4, 127.

[23] For a discussion of Bacon's concept of eliminative induction and its limits, see Gaukroger, *Francis Bacon and the Transformation of Early-Modern Philosophy*, 148–53.

[24] For this theme in Bacon and Boyle, see Knight, 'Boyle's Baconianism', 15.

unbroken ascent, so that it arrives at the most general axioms last of all'.[25] The Baconian inductive process thus represents an ascent from the particularities of empirical observation to a plausible set of broader generalizations.

So how is reliable human knowledge to be acquired? Bacon's starting point is to recognize how human thinking can be skewed and distorted by inherited patterns of thought. This idea resonated with a theological concern of the period—namely, the belief that original sin contaminated or distorted the human capacity for reasoning, thus leading to unreliable conclusions. While this concern could be restated in non-theological terms, it emerged as a significant issue in the late sixteenth century.[26] Kepler, for example, followed the leading Lutheran theologian Philip Melanchthon in arguing that mathematics and geometry were not compromised in this way by sin, and could both be grasped by the 'natural light' of human reason and used to mirror the structures of the universe.[27]

Bacon claims that four internal impediments to knowledge, which he terms 'Idols of the Mind', can lead to significant misreadings and misunderstandings of nature. Bacon's intention here is to highlight how we can become trapped within a web of false productions of the human imagination, which arise from the 'crooked mirror' of the human mind.[28] 'I call the first, Idols of the Tribe; the second, Idols of the Cave; the third, Idols of the Market Place; and the fourth, the Idols of the Theatre.'[29] Each of these 'Idols' subverts a true understanding of the world by 'besetting' the mind through conventional misreadings of the world, cultural and social pressures, or 'attachment to preconceived ideas'—such as the idea that planetary orbits are necessarily circular. A method of inquiry is necessary that is able to neutralize or overwhelm these natural cognitive deficiencies. For Bacon, 'the proper remedy to be applied for the keeping off and clearing away of idols' is the 'formation of ideas and axioms by true induction'.[30] A natural philosophy thus corrects the failings of human reason, when left to its own devices.

For Bacon, induction is both empirical and rational, involving the observation and interpretation of nature, coupled with an experimental method

[25] Bacon, *Works*, vol. 4, 50.

[26] See Harrison, *The Fall of Man and the Foundations of Science*, especially 186–244.

[27] On Melanchthon's views on natural philosophy, see Kusukawa, *The Transformation of Natural Philosophy*. Kepler would have encountered these ideas at Tübingen, probably through the Lutheran theologian Jacob Heerbrand: Methuen, *Kepler's Tübingen*, 136–7.

[28] Brandt, 'Francis Bacon, Die Idolenlehre'.

[29] Bacon, *Works*, vol. 4, 53. For a good account of these 'Idols', see Gaukroger, *Francis Bacon and the Transformation of Early-Modern Philosophy*, 122–7.

[30] Bacon, *Works*, vol. 4, 54.

devised to resolve points of ambiguity or uncertainty. Bacon's central concern is to be able to go beyond merely noting and documenting the surface appearances of reality, and to penetrate to the reality that lies behind them.[31] After considering the possible methods of investigation of nature, he concluded that the best method is the gathering of observations or facts about nature and proceeding to the detection of patterns within nature. His overall strategy is to begin with experience of the world, and infer certain first principles on this basis, and then inferring from these first principles what experimental methods might be devised to confirm or extend this knowledge. Gaukroger summarizes Bacon's approach as follows:

> [Bacon's] aim is a twofold one, reminiscent of the *regressus* theory that had dominated discussions of method in the sixteenth century: to move inferentially from experience to first principles, and then to move inferentially from these first principles to new experiments.[32]

To reiterate a point made earlier, Baconian induction is not to be seen as a naïve empiricism, the mere gathering of facts, followed by an attempt to develop an orderly presentation or tabulation of these observed particulars.[33] This is unquestionably part of Bacon's method; yet it is only its starting point, not its conclusion. Bacon's fundamental concern is to explain why *these* patterns, and not others, are observed in nature. To answer this question, we need to explore 'the remoter and more hidden parts of Nature' through experimentation to understand the patterns of behaviour or organization that we observe in nature.[34] It is through experiments that we are able to remove 'the mask and veil from natural objects, which are commonly concealed and obscured under the variety of shapes and external appearances'.[35] (This concern to penetrate beneath the *visibilia* of the natural world may well help us understand the importance attached to the invention of the microscope, and its use by Robert Hooke.[36])

[31] For detailed analysis of his approach, see Gaukroger, *Francis Bacon and the Transformation of Early-Modern Philosophy*, 132–65; Giglioni, 'Learning to Read Nature'; Serjeantson, 'Francis Bacon and the "Interpretation of Nature" in the Late Renaissance'.

[32] Gaukroger, *Francis Bacon and the Transformation of Early-Modern Philosophy*, 142.

[33] A concern noted in Sargent, 'Baconian Experimentalism', 311–13.

[34] Bacon, *Works*, vol. 4, 18.

[35] Bacon, *Works*, vol. 4, 257.

[36] On which see Wilson, *The Invisible World*. Wilson's useful analysis fails to do justice to the importance of the interpretation of instrumental observation, a point we noted earlier in relation to Galileo's observations of mountains and valleys on the moon (pp. 57–8): on this point, see Baigrie, 'Catherine Wilson's *The Invisible World*'.

Bacon was thus clear that the accumulation of facts was an important and necessary prelude to their *interpretation*. In his *Novum Organum* Bacon sets out a set of aphorisms contrasting those who merely collect facts and observations and those who generate their ideas without any reference to the external world. Bacon, it must be made clear, respected those who constructed 'histories' (collections of experimental results and accurate observations); his criticism was that these facts required *interpretation*. Insisting that the human mind must generate ideas in response to what is observed in the natural world, rather than invent or inherit such ideas and force nature to conform to them, Bacon suggests that the humble bee might offer a model for a 'true [natural] philosophy'.

> Those who have handled science have been either men of experiment or men of dogmas. The men of experiment (*empirici*) are like the ant, they only collect and use; the reasoners (*rationales*) resemble spiders, who make cobwebs out of their own substance. But the bee takes the middle course: it gathers its material from the flowers of the garden and field, but transforms and digests it by a power of its own (*propria facultate vertit et digerit*). Not unlike this is the true business of philosophy.[37]

Reflection on observations of the natural world thus leads not merely to the generation of theories concerning the world, through the intellectual digestion and assimilation of what is observed, but to the *correction* of views that are discovered to be empirically contestable. For Bacon, an experimental natural philosophy possesses an internal capacity for self-correction without a corresponding parallel in philosophy or theology.

The forms of experimental natural philosophy which developed from Bacon's approach held that observation and experiment, understood in general terms as activities that produced 'facts', had epistemic priority over hypotheses and general principles in the acquisition of knowledge.[38] These 'facts' might arise from direct sensory observation or from the manipulation of observed objects by the use of instruments. This new approach involved an invasive 'penetrating interrogation of nature' through various means, rather than the 'respectful observation' which was characteristic of Aristotelian

[37] Bacon, *Works*, vol. 4, 92–3.

[38] Anstey, 'Philosophy of Experiment in Early Modern England'. For the suggestion that Bacon's experimental methods are modelled on procedures developed by goldsmiths, see Pastorino, 'Weighing Experience', 556–7.

science.[39] Yet these facts might take two different forms: *experimenta lucifera* ('luciferous experiments') aiming to discover (or cast light on) underlying causes of phenomena; and *experimenta fructifera* ('fructiferous experiments'), which may yield practical or economic outcomes, but do not advance an understanding of the natural world.[40] While Bacon's agenda of human dominion over nature led him to see fruitful experiments positively, he considered them to lack the intellectual significance of illuminating experiments. Dominion over nature ultimately required an understanding of nature and its processes, in order that these might be redirected for the public good.

Bacon's legacy is contested, both in terms of how 'Baconism' is to be understood, and what its significance has been, particularly in relation to the emergence of the scientific method. In his recent tendentious account of the development of the natural sciences, Steven Weinberg treats him as something of an irrelevance.[41] Edward O. Wilson, however, sees Bacon as being something of an intellectual colossus, shaping the future development of western thought into the period of the Enlightenment.[42] Bacon's views on the experimental method are clearly underdeveloped; they could, however, be extended and refined. This process is probably best seen in the works of Robert Boyle, widely regarded as the epitome of a 'natural philosopher', who we shall consider in the next section.[43]

Robert Boyle: Interpreter of a Baconian Natural Philosophy

Born in south-east Ireland, Boyle made his way to Oxford in 1655, where he established himself in some rented rooms close to University College.[44] While there are no good reasons for supposing that he ever held a university or college appointment during his time at Oxford, Boyle clearly found it to be an

[39] Pesic, 'Proteus Unbound', 428. Some scholars have expressed concern about Bacon's reference to experimentation as the 'vexation of nature', interpreting this to mean the 'torture' of nature. For a critique of this misreading, see Pesic, 'Proteus Rebound'.

[40] For the distinction, see Bacon, *Works*, vol. 4, 95. See also Ducheyne, 'The Status of Theory and Hypotheses'.

[41] Weinberg, *To Explain the World*, 201–2, 212–14. Weinberg's modernizing narrative shows little interest in exploring the way in which Bacon's work laid the foundations for Robert Boyle's scientific projects.

[42] Wilson, *Consilience*, 23–8.

[43] For Boyle's strong continuity with Bacon's approach, see Hunter, 'Robert Boyle and the Early Royal Society'; Anstey, 'Philosophy of Experiment in Early Modern England'; Knight, 'Boyle's Baconianism'.

[44] The best biography is now Hunter, *Boyle: Between God and Science*, which takes full account of the changing scholarly understanding of this period, including the relationship between chemistry and alchemy, for a proper understanding of Boyle's intellectual development and significance.

excellent base for developing his natural philosophy, of which chemistry was an important part.[45]

Although Boyle appears to have been well-read in the Greek and Latin classics, he was strongly critical of 'speculative' approaches to natural philosophy that he found in Aristotle and the classical tradition.[46] Boyle is generally agreed to have played an important role in the gradual displacement of Aristotelianism by a mechanical natural philosophy, in which natural phenomena were explained on the basis of the motion of inert particles of matter.[47]

Boyle argued for an approach which aimed to penetrate beneath the surface of the natural world through careful and detailed observational analysis rather than being constrained by preconceived interpretations of nature, such as that of Aristotle.[48] Observation of nature leads to intentional experimentation upon nature to test (Boyle regularly uses the term 'prove' in the sense of 'put to the test') the reliability of certain hypotheses. Boyle makes this point in commending the *virtuosi* who

> consult Experience both frequently and heedfully; and, not content with the *Phaenomena* that Nature spontaneously affords them, they are solicitous, when they find it needful, to enlarge their Experience by Tryals purposely devis'd; and ever and anon Reflecting upon it, they are careful to Conform their Opinions to it; or, if there be just cause, Reform their Opinions by it. So that our *Virtuosi* have a peculiar Right to the distinguishing Title that is often given them, of *Experimental Philosophers*.[49]

Following Bacon, Boyle stressed the importance of accumulating observational and experimental data, seeing the harvesting of these facts not as an

[45] For discussion, see Hunter, 'Robert Boyle and the Early Royal Society'; MacIntosh, 'Boyle's Epistemology'; Sargent, *The Diffident Naturalist*; Wojcik, *Robert Boyle and the Limits of Reason*. For Boyle's complex relationship with alchemy, see Principe, *The Aspiring Adept*. For the development of Boyle's understanding of chemistry, see Clericuzio, 'Boyle's Chemistry'.

[46] For comment, see Levitin, 'The Experimentalist as Humanist', 152–9. See also Anstey and Vanzo, 'The Origins of Early Modern Experimental Philosophy'.

[47] For Boyle's views on this topic and their reception, see Cook, 'Divine Artifice and Natural Mechanism'. Many at the time hoped that the development of microscopes would allow these 'corpuscules' of matter to become visible objects of study. For this question in Locke's account of the visibility of corpuscular particles of matter, see Downing, 'Are Corpuscles Unobservable in Principle for Locke?'

[48] Boyle's later works are severely and perhaps unfairly critical of Aristotle: for some suggestion of what lies behind this animus, see Levitin, 'The Experimentalist as Humanist', 172–81.

[49] Boyle, *Works*, vol. 11, 292.

end in itself, but in order to generate understanding through the proposal and subsequent testing of hypotheses and theories.[50]

> It was not my chief Design to establish Theories and Principles, but to devise Experiments, and to enrich the History of Nature with Observations faithfully made and deliver'd; that by these, and the like Contributions made by others, men may in time be furnish'd with a sufficient stock of Experiments, to ground *Hypotheses* and *Theories* on.[51]

Boyle's forty-three air-pump experiments of 1660 indicate the scope and value of this experimental method, particularly in relation to clarifying aspects of the respiratory process.[52]

In an early work entitled 'Of the Study of the Booke of Nature', Boyle spoke of his decision to supplement this study by engaging two other 'Books'—the Bible, and Conscience.[53] Boyle does not appear to have pursued this idea of 'Three Books'; like many writers of this age, Boyle settled on the established Renaissance metaphor of 'God's Two Books' to establish a conceptual and semiotic link between his natural philosophy and the world of religion. Although the origins of this metaphor can be traced back to the early Middle Ages, it became increasingly important during the sixteenth and seventeenth centuries.

The notion of a 'Book of God's Works', which expands or enacts the 'Book of God's Words', offered an intellectual imaginary of particular significance in the early modern period, that held together two distinct understandings of the natural world, with the potential for convergence on the one hand, or divergence on the other—the world as God's creation, and the world as an object of scientific study. As Thomas Burnet pointed out, truths arising from these divergent disciplines could at least be held together, if not woven together: 'We are not to suppose that any truth concerning the natural world can be an enemy to religion.'[54] Sir Thomas Browne, an important representative of early modern natural philosophy in the generation before Boyle,[55]

[50] Ben-Chaim, 'The Value of Facts in Boyle's Experimental Philosophy'.

[51] Boyle, *Works*, vol. 3, 12.

[52] For the experiments, see West, 'Robert Boyle's Landmark Book of 1660 with the First Experiments on Rarified Air'. For discussion of their significance, see Gaukroger, *The Emergence of a Scientific Culture*, 368–79.

[53] Boyle, *Works*, vol. 13, 147.

[54] Thomas Burnet, *Sacred Theory of the Earth* (1684); cited Gaukroger, 'The Challenges of Empirical Understanding in Early Modern Theology', 566.

[55] For Browne's contributions, see Preston, *Thomas Browne and the Writing of Early Modern Science*; Hughes, 'The Medical Education of Sir Thomas Browne'.

considered this image to be foundational in holding together a coherent philosophical and religious account of the world.

> Thus there are two books from whence I collect my Divinity; besides that written one of God, another of his servant Nature, that universal and public Manuscript, that lies expans'd unto the eyes of all; those that never saw him in the one, have discovered him in the other.[56]

Boyle was one of many to deploy this metaphor. 'The World is the great Book, not so much of Nature, as of the God of Nature, which we should find ev'n crowded with instructive Lessons, if we had but the Skill, and would take the Pains, to extract and pick them out.'[57] In an early reflection, Boyle remarked that he would spend his Sundays studying the 'Booke of the Creatures', and occasionally 'trying those Experiments that may improve my Acquaintance with her'.[58] Yet Boyle is clear that these Books are different, requiring different modes of 'reading', and leading to different outcomes, even if these were capable of correlation or even integration. Both natural philosophy and theology speak about God and the world; but they use different means of representation in doing so, and hence demand different interpretative approaches.[59] Boyle thus advocated a 'physico-theological' reading of nature, seeing this as a spiritual exercise which moved away from a theoretical detachment from nature towards the cultivation of a reverential attitude towards the natural order, on account of what it represents.[60]

Boyle on the Purposes of Natural Philosophy

So what benefits did natural philosophy, as Boyle understood it, bring to those *Virtuosi* who practised it?[61] England was severely shaken by the Civil War and its aftermath in the middle of the seventeenth century. It became clear that Christians might read the same Bible, yet draw different, and often radically divergent, theological and political conclusions from their reading—without being able to agree upon any ultimate authority by which such

[56] Browne, *Religio Medici*, 17.

[57] Boyle, *Works*, vol. 5, 39.

[58] Hunter, *Robert Boyle (1627–91)*, 29. Hunter's detailed account of 'how Boyle became a scientist' merits close study: Hunter, *Robert Boyle (1627–91)*, 15–57.

[59] Osler, 'Mixing Metaphors', 102–3.

[60] Corneanu, *Regimens of the Mind*, 114–41.

[61] This term is used in the 1690 work *The Christian Virtuoso: Shewing, That, by being addicted to Experimental Philosophy a man is rather assisted than indisposed to be a good Christian.*

interpretative disputes might be resolved.[62] Three broad strategies may be discerned as emerging from this period of social and intellectual confusion. The first, famously advocated by John Locke in his 'Letter Concerning Toleration' (1689), was that since such religious disagreements could not be resolved consensually by any existing authority, it was necessary to find ways of tolerating such disagreement.[63] The second was to suggest that human reason stood above the ambiguities and uncertainties of biblical interpretation experienced by religion. By recognizing the supreme authority of human reason, such seemingly insoluble disputes could be resolved, if they could not be avoided.[64]

The third approach is Boyle's: engaging with the natural world allows a clarity of thought and the application of processes of confirmation which minimize disagreement. As Stephen Gaukroger points out, Boyle argued that 'an empirical understanding of the natural world can be promoted as revealing God's purposes in a way that avoids sectarianism.'[65] Boyle clearly saw natural philosophy as a way of engaging theological questions that was free of the confessional religious commitments and tensions that had arisen in England since the English Civil War. Natural philosophy was a means of doing theology, shorn of the inconclusive debates of the age concerning ecclesiastical authority and biblical interpretation. From about 1690, 'physico-theology' (from the Greek term *physikos*, 'natural') came to be seen as a form of theology which arose from the application of the methods of natural philosophy.

Engaging the world of nature thus transcended religious partisanship—a point that is sometimes linked with suggestions of 'Latitudinarianism' within the Royal Society in the late seventeenth century.[66] Its focus on pursuing a natural philosophy gave a coherence to the institutional identity of the Royal Society, particularly during the period of religious and political instability preceding the 'Glorious Revolution' of November 1688. Although Boyle did not, to my knowledge, develop this point, the pursuit of a natural philosophy

[62] For Boyle's own experiences of and concerns about such 'sectarianism' in religious and other contexts, see Hunter, *Robert Boyle (1627–91)*, 51–7.

[63] Marshall, *John Locke, Toleration and Early Enlightenment Culture*, 469–535. The rise of 'Deism', which began during Boyle's time, is often portrayed as an accommodation of religion to an increasingly rational and scientific culture. See Harrison, 'Natural Theology, Deism, and Early Modern Science'; Gaukroger, *The Collapse of Mechanism and the Rise of Sensibility*, 40–54.

[64] This approach was rendered problematic by the growing realization that human beings reasoned in different ways in different historical and cultural locations: see McGrath, *The Territories of Human Reason*, 19–35.

[65] Gaukroger, 'The Challenges of Empirical Understanding in Early Modern Theology', 567.

[66] For a critical assessment of this suggestion, see Hunter, 'Latitudinarianism and the "Ideology" of the Early Royal Society'.

clearly enabled the emergence of an international scientific 'Republic of Letters', in which national and religious concerns were seen as subordinate to a more universal understanding of the natural world, resting on certain agreed methods of investigation and reasoning.[67]

Yet Boyle's most interesting reflections on the utility of natural philosophy concern the quality of the human engagement with nature it facilitates, which is clearly understood to be affective, not merely cognitive, eliciting admiration and wonder as well as understanding.[68]

> The Book of Nature is to an ordinary Gazer, and a Naturalist, like a rare Book of Hieroglyphicks to a Child, and a Philosopher: the one is sufficiently pleas'd with the Odnesse and Variety of the Curious Pictures that adorne it; whereas the other is not only delighted with those outward objects that gratifie his sense, but received a much higher satisfaction in admiring the knowledge of the Author, and in finding out and inriching himself with those abstruse and vailed Truths dexterously hinted in them.[69]

The image of hieroglyphics immediately points to the need for an observer to be trained in the symbols or language of nature, if it is to be understood.[70] The natural philosopher is privileged and educated, and thus able to see aspects of the natural world that would not be accessible to the unlearned—a theme we noted in the previous chapter (pp. 61–2).

Boyle can be seen a transitional figure between an anthropocentric view of the natural world, which holds that it was created for human benefit, and a more expansive view hinted at in the writings of John Ray in the 1690s, and stated more explicitly two decades later by writers such as William Derham, who remarked that an increasing appreciation of the vastness of the universe offered a 'far more extensive, grand, and noble view of God's works' than that of the 'old vulgar opinion, that all things were made for man'.[71] Hints of this

[67] Pollock, 'The Voyage Account, the Royal Society and Textual Production, 1687–1707'. For Kepler and the 'Republic of Letters', see Grafton, 'Chronology, Controversy, and Community in the Republic of Letters'.

[68] For a good account of this attitude, including discussion of Boyle, see Smith, *Empiricist Devotions*. There is also some excellent analysis in Mandelbrote, 'The Uses of Natural Theology in Seventeenth-Century England'.

[69] Boyle, *Some Considerations Touching the Vsefulnesse of Experimental Naturall Philosophy*, 4. See also Clody's comments on Bacon's 'alphabet of nature': 'Deciphering the Language of Nature', 122–4.

[70] For the theme of breaking nature's code in Bacon, see Clody, 'Deciphering the Language of Nature'.

[71] This transition is discussed in more detail in Brooke, 'Wise Men Nowadays Think Otherwise'. For the quotation from Derham's *Astro-Theology* (1715), see Brooke, 'Wise Men Nowadays Think Otherwise', 209.

older view are still found in Boyle's works; it is, however, clear that a process of recalibration of both the status and responsibilities of humanity within the natural world was in the process of emerging.

The forms taken by the 'natural philosophy' of the seventeenth century cause some difficulties for those historians of the natural sciences, anxious to trace the historical origins of their disciplines, yet who struggle to cope with the obvious interest shown by writers such as Kepler, Galileo, Boyle, and Newton in questions of theology, which they clearly considered to be part of their 'natural philosophy'.[72] These cannot be dismissed simply as deferential gestures to a religious culture, in that at multiple points theology is allowed to *inform* the interpretation of nature—for example, in Kepler's grounding of mathematical archetypes in the mind of God.[73] Newton himself was clear that God had to be included within the purview of any responsible natural philosophy. His *General Scholium* declared that 'discoursing' of God 'from the appearances of things, does certainly belong to Natural Philosophy'.[74] This naturally leads us to consider the importance of Newton for the shaping of early modern natural philosophy in greater detail.

Newton's Integrative Natural Philosophy

Isaac Newton's *Philosophiae Naturalis Principia Mathematica* ('The Mathematical Principles of Natural Philosophy', 1687), is probably the best-known work of natural philosophy, and continues to attract close attention from historians and philosophers. Yet both this work, and Newton's intellectual achievements in general, still tend to be read through the distorting prism originating from the French Enlightenment, which adopted Newton as a mascot for its own naturalist cultural projects.[75] Phrases such as 'Newtonian scientific worldview' or 'clockwork universe' are still widely used to refer to this modernist, secular, and naturalistic interpretation of Newton's natural philosophy.[76] Newton certainly came to be associated in popular writing with the idea of a

[72] For this concern, see Lüthy, 'What to Do with Seventeenth-Century Natural Philosophy?'

[73] Hon, 'Kepler's Revolutionary Astronomy'.

[74] Snobelen, '"God of Gods, and Lord of Lords"'. For the wider issue of the explicit and implicit theology of the *Principia*, see Snobelen, 'The Theology of Isaac Newton's *Principia Mathematica*'.

[75] Shank, *The Newton Wars and the Beginning of the French Enlightenment*, 1–3. For the wider reception of Newton, see Boran and Feingold, eds., *Reading Newton in Early Modern Europe*.

[76] For recent studies of Newton which help remove ideologically driven accounts of Newton's natural philosophy, see Shank, 'Between Isaac Newton and Enlightenment Newtonianism'; Iliffe, *Priest of Nature*.

'clockwork universe', despite the fact that he did not in fact develop such a notion, and the lines of argument he developed actually point against it.[77]

But how did this misreading of Newton take place? It is relatively easy to understand how Newton was adopted as a figurehead by movements with their own ideological agendas, anxious to draw upon his unparalleled international reputation as a genius.[78] Yet the development of Newton scholarship in the twentieth century also contributed to this partial and inadequate account of Newton's overall intellectual vision, including his natural philosophy. As Robert Iliffe points out,[79] early twentieth-century studies of Newton focused on his mathematics, physics, and chemistry. His 'alchemical' and theological writings were at this stage often seen as peripheral to any serious study of scientific progress, if not deplorable aberrations from scientific orthodoxy, and his anti-Trinitarian theological views treated as if they were anticipations of the Enlightenment criticisms of the perceived irrationality of the doctrine of the Trinity.[80] However, intense recent scholarly activity has led to a revision of our understanding of this period.[81]

By the late seventeenth century, natural philosophy had bifurcated into two distinct strands.[82] On the one hand was a *speculative* natural philosophy, exemplified by Descartes's vortex theory of planetary motions, which was characterized by the formulation of speculative hypotheses without attempting to ground these in experiment or observation, even if it was conceded that such hypotheses would aim to 'save the phenomena'.[83] On the other hand was an *experimental* natural philosophy, characterized by its commitment to observation and experiment, and hostile to the use of speculative hypotheses.

Although Newton was initially sympathetic to the form of natural philosophy set out in Descartes's *Principia Philosophiae* (1644), he clearly moved away from such an approach. In his letter of 10 June 1672 to Henry Oldenburg,

[77] Snobelen, 'The Myth of the Clockwork Universe'. While the idea is found in later 'Newtonian' writers, it cannot be retrojected onto Newton himself: Davis, 'Newton's Rejection of the "Newtonian World View"'.

[78] For this reputation and its social functions, see Iliffe, '"Is He Like Other Men?"' Kant's philosophical resistance to recognizing Newton as a genius seems to rest on a misunderstanding of his 'logic of discovery': see Hall, 'Kant on Newton, Genius, and Scientific Discovery'.

[79] Iliffe, 'Abstract Considerations', 428–9.

[80] Mandelbrote, 'Eighteenth-Century Reactions to Newton's Anti-Trinitarianism'.

[81] E.g., Nummedal, 'Alchemy and Religion in Christian Europe'; Parshall et al., eds., *Bridging Traditions*; Principe, 'Transmuting Chymistry into Chemistry'.

[82] For what follows, see Anstey, 'Experimental versus Speculative Natural Philosophy'; Anstey and Vanzo, 'The Origins of Early Modern Experimental Philosophy'.

[83] Gaukroger, *Descartes' System of Natural Philosophy*, especially 142–6. This idea is found in two works read carefully by Newton: Thomas Streete's *Astronomia Carolina* (1661) and Vincent Wing's *Astronomia Britannica* (1669).

the first Secretary of the Royal Society, Newton made clear his own commitment to an experimental natural philosophical method:

> The best and safest method of philosophizing seems to be this: first to search carefully for the properties of things and establish them by experiments, and then more cautiously to assert any explanatory hypotheses. For hypotheses should be adapted to the properties of things which require to be explained (*ad explicandas rerum proprietates tantum accommodari debent*), and are not to be used to determine them (*non ad determinandas usurpari*).[84]

Newton restated his commitment to such an experimental philosophy in a late reflection of 1715, in which he summarized his overall method thus:

> It is not the Business of Experimental Philosophy to teach the Causes of things any further than they can be proved by Experiments. We are not to fill this Philosophy with Opinions which cannot be proved by Phaenomena. In this Philosophy Hypotheses have no place, unless as Conjectures or Questions proposed to be examined by Experiments.[85]

But did such reflection on phenomena lead to understandings of the world that were probably true, or perhaps even certainly true? In the 1660s, there was intense cultural suspicion within the circles of the Royal Society of the dogmatic certainties of religious 'enthusiasts', which had caused such social and intellectual disruption during the recent English Civil War. For most members of the Royal Society, it was important to remain 'modest in opinions, avoiding both peremptory dogmatism and an unwarrantable Scepticism'.[86] As Niccolò Guicciardini points out, the intellectual ethos of the Royal Society was 'probability through patient collection of facts', accompanied by a 'sceptical avoidance of dogmatic certainty'.[87] Joseph Glanvill's *Vanity of Dogmatizing* (1661) reflects this distaste for ideological dogmatism and personal arrogance, which led to suspicions of the idea that the intellectual certainties of geometry had any parallel in an experimental natural philosophy.[88] While Newton was aware of the limits of offering mathematical

[84] *The Correspondence of Isaac Newton*, vol. 1, 164.
[85] Newton, 'An Account of the Book Entituled *Commercium Epistolicum*', 222.
[86] Corneanu, *Regimens of the Mind*, 81.
[87] Guicciardini, 'Reconsidering the Hooke–Newton Debate on Gravitation', 514.
[88] For the debates about the mathematization of nature around this time, see Cormack, 'The Role of Mathematical Practitioners and Mathematical Practice in Developing Mathematics as the Language of Nature'; Mahoney, 'Changing Canons of Mathematical and Physical Intelligibility in the Later

accounts of the natural world, his emphasis on the mathematization of nature placed him somewhat on the periphery of the prevailing consensus of the Royal Society.

This method was put into practice in his *Philosophiae Naturalis Principia Mathematica* (1687). To explore Newton's approach, we shall consider what is probably the best-known instance of its application—the formulation of the principle of universal gravitation, and its application as a unifying principle in correlating terrestrial and celestial mechanics.[89] The background to this is an explanatory difficulty that remained unresolved concerning Kepler's theory of planetary orbits: why did the planets orbit in ellipses around the sun? In fact, why did they orbit around the sun at all? What was the physical basis for Kepler's geometrical summaries (they were not yet described as 'laws') of the observed orbital motion of a planet around the sun?[90] Newton was aware of the main themes of Kepler's account of planetary motion, having encountered these indirectly (probably in 1685 or 1686) through Ishmaël Boulliau's *Astronomia Philolaica* (1645).

During the 1660s, Robert Hooke began to frame a possible explanation, based on the notions of inertia and gravitation. The orbital motion of a planet, he suggested, was the consequence of 'compounding' its inertial motion along a straight line with an attractive motion towards the sun, resulting from its gravitational attraction.[91] Yet Hooke experienced difficulties in expressing his physical principles of celestial mechanics *mathematically*, and in November 1679 wrote to Newton asking for his assistance. Scholarship remains divided on whether the eighteenth-century French mathematician Alexis-Claude Clairaut was right when he suggested that Hooke's analysis makes it clear that there is a considerable distance separating 'a truth that is glimpsed and a truth that is demonstrated'.[92]

Seventeenth Century'; Dear, *Discipline & Experience*; Gorham and Hill, eds., *The Language of Nature*. It should also be recalled that part of the Aristotelian legacy concerning natural philosophy saw ideal human knowledge as demonstrative and certain.

[89] For accounts of this process, see MacDougal, *Newton's Gravity*, 127–64; Harper, *Isaac Newton's Scientific Method*, 220–56. Iliffe's argument that Newton's explanatory success rested on blurring the traditional distinction between mechanics and mathematics is important here: Iliffe, 'Abstract Considerations', 430–4.

[90] For the issues, see Katsikadelis, 'Derivation of Newton's Law of Motion from Kepler's Laws of Planetary Motion'.

[91] For details, see Nauenberg, 'Robert Hooke's Seminal Contribution to Orbital Dynamics'. Hooke set out this theory in a lecture of 23 May 1666, entitled 'A Statement of Planetary Movements as a Mechanical Problem'.

[92] Nauenberg, 'Robert Hooke's Seminal Contribution to Orbital Dynamics', 11. See further Dittrich, 'The Eastward Displacement of a Freely Falling Body on the Rotating Earth'.

Despite some questions which remain about the originality of Newton's concept of universal gravitation, there is little doubt that Newton was able to mathematize this notion—and in doing so, demonstrated the potential of his natural philosophy to explain the deep structure of the solar system.[93] The first two books of the *Principia* offered an essentially mathematical account of motion, without any commitment to metaphysical or ontological entailment. In the third book, Newton turns to consider how the mathematical and physical realms are to be correlated.

Two points may be highlighted. First, Kepler had established three general principles underlying the orbits of the planets; he was, however, unable to explain these principles. Newton was able to show that all three of what he now named Kepler's 'rules' (*regulae*) of planetary motion could be explained on the basis of his theories of inertia and universal gravitation. Kepler had identified some important patterns; Newton demonstrated why those patterns existed in the first place, at the same time offering a predictive account of the future movements of the planets. He also extended their scope, asking his astronomer friend John Flamsteed to make precise observations of the movements of the four moons of Jupiter in December 1684, which he was able to show conformed to Kepler's orbital rules.[94] Moons orbiting planets beyond the earth thus obeyed the same laws of motion governing the moment of the planets round the sun. The forces that governed motion on earth seemed to work everywhere, in the same way. For example, Newton was able to show that celestial motion under an inverse square law of gravitational attraction would take the form of an ellipse (observed in the orbits of planets) or a parabola (observed in the orbits of certain comets).[95]

Second, Newton showed that the same fundamental rules that applied to motion below the moon (i.e., on earth) also applied beyond the moon (i.e., in the case of the planets). The importance of this point is easily overlooked today, when the notion of the 'sublunary' world no longer has the imaginative significance that it did for many in the seventeenth century (see pp. 53–4). Newton's analysis demonstrated that the same fundamental principles seemed to operate throughout the known universe of his day. This is an important example of 'theory unification', demonstrating how terrestrial phenomena

[93] For a good account of Newton's explanation of the *regula Kepleriana*, and reflections on the later reception of Newton's theories, see Wilson, 'Newton and Celestial Mechanics'. Yet as Mahoney points out, the full impact of the universal mechanics of Newton's *Principia* could only be attained once its 'essentially geometrical style' had been transposed into the symbolic algebra of Leibniz's calculus: see Mahoney, 'The Mathematical Realm of Nature', 703.

[94] For Newton's correspondence with Flamsteed, see Newton, *Correspondence*, vol. 2, 404, 407–8.

[95] For the theory see Cook, 'The Inverse Square Law of Gravitation'.

such as the tides of the sea (and the falling of apples) and celestial phenomena such as the movements of the moon and planets could all be accounted for on the basis of the same principle of universal gravitation.[96] What had once seemed to be disconnected could now be held together within the scope of a grander theoretical perspective, suggesting that the universe was more ordered and coherent than some had feared. Newton's international reputation as a genius often rested on this seemingly preternatural capacity to make connections between apparently disconnected phenomena.[97]

Newton's demonstration of cosmic coherence showed that certain fundamental forces and principles seemed to apply throughout the observable universe, and even to be transferable to the social and political orders.[98] Alexander Pope's tribute to Newton following his death in 1727 is well known, taking its cue from Newton's remarkable illumination of the nature of light, set out in his *Opticks*:[99]

> All Nature and her Laws lay hid in Night.
> God said, *Let Newton be*! and All was *Light*.

Other poetic celebrations of his genius went further, often suggesting a link between Newton as a natural philosopher and the enhanced appreciation of the natural world that resulted from the penetrating cosmic vision he created. Significantly, at this stage Newton is often presented as a sage or visionary who empowers the poetic imagination, not someone who suppresses that imagination and restricts its scope.[100] Yet it has to be conceded that Newton's emphasis on the regularity and order of the natural world was too easily reconceptualized in mechanical metaphors, such as a 'clockwork universe' or God as a 'watchmaker'—an image that was artfully redeployed by William Paley in his *Natural Theology* (1802) to reaffirm the rationality of religious belief.

[96] Morrison, *Unifying Scientific Theories*, 7–34.

[97] For example, see the 1782 comment of Louis-Sebastien Mercier, cited by Shank, *Newton Wars*, 4: 'Newton saw an apple drop and after meditating on it conceived the system of universal gravitation. Another, lacking the ability to see the ties that bind the planets to their orbits, would simply have grabbed the apple and eaten it.'

[98] Striner, 'Political Newtonianism'.

[99] Pope's tribute to Newton focused on his natural philosophy; his official memorial in Westminster Abbey celebrated him in a more comprehensive manner: *Naturæ, Antiquitatis, S. Scripturæ sedulus, sagax, fidus Interpres* ('Of Nature, Antiquity, and of Holy Scripture a diligent, wise, and faithful Interpreter'): Fara, *Newton*, 41.

[100] A point explored by Ketcham, 'Scientific and Poetic Imagination in James Thomson's "Poem Sacred to the Memory of Sir Isaac Newton"'.

Physico-Theology: The Religious Dimension of Natural Philosophy

We have already noted how the early modern concept of natural philosophy wove together domains of thought that today would be understood to belong to quite distinct disciplines—such as physics, biology, philosophy, mathematics, theology, and music. Yet more needs to be said about the form of natural theology that emerged in this period, and its importance for any understanding of the focus and limits of natural philosophy. Like many in this age, Boyle saw a seamless connection between the study of nature and religious belief. He considered that 'the Discoveries [natural philosophers] make in the Book of Nature' would lead them and others to 'be excited and qualifi'd the better to admire and praise the Authour, whose Goodness does so well match the Wisdom they celebrate'.[101] This recognizable variant of the 'Priest in the Temple of Nature' motif emphasizes the importance of the natural philosopher's relationship with nature, seeing it not simply as an object for study, but as a potential source of wisdom.[102]

Religious concerns are an integral part of any account of the development of early modern natural philosophy, not least in their potential to offer a motivation for the attentive observation of the natural order.[103] Galileo may have had differences with the institutional church of his day; this did not stand in the way of his productive engagement with questions of theological interpretation. Newton himself appears to have carefully demarcated his views about God arising from his natural philosophy (such as the concepts of space and action at a distance) from his more explicitly theological views, based on a reading of the Bible.[104] Others, however, recognized that religious concerns offered a powerful motivation for the detailed study of nature as creation.

For example, Boyle notes that many philosophers, irrespective of their religious commitments, 'have been, by the contemplation of the world, moved to consider it under the notion of a temple'.[105] This motif of a 'priest in the temple of nature' should not be seen as a reversion to some natural religion, but rather to highlighting the privileged role of the natural philosopher, who having discerned the beauty and complexity of the natural world, feels obliged to

[101] Boyle, *The Excellency of Theology*, 220.
[102] Harrison, '"Priests of the Most High God"'.
[103] Vassányi, *Anima Mundi*, 103–23.
[104] Greenham, 'Clarifying Divine Discourse in Early Modern Science'.
[105] Boyle, *Works*, vol. 5, 31–2. See further Fisch, 'The Scientist as Priest'; Harrison, *The Bible, Protestantism and the Rise of Natural Science*, 198–9.

revere, respect, and tend nature—and encourage others to share in this worship.[106] Boyle was clear that an attentive engagement with the natural world was both cognitive and affective, changing both thought and behaviour.[107] As Scott Mandelbrote has pointed out, Boyle and others held that 'a cause and a consequence of the impulse to natural theology' was the desire to worship the creator through 'a proper appreciation of the form of his creation'.[108]

The category of 'physico-theology' dates from this period, and first makes its explicit appearance in Walter Charleton's *The Darknes of Atheism Dispelled by the Light of Nature: A Physico-Theologicall Treatise* (1652).[109] The prefix '*physico-*' derives from the Greek term *physis*, 'nature'. Some have interpreted physico-theology as anticipating the agendas of the Enlightenment; others have seen it as offering a defence of the rationality of Christianity in an age of growing secularization.[110] In general terms, however, 'physico-theology' appears to be have been understood as a form of theology developed on the basis of the methods of natural philosophy—a knowledge of God drawn from reflection on the 'Book of Nature', rather than a knowledge of God based on reflection on the 'Book of Scripture'.[111] From the standpoint of orthodox Christian theology, such an approach would focus on the doctrine of creation, but would have nothing to say about the more controversial matters of Christology or the doctrine of the Trinity—two matters on which some natural philosophers, such as Newton, were reluctant to comment publicly.[112]

This might suggest a parallel between physico-theology and Deism, an approach to religious belief which focuses on the notion of God as creator and lawgiver. It is certainly true that Newton's ideas were a catalyst to the

[106] For physico-theology as a religiously transformative engagement with nature, see Corneanu, *Regimens of the Mind*, 169–72. For a good account of Boyle's 'endeavor to subordinate the study of nature to the worship of God', see Ben-Chaim, 'Empowering Lay Belief', 54–8.

[107] Wragge-Morley, *Aesthetic Science*, 22–46.

[108] Mandelbrote, 'The Uses of Natural Theology in Seventeenth-Century England', 468, referencing Boyle and John Ray.

[109] For excellent studies of 'physico-theology', see Harrison, 'Physico-Theology and the Mixed Sciences'; Blair and von Greyerz, eds., *Physico-Theology*. This should not be confused with later understandings of natural theology, particularly those which frame its objective as the demonstration of God's existence by pure reason. See McGrath, *Re-Imagining Nature*, 6–25; McGrath, 'Natürliche Theologie'.

[110] For an excellent account of the role of natural theology at this time, see Mandelbrote, 'The Uses of Natural Theology in Seventeenth-Century England'.

[111] Vidal and Kleeberg, 'Knowledge, Belief, and the Impulse to Natural Theology', 381. Note also the comments of Cunningham, 'How the *Principia* Got Its Name', 387: physico-theology is about 'arguing *from* the findings of natural philosophy *to* the existence and attributes of God'.

[112] We have already noted Newton's anti-Trinitarianism. For the heated debates associated with this doctrine at this time, see Dixon, *Nice and Hot Disputes*. In terms of his Christology, Newton is often considered an Arian: see Fallon, 'Milton, Newton, and the Implications of Arianism'; Trubowitz, 'Reading Milton and Newton in the Radical Reformation'. For the nature of Arianism, including reflection on Newton's views, see Wiles, *Archetypal Heresy*, 77–92.

emergence of both these movements;[113] yet the nature of that catalysis was different in each case. The term 'Deism' designates a broad spectrum of individual opinions, rather than a well-defined coordinated movement, characterized by varying degrees of scepticism concerning both the rationality and utility of traditional religious beliefs. Deism takes too many historical forms to be plausibly enfolded within a single definition or history, even though it is helpful to think of it as a family of beliefs and attitudes.[114]

The intellectual origins of Deism, however, are perhaps better understood not to lie in 'natural philosophy', but rather in forms of rationalism that were critical or suspicious of certain traditional Christian doctrines, such as the incarnation and Trinity.[115] Deism was thus framed antagonistically, in that it defined itself in relation to what it deemed to be incorrect. Deism and physico-theology may converge in at least some of their conclusions; yet these are attained by divergent intellectual pathways, and have different fundamental motivations.

It remains unclear whether physico-theology is to be distinguished from natural theology, or whether both represent potentially overlapping domains of a spectrum of knowledge. Some have suggested that physico-theology was 'a distinctive intellectual project', to be distinguished from natural theology, understood as an attempt to gain knowledge of God 'by human reason alone'. Physico-theology should thus be seen as an essentially empirical approach to the natural world, which sought 'to establish the compatibility of the new science with the biblical narrative'.[116] While there are some difficulties with this view, not least the intellectual heterogeneity of the concept of natural theology,[117] a case can certainly be made for suggesting that the introduction of the new term 'physico-theology' reflects a new and more empirically based approach to the relation of religion and the natural world.

This kind of approach offered a means of engaging religious questions without becoming involved in partisan and sectarian debates about the sources and norms of theology, which had become particularly significant in the fraught political and religious atmosphere following the English Civil War.

[113] On Deism, see Force, 'Biblical Interpretation, Newton and English Deism'.

[114] The best discussion remains Barnett, *The Enlightenment and Religion*, 11–44, 81–129. See also Pocock, 'Historiography and Enlightenment'; Wigelsworth, *Deism in Enlightenment England*, 1–13.

[115] Rogers, 'Stillingfleet, Locke and the Trinity'; Antognazza, *Leibniz on the Trinity and the Incarnation*; Lim, *Mystery Unveiled*, 69–123.

[116] This is the view developed by Blair and von Greyerz in their recent revisionist account of physico-theology: see Blair and von Greyerz, eds., *Physico-Theology*, 1–2.

[117] See the points made in McGrath, 'Natürliche Theologie'; Pickering, 'New Directions in Natural Theology'.

For Boyle, an early advocate of physico-theology, a theological understanding of the world may lie beyond the scope of natural philosophy; such a religious framework nevertheless offers an explanation of why such a natural philosophy is possible in the first place. In developing this general approach, Boyle sets out an important (and much-cited) distinction between 'things that are above reason' and 'things that are contrary to reason'.[118] This distinction is important in relation to Boyle's concern that implications of excessively rationalistic views about nature impoverished human appreciation of its deeper meaning. Although Boyle's caution about physico-theology partly reflects some theological concerns, his views fit in with the wider views of his culture—for example, echoed in Locke's belief that God 'providentially designs us with limited epistemic capacities in order to check our pride and to motivate us to seek perfection in God'.[119]

Margaret Osler has pointed out that scholarship has used two main metaphors to correlate Boyle's natural philosophy with his theological interests.[120] Some consider him to see theology and natural philosophy as two distinct and self-contained domains of thought which are nevertheless in harmony with each other; others suggest that he appears to use a method of a process of 'appropriation and translation', adapting concepts that originated from within a theological context to serve new purposes for natural theology. For Margaret Cook, this second model has the advantage of recognizing the 'permeable, historically contingent boundaries between the two deeply connected discursive domains of theology and natural philosophy',[121] and perhaps helps us to understand why there are so many logical, empirical, and theological loose ends in the accounts of physico-theology of this age.

Newton did not himself contribute to the writing of such 'physico-theological' works; his natural philosophy, however, was widely taken up by others with such interests, and incorporated into tracts and sermons affirming the fundamental rationality of the Christian faith, which seemed to be confirmed by Newton's dazzling demonstration of the ordering of the universe, grounded in its origins in the mind of God. The Boyle Lectures, established under the terms of the will of Robert Boyle and delivered annually over the forty-year period 1692–1732, represent perhaps the most sustained adoption of Newtonian natural philosophy in the service of theism in the

[118] For two divergent accounts of this distinction, and an assessment of its influence, see Wojcik, *Robert Boyle and the Limits of Reason*; Holden, 'Robert Boyle on Things above Reason'.

[119] Rossiter, 'Locke, Providence, and the Limits of Natural Philosophy', 218.

[120] Osler, 'Mixing Metaphors', 97–104.

[121] Cook, 'Divine Artifice and Natural Mechanism', 136.

early modern period.[122] Poets such as James Thompson developed Newton's natural philosophy as a tool for gaining an enhanced appreciation of the natural world, and coping with its seeming incoherence and instability.[123] A good example of this is found in Thompson's comparison of a 'vulgar' reading of the world as chaotic, dangerous, and threatening, with a superior reading through Newton's 'philosophic eye'. While others see disorder and lack of purpose, Newton allows us to see the harmony and order beneath the surface.[124]

Writing in 1611, before the dramatic interventions of the natural philosophy of Kepler, Galileo, and Newton, John Donne had expressed alarm at the implications of a 'New Philosophy' which 'calls all in doubt', seeming to rob the world of meaning and purpose. Donne's 'First Anniversarie' had balefully proclaimed the fragmentation of a unified view of the world. ''Tis all in pieces, all coherence gone'.[125] Newton, it seemed to many, had restored coherence to the world, showing its fundamental harmony and order.[126] Like Kepler before him, he seemed to have achieved a harmonious synthesis, not merely of celestial and terrestrial mechanics, but of theology and natural philosophy.

Yet Newton's legacy proved to be open to other interpretations. Newton had indeed highlighted the ordering and regularity of the universe; this subsequently led others—though not Newton himself—to compare the universe to a clockwork machine.[127] And so what Herbert Odom nicely described as the gradual 'estrangement of celestial mechanics and religion' began to set in.[128] By the end of the eighteenth century, Newton's natural philosophy was associated with the notion of the created order as an autonomous mechanism which did not require divine support or intervention for its continuation.

In this chapter, we have considered what many regard as the 'Golden Age' of empirical natural philosophy. Yet the personal syntheses of Boyle and Newton began to fragment in the eighteenth century, and fall apart completely in the nineteenth. In the following chapter, we shall document this development, and reflect on its implications.

[122] Dahm, 'Science and Apologetics in the Early Boyle Lectures'; Re Manning, *God's Scientists*.

[123] See especially the discussion of the 'Literature of Physico-Theology' in Connell, *Secular Chains*, 177–209. On Thompson, see especially Ketcham, 'Scientific and Poetic Imagination in James Thomson's "Poem Sacred to the Memory of Sir Isaac Newton"'.

[124] Connell, *Secular Chains*, 194–5.

[125] For Donne's preoccupation with this concern, see Grady, *John Donne and Baroque Allegory*, 137–69, 170–206.

[126] For the importance of this theme for discussion of morality and divine rationality in the early modern period, see Harrison, *The Bible, Protestantism and the Rise of Natural Science*, 185–204.

[127] For this comparison, see Dolnick, *The Clockwork Universe*, 182–3, 310–13. For a correction of this account, see Snobelen, 'The Myth of the Clockwork Universe'.

[128] Odom, 'The Estrangement of Celestial Mechanics and Religion'.

5

The Parting of the Ways

From Natural Philosophy to Natural Science

The questions of precisely when, why, and how the 'natural sciences' became distinguished and then detached from 'natural philosophy' remain unresolved, partly on account of difficulties in definition and an absence of clear boundary-markers along the way. For many scholars, this transition happened episodically, perhaps over the eighteenth century, without any specific crisis or revolution that may be said to have precipitated it. John Herschel's *Preliminary Discourse on the Study of Natural Philosophy* (1830) is perhaps the last statement of a 'natural philosophy' offering a comprehensive framework of engagement of nature, which embraced the question of humanity's moral relation with the natural world and practices that were appropriate for the meaningful inhabitation of nature.[1]

Herschel's work achieved the status of a minor classic, and was read with approval by many, including the young Charles Darwin as a final-year undergraduate at Cambridge University.[2] Darwin later recorded his particular appreciation of Herschel's reflections on the origins of species, which would motivate his own biological reflections: 'Herschel calls the appearance of new species the mystery of mysteries, & has a grand passage upon the problem!'[3] The proper study of nature, for Herschel, was religiously and spiritually fulfilling, and opened the way to reflection on how to enable the 'collective wisdom' of humanity to resolve ongoing social problems, by dealing with 'those obstacles which individual short-sightedness, selfishness, and passion, oppose to all improvement'.[4]

Yet Herschel increasingly seemed to represent an outlook that was receding into the past. Concerns that studying the natural world could not generate or sustain moral values had gained traction during the later eighteenth century,

[1] For accounts of Herschel's significance for the shaping of early nineteenth-century natural sciences, see Cannon, 'John Herschel and the Idea of Science'; Snyder, *The Philosophical Breakfast Club*.

[2] Darwin, *Autobiography*, 67–8. See further Warner, 'Charles Darwin and John Herschel'.

[3] Cited Warner, 'Charles Darwin and John Herschel', 436.

[4] Herschel, *Preliminary Discourse*, 74.

largely on account of David Hume's distinction between 'fact' and 'value'.[5] Although Hume's criticism of natural theology or natural religion has attracted much scholarly attention,[6] he also raised significant concerns about the idea that the respectful and attentive observation of nature could be morally informative and improving, without the interposition of some bridging theory, which was not itself empirical in its origins.

On the basis of Hume's analysis, there are two possible ways of understanding the relation of the moral and non-moral realms.[7]

1. It is not possible to move from non-moral premises to moral conclusions by purely logical means.
2. It may be possible to move from non-moral premises to moral conclusions through the use of bridging arguments.

It was, of course, possible to draw on a theistic framework as a bridging framework between the *observation* of nature and the *ethical inhabitation* of nature. As Tamás Demeter points out, the conviction that 'natural philosophers were studying God's creation provided the basic ideological framework of early modern science'.[8] As we shall see, the nineteenth century saw a growing trend to move away from theistic approaches to nature characteristic of the early modern period to the more 'naturalist' approaches associated with Thomas H. Huxley. We shall discuss this point further later in this chapter.

Yet the traditional category of 'natural philosophy' was undermined by more than a shifting ideological context; the propriety of continuing to use this term to refer to an essentially empirical discipline became increasingly contested. On at least two occasions during the 1830s, William Whewell suggested (anonymously) replacing 'natural philosopher' with the single word 'scientist'.[9] By 1840, Whewell had gone public, and openly advocated the use of this new term. 'We need very much a name to describe a cultivator of science in general. I should incline to call him a Scientist. Thus we might say, that as an Artist is a Musician, Painter, or Poet, a Scientist is a Mathematician,

[5] For Hume's views and their contemporary interpretation, see the material collected in Pigden, ed., *Hume on 'Is' and 'Ought'*.

[6] Demeter, 'Natural Theology as Superstition'; Graham, 'Hume and Smith on Natural Religion'.

[7] Pigden, 'Logic and the Autonomy of Ethics'. The most significant challenge to Hume's conclusions is Prior, 'The Autonomy of Ethics'; for the view that Prior merely demonstrates a 'shallow natural–normative entanglement', which is a consequence of 'simple logical relationships' that hold between just about any two subject matters, see Oddie, 'Non-Naturalist Moral Realism, Autonomy and Entanglement'.

[8] Demeter, 'Natural Theology as Superstition', 177.

[9] Ross, 'Scientist: The Story of a Word', especially 71–5.

Physicist, or Naturalist.'[10] The term 'scientist' resonated with the new cultural mood, and gradually displaced the older vocabulary.

A second development of importance was the emergence of the term 'science' *in the singular*, which appears to be linked with the establishment of the British Association for the Advancement of Science in 1831.[11] (The French and German equivalents—the Académie des Sciences, founded in 1666, and the Königlich Preussischen Akademie der Wissenschaften zu Berlin, founded in 1700—used the term 'science' *in the plural*.) At that time, a number of distinct scientific disciplines were in the process of emerging in Great Britain, giving rise to the possibility that these might be seen as unrelated fields of studies, and hence diminish the social and cultural authority of 'science' in British culture. 'The fragmentation of the field of scientific inquiry called forth a vision of its potential unification.'[12] In their understandable desire to avoid the isolation of specific individual sciences, the founders of the British Association chose to speak simply of 'science' *in the singular*—and in doing so, may have created an impression that every natural science uses a singular and identical 'scientific method', thus failing to take account of the fact that individual sciences develop research methodologies which are specifically adapted to their own objects of study.[13]

These two related developments gradually changed the cultural landscape of the academic and professional world in the final quarter of the nineteenth century, with older terms—above all, 'natural philosophy'—increasingly being set to one side as the new vocabulary achieved acceptance. While aspects of the moral agendas of natural philosophy would initially remain embedded within the emerging concept of 'natural science',[14] its focus now lay elsewhere. In 1879, Thomas H. Huxley set out his own approach to the relation of science and philosophy, which avoids making any reference to the now outdated notion of 'natural philosophy', as follows:

> Fundamentally, then, philosophy is the answer to the question, What can I know? and it is by applying itself to this problem, that philosophy is properly distinguished as a special department of scientific research. What is commonly called science, whether mathematical, physical, or biological,

[10] Whewell, *Philosophy of the Inductive Sciences*, vol. 2, 560. See also Yeo, *Defining Science*, 5.

[11] For the historical context, see Orange, 'The Origins of the British Association for the Advancement of Science'.

[12] Golinski, 'Is It Time to Forget Science?', 20.

[13] McGrath, *Territories of Human Reason*, 75–92.

[14] Think, for example, of the subtle moral agendas informing Darwin's evolutionary theories: Richards, 'Darwin's Theory of Natural Selection and Its Moral Purpose'.

> consists of the answers which mankind have been able to give to the inquiry, What do I know? They furnish us with the results of the mental operations which constitute thinking; while philosophy, in the stricter sense of the term, inquires into the foundation of the first principles which those operations assume or imply.[15]

'Natural philosophy', once seen as a unified field of study, now bifurcated into 'science' and 'philosophy'.[16] This led to the gradual erosion of the moral agendas originally associated with the study of nature. This loss of a moral focus on nature was catalysed by the English Industrial Revolution, which was based on new technologies that led to radical degradation of the natural environment and significant social dislocation.[17] John Ruskin's 1884 lectures on the 'Storm Cloud' may well represent one of the earliest recognitions of the phenomenon of climate change resulting from industrial activity; yet Ruskin's concerns are deeper than this, hinting at some of the lost concerns of an older natural philosophy.[18] The Industrial Revolution seemed to Ruskin to abandon a respectful engagement with the natural world, leading him to develop his own way of bringing the arts and sciences into contact and constructive dialogue.[19] A new vision of nature, engaging both arts and sciences, seemed to be required in the face of a 'demonstrable threat posed by rapacious industrial capitalism to the lives of people, animals and plants'.[20]

The Enlightenment Reformulation of Newtonian Natural Philosophy

The scholarly understanding of the phenomenon usually styled 'the Enlightenment' has undergone significant change since the 1970s. While often portrayed as an intellectually homogeneous movement, critical historical scholarship has led many to conclude that the 'Enlightenment' was

[15] Huxley, *Hume*, 48–9. Huxley is commenting on Hume in this volume, but here sets out his own ideas before moving on to discuss Hume's.

[16] Maxwell, 'In Praise of Natural Philosophy', 705–9.

[17] A good example of this trend can be seen in John Banks (1740–1805), who vigorously advocated the mechanical exploitation of nature: Soares, 'John Banks'. For further documentation and comment on such trends, see Jacob, *Scientific Culture and the Making of the Industrial West*; Jacob and Stewart, *Practical Matter*; Hankins, *Science and the Enlightenment*.

[18] For the difficulties in interpreting this metaphor, see Ford, 'Ruskin's Storm-Cloud'.

[19] For this agenda, and its links with the Oxford University Museum, see Holmes and Smith, 'Visions of Nature'.

[20] Holmes and Smith, 'Visions of Nature', 8.

essentially a polemical term devised in the nineteenth century to make sense of what had happened in the eighteenth, and locate it within certain privileging narratives of progress.[21] It is perhaps best to think of the Enlightenment as a competing series of visions of rationality, rather than a single coherent movement, whose agendas and concerns were often shaped by local concerns, while being informally linked through various networks.[22] Yet scholars of the Enlightenment have perhaps not given due account of the importance of Newton's natural philosophy, which seemed to many of his European admirers to demonstrate the power of human reason to penetrate the mysterious workings of the cosmos.[23]

Populist accounts of the Enlightenment often portray it as a movement grounded on a culturally and historically invariant reason, which could serve as both the foundation and criterion of reliable human knowledge.[24] Yet the precision and persuasiveness of Newton's natural philosophy led many to adopt the Baconian method of acquiring knowledge inductively through the close observation of nature, which was seen as circumventing the interminable and sterile conflicts among philosophers regarding the meaning and validity of first principles of reason. Rationalist confidence in humanity's intellectual powers was not primarily affirmed in terms of *a priori* reasoning, but rather in terms of the *a posteriori* forms of inductive reasoning developed by Bacon, and so successfully applied by Newton.

While it is true that some leading Enlightenment writers—such as Christian Wolff—attempted to reason from first principles, the outcomes of this process were generally seen to lack the coherence of those offered by Newtonian natural philosophy.[25] This was particularly the case with the French Enlightenment, which shows clear lines of influence from Newtonian

[21] Clark, 'Providence, Predestination and Progress'. See further Hunter, *Rival Enlightenments*, 33–92; Schmidt, 'Enlightenment as Concept and Context'.

[22] For the diversity of the Enlightenment, see Carey, *Locke, Shaftesbury, and Hutcheson*; Schmidt, 'Inventing the Enlightenment'; Withers, *Placing the Enlightenment*, 25–41; McGrath, *The Territories of Human Reason*, 75–92. For the intellectual networks which gave the movement some sense of coherence, see Withers, *Placing the Enlightenment*, 42–61; Gaukroger, *The Collapse of Mechanism and the Rise of Sensibility*, 232–47; Hobson, *Diderot and Rousseau*; Gies and Wall, *The Eighteenth Centuries*.

[23] For important exceptions to this criticism, see Ducheyne, 'Reid's Adaptation and Radicalization of Newton's Natural Philosophy'; Shank, *The Newton Wars*, 1–35; Demeter, *David Hume and the Culture of Scottish Newtonianism*; Boran and Feingold, eds., *Reading Newton in Early Modern Europe*. On the Enlightenment's adaptations of Newton, see Shank, 'Between Isaac Newton and Enlightenment Newtonianism'. It should be noted that some Enlightenment writers disliked the mathematized, mechanical understanding of natural philosophy they found in Newton's *Principia*, and preferred the suggestions of various principles of natural self-activity found in Newton's *Opticks*: Demeter, *David Hume and the Culture of Scottish Newtonianism*, 134.

[24] Flyvbjerg, *Rationality and Power*, 1–8.

[25] For a good survey, see Gaukroger, *The Collapse of Mechanism and the Rise of Sensibility*, 97–225.

sources.[26] Voltaire spent some thirty-three months in self-imposed exile in England during the 1720s, and appears to have absorbed the new confidence in rational investigation of nature arising from Newton's legacy.[27]

Yet an unfortunate outcome of the Enlightenment adoption and partial assimilation of Newton's natural philosophy was the perceived alignment of Newton's intellectual achievement with modernist narratives of modernization and progress. Newton was thus read through the prism of the *Principia*, and was seen as 'the heroic founding father of secular scientific modernity',[28] rather than as the advocate of a rich and complex natural philosophy. In portraying Newton in this way, Enlightenment writers such as Pierre-Simon Laplace were reading him selectively, assimilating him to their own agendas, and—perhaps most significantly, for our purposes—detaching what would now be seen as Newton's *scientific* ideas from their informing context. Newton's 'science' (to use this word anachronistically) was embedded within his wider natural philosophy; his later interpreters either divorced them from this context, or interpreted his 'natural philosophy' simply as an older term for 'natural science', which could be displaced without intellectual loss or impoverishment.

The Romantic Reaction against Popular Newtonianism

This historical observation underlies the growing critique of Newtonianism as an intellectual system that was inattentive to the complexity of the natural world, or to the place of humanity as both an observer of and a participant in nature. Such concerns were articulated in the closing years of the eighteenth century in Germany by writers such as Friedrich Wilhelm Joseph von Schelling, who developed a 'natural philosophy' (*Naturphilosophie*) which took full account of humanity's location within the natural order.[29] Nature was to be seen as 'a self-productive organic whole', resistant to dissection and rational analysis.[30] Goethe espoused a strongly holistic view of nature,

[26] This point was highlighted by Jacob, *The Newtonians and the English Revolution 1689–1720*. For an alternative perspective, see Edelstein, *The Enlightenment*, 24–30.

[27] Note especially Voltaire's *Éléments de la philosophie de Newton* (1738). For Voltaire's high estimation of both Locke and Newton, see Edelstein and Kassabova, 'How England Fell Off the Map of Voltaire's Enlightenment'. On his return to France, Voltaire made the study of Newton one of his primary concerns. Voltaire was present in England at the time of Newton's death on 31 March 1727.

[28] Shank, 'Between Isaac Newton and Enlightenment Newtonianism', 78.

[29] Guilherme, 'Schelling's *Naturphilosophie* Project'; Grant, *Philosophies of Nature after Schelling*, 1–25.

[30] Nassar, 'From a Philosophy of Self to a Philosophy of Nature', 315.

insisting upon the unity of nature despite its many individual aspects.[31] Although Max Weber's notion of the 'disenchantment' (*Entzauberung*) of nature represents a twentieth-century interpretation of the changing understanding of nature during the eighteenth century,[32] the language of 'magic' was used at this time in Germany to refer to a sense that something had been irretrievably lost. The later Romantic poet Joseph von Eichendorff, for example, spoke about nature being able to 'sing', if only the 'Magic Word' (*Zauberwort*) could be found to unlock its secrets.[33] Yet many would argue that the most perceptive critical assessment of Newtonianism took place within England, mounted initially by poets who believed that the natural world had been impoverished by its reductionist agenda.

William Blake, for example, was strongly critical of Newtonianism, which he considered to be imaginatively and aesthetically emaciated.[34] Blake singled out Bacon, Locke, and Newton—the three intellectual archetypes of an experimental natural philosophy—for criticism, apparently on the basis of his mistaken assumption that all three are to be seen as proponents of a mechanical worldview. Blake's view is probably best known from the familiar line from *Jerusalem* (1804–20), 'Bacon and Newton, sheathed in dismal steel',[35] but is set out more fully in these lines lamenting the captivation of the intellectual and cultural mindset of Europe by a mechanical philosophy, in which nature is understood as a machine:

> I turn my eyes to the schools and universities of Europe
> And there behold the Loom of Locke, whose woof rages dire,
> Wash'd by the Water-wheels of Newton: black the cloth
> In heavy wreaths folds over every nation: cruel works
> Of many Wheels I view, wheel without wheel, with cogs
> tyrannic
> Moving by compulsion each other, not as those in Eden, which,
> Wheel within wheel, in freedom revolve in harmony
> and peace.[36]

[31] For a full discussion, see Bortoft, *The Wholeness of Nature*, 8–9.

[32] Joas, *Die Macht des Heiligen*; Münch, 'Max Webers These der Entzauberung der Welt'. Weber's term 'Entzauberung' was borrowed from Schiller.

[33] McGrath, '"Schläft ein Lied in allen Dingen"?' The 'Magic Word' is best understood as a hermeneutical principle that discloses nature in its true light, as opposed to rival reductionist accounts.

[34] See especially Ault, *Visionary Physics*; Gilpin, 'William Blake and the World's Body of Science'.

[35] Blake, *Complete Poetry and Prose*, 157, line 11. This work, subtitled 'The Emanation of the Giant Albion', is not to be confused with the short poem of the same name, later set to music by Hubert Parry.

[36] Blake, *Complete Poetry and Prose*, 157, lines 14–20.

Blake's decidedly negative evaluation of Newton's natural philosophy has often been asserted to arise from his direct engagement with Newton's writings, especially the *Principia*.[37] Yet this seems improbable. Newton's *Principia* is notoriously difficult to interpret for those unable to understand its mathematical language. Newton himself declined to offer a popular account of his theories, and thus created a market for popular interpretations of 'Newtonianism'.[38] As a result, discussion of 'Newtonianism' was relocated from the academic world, being widely disseminated through public lectures and cheap pamphlets, with the result that 'coffeehouses formed the matrix of Newtonian persuasion'.[39] It is therefore important to note that recent scholarly examination of Blake's writings has called into question whether Blake had any *direct* familiarity with Newton's works, and points rather to Blake being familiar with popular Newtonianism, rather than with Newton's own works.[40] If this is so, Blake's criticisms are really directed against certain culturally influential readings of Newton, not necessarily Newton himself.

While Blake's concerns about Newton's legacy focused on his *Principia*, other criticisms of this period concerned his *Opticks*.[41] In his rambling 1820 poem 'Lamia',[42] John Keats was particularly critical of the 'unweaving' of the rainbow, protesting against the conquering of a rich and mysterious reality, which reduced it to the realm of the dull and common.

> Do not all charms fly
> At the mere touch of cold philosophy?
> There was an awful rainbow once in heaven:
> We know her woof, her texture; she is given
> In the dull catalogue of common things.
> Philosophy will clip an Angel's wings,
> Conquer all mysteries by rule and line,
> Empty the haunted air, and gnomed mine –
> Unweave a rainbow.

[37] Ault, *Visionary Physics*, 3.

[38] Wigelsworth, 'Competing to Popularize Newtonian Philosophy'; Feingold, *The Newtonian Moment*, 142–67. A good early example of this genre is Henry Pemberton's *View of Sir Isaac Newton's Philosophy* (1728).

[39] Stewart, *The Rise of Public Science*, 146–7.

[40] See the nuanced discussion of this important point in Cooper, 'William Blake's Aesthetic Reclamation'. For a good account of such forms of 'popular Newtonianism', see Smith, *Empiricist Devotions*, 69–105.

[41] Goethe's evaluation of Newton's theory should be noted here: see Heisenberg, 'Die Goethe'sche und Newtonische Farbenlehre im Lichte der modernen Physik'.

[42] Keats, *Poetry and Prose*, 412–29.

Although there is no explicit reference to Newton here, some have made a connection with reports of Keats attending a dinner at which Newton was criticized for having 'destroyed all the Poetry of the rainbow by reducing it to a Prism'.[43] Yet once more, questions must be raised about whether Keats had actually read Newton, or whether he was responding to an uninterrogated popular cultural perception about Newton.

Some have interpreted Keats's protest against a 'cold philosophy' to refer to philosophy in the modern sense of that term.[44] It is, however, important to appreciate that in 1820 the term was often used as a synonym for '*natural* philosophy'. The fourth edition of the authoritative *Encyclopaedia Britannica* (1810) offers this descriptive account of the term 'philosophy' as it was then used:

> The chief object of the philosopher is to ascertain the *causes* of things; and in this consists the difference between his studies and those of the natural historian, who merely enumerates phenomena and arranges them into separate classes.[45]

Yet from this definition of philosophy, along with the frequent references to the work of Francis Bacon, it is clear that the term 'philosophy' approximates to 'natural philosophy'. This sense of the term, when applied to Keats's concerns in 'Lamia', make far more sense of his remarks than the notion of philosophy as critical reasoning.[46]

Perhaps Keats's real concern, echoed by many within and beyond the Romantic movement, is the increasing emphasis placed upon objectivity in the study of nature.[47] Given the importance of this theme for our later reflections, its historical development merits discussion at this point.

Objectivity: A Disengaged Account of Nature

Such concepts as 'objectivity' and 'impartiality' attained the status of epistemological virtues in the late seventeenth and eighteenth centuries, encouraging the view that nature could 'speak for itself', without fear of intellectual

[43] Motion, *Keats*, 219.

[44] This trend can be seen, for example, in Lindholm, '"At the Mere Touch of Cold Philosophy"'.

[45] 'Philosophy', in *Encyclopaedia Britannica*, vol. 16, pt. 1, 369–70. There is a separate entry for 'Natural Philosophy': vol. 14, pt. 2, 641.

[46] Wordsworth read Newton's *Opticks* at Hawkshead Grammar School while preparing for Cambridge University admission examinations in December 1789. For Wordsworth's reflections on natural philosophy, see Hamilton, 'Deep History'.

[47] For the roots of this development in Boyle and Newton, see Markley, 'Objectivity as Ideology'; Zagorin, 'Francis Bacon's Objectivity and the Idols of the Mind'.

distortion arising from partisan concerns or cultural precommitments.[48] This emphasis on the possibility of an objective account of nature was important within the context of a culture shaped by memories of social tensions and disruptions associated with earlier European Wars of Religion. It is easy to understand the appeal at this time of such an external objective standard of rationality and morality,[49] in that it locates epistemic authority beyond human agency. Part of the appeal of a 'natural philosophy' was that it represented a set of ideas and values that was grounded in something lying beyond the control and manipulation of humanity. Wordsworth's lines might be cited in support of this view:

> To the solid ground
> Of nature trusts the Mind that builds for aye.[50]

Indeed, such was the appeal of these lines that they were adopted as the motto for the new scientific journal *Nature*, first published in November 1869.[51]

Yet it is one thing to be an *unbiased interpreter* of nature; it is quite another to be a *detached observer* of the natural world. The 'subjective turn' evident in British Romantic writers such as Coleridge and Wordsworth is often understood as resulting from discontentment with 'the passivity of consciousness in empiricist accounts' of nature on the one hand, and a growing interest in 'newer physiological theories of sensation, of subjectivity as constituted by the creative perceptual activity of imagination' on the other.[52] For Wordsworth and Coleridge, the subjectivity of the observer had to be enfolded within any account of the observation of nature. Every 'act of existential perception' thus becomes 'an act to which the perceiver contributes an essential ingredient of his own knowledge'.[53] Yet while Romantic writers drew on multiple cultural and intellectual resources in framing and exploring the subjective appreciation of the natural world, the shadow of a 'scientific other' appears to hover over these subjective reflections.

While Romantic poets might differ in how they grounded and expressed their subjective concerns, they were clearly concerned at the loss of an

[48] As argued by Daston, 'Objectivity and the Escape from Perspective'; Murphy and Traninger, eds., *The Emergence of Impartiality*; Solomon, *Objectivity in the Making*, 1–8; Daston and Galison, *Objectivity*, 27–35.

[49] Adam, *Theoriebeladenheit und Objektivität*, 179–244.

[50] Wordsworth, *Last Poems*, 24.

[51] They were quietly removed from the journal's masthead in 1963: Owens, '*Nature*'s Motto'.

[52] Keymer, 'The Subjective Turn', 320. For a significant attempt to 'historicize Romantic subjectivity', see Henderson, *Romantic Identities*.

[53] Shaw, 'The Optical Metaphor', 295.

affective relational dynamic between the human observer and the natural world. Other voices of the age tried to find ways of reconnecting affectively and imaginatively with the natural world. John Ruskin, for example, was critical of the 'cold, merely materialist science' of the middle Victorian period, and sought to recapture the spirit of earlier natural philosophies by explaining 'what natural forms had meant, intellectually and morally, to previous generations'.[54] Yet such attempts at reconnection ultimately seem to have failed to halt or reverse the growing professional and intellectual divergences between the emerging natural sciences and the more traditional arts and humanities.

The Transition from 'Theistic Science' to 'Naturalistic Science'

A conceptually important, though historically extended, transition of importance to our reflections on the changing shape and fortunes of natural philosophy concerns the notions of 'theistic science' and 'naturalistic science' in the nineteenth century.[55] Two major schools of thought found themselves competing for public support in Britain during the nineteenth century: 'theistic science', comparable in many ways to earlier forms of natural philosophy, and an emerging 'naturalist science'.[56] Crucially, historians have noted how these two approaches rested on very similar core beliefs (such as the uniformity of nature) and working methods and norms.[57]

While both theistic and naturalistic approaches to science assumed (and ultimately depended upon) the principle of the uniformity of nature, the former grounded this in an objective divine act of creation, and the latter through an emerging account of the metaphysics of nature. As Huxley pointed out, critical assumptions underlying the sciences—such as the uniformity of nature—are 'neither self-evident nor are they, strictly speaking, demonstrable'.[58] Both theistic and naturalist approaches to science rested on assumptions about the natural order that could not be proved to be correct.

[54] O'Gorman, 'Religion', 152–3.

[55] For some significant recent studies of the motivations underlying this transition, and its historical development, see Stanley, *Huxley's Church and Maxwell's Demon*, 242–63; Lightman and Reidy, eds., *The Age of Scientific Naturalism*; Dawson and Lightman, eds., *Victorian Scientific Naturalism*.

[56] This conflict can, of course, be framed within a 'secularizing' grand narrative. For the problems with such a narrative, see Clark, 'Secularization and Modernization'.

[57] For detailed discussion of this point, see Stanley, *Huxley's Church and Maxwell's Demon*, 34–79.

[58] Huxley, 'The Progress of Science', 61. Huxley continues: 'The justification of their employment, as axioms of physical philosophy, lies in the circumstance that expectations logically based upon them are verified, or, at any rate, not contradicted, whenever they can be tested by experience.'

Scientific 'naturalism' is not a 'natural' category, but is better understood as an 'actor's category', the meaning of which 'was forged and fought over alongside major developments in the history of science and in British history more generally'.[59] This observation helps us understand how Huxley's 'descriptive binary' could be reframed polemically as the 'prescriptive corollary' of the 'warfare' model of science and religion.[60] Yet despite these very different understandings of the metaphysical foundations of natural science, both theistic and natural approaches led to the same theoretical and practical outcomes, using the same norms and methods. Any adjudication concerning which was to be preferred seemed to lie beyond the scope of an inductive method.

With this point in mind, we may turn to what is now widely seen as one of the most significant statements of a 'naturalistic' approach to biological science: Charles Darwin's *Origin of Species* (1859). Darwin's assessment of the relative merits of theistic and naturalist approaches to science focused on which appeared to offer the better and more persuasive inductive explanation of certain otherwise puzzling observations of the distribution of biological species.

Darwin: Natural Philosophy and the Status of Humanity within Nature

Darwin did not see himself called upon to prove the existence of biological evolution: that idea had gained credibility since the publication of his grandfather Erasmus Darwin's *Zoonomia* (1794–6), and was further reinforced by Jean-Baptiste Lamarck's *Histoire naturelle des animaux sans vertèbres* (1815–22). Although some influential voices—such as the geologist Charles Lyell—resisted the idea, Lamarck's transformist account of evolutionary change was being widely discussed in English scientific circles during the 1850s.[61] In 1858, the year before the publication of Darwin's *Origin of Species*, Thomas H. Huxley delivered a lecture on the 'progressive development of animal life in time', arguing that it was not the evolution of species that was in doubt; the question that remained to be resolved was how this process of progressive development took place.

[59] Cowles, 'History Naturalized', 108. [60] Cowles, 'History Naturalized', 110.

[61] Corsi, *The Age of Lamarck*; Corsi, 'Before Darwin'; Galera, 'The Impact of Lamarck's Theory of Evolution before Darwin's Theory'.

> There is no doubt that the living beings of the past differed from those of the present period; and again, that those of each great epoch have differed from those which preceded, and from those which followed them. That there has been a succession of living forms in time, in fact, is admitted by all; but to the inquiry – What is the law of that succession? different answers are given; one school affirming that the law is known, the other that it is for the present undiscovered.[62]

This was the main question that Darwin addressed: identifying the natural process that lay behind, and thus explained, the phenomenon of biological evolution.

Although Darwin is often described as a 'scientist', his working method actually places him within the Baconian inductive tradition of natural philosophy.[63] Darwin was not an experimentalist; rather, he had accumulated a series of observations while sailing round the world on H.M.S. *Beagle* which seemed difficult to explain using existing understandings of the origins and development of biological life. Darwin's own account of his discovery of 'natural selection'[64] fits—as he himself suggests—within a Baconian inductive paradigm. Perhaps the best place to start is with the opening paragraph of the *Origin of Species*:

> When on board H.M.S. *Beagle*, as naturalist, I was much struck with certain facts in the distribution of the inhabitants of South America, and in the geological relations of the present to the past inhabitants of that continent. These facts seemed to me to throw some light on the origin of species—that mystery of mysteries, as it has been called by one of our greatest philosophers.[65]

[62] Huxley, 'On Certain Zoological Arguments', 301.

[63] In his *Autobiography*, Darwin declared that he proceeded 'on true Baconian principles and without any theory collected facts on a wholesale scale': Darwin, *Autobiography*, 119. Although the term 'natural philosophy' is absent from *The Origin of Species*, the terms 'science' and 'scientific' (though not 'scientist') are regularly encountered in Darwin's text. The term 'natural philosopher' is used once in the sixth edition of the *Origin*, in discussing some problems relating to the inductive method: Darwin, *Origin of Species*, 6th edition, 444.

[64] There has been extensive discussion as to whether Darwin's personal account of his discovery of this principle can actually be accommodated with the documentary evidence. In part, the difficulty lies in Darwin's occasionally contradictory statements concerning the dating of his evolutionary ideas: see, for example, Ayala, 'Darwin and the Scientific Method', 10033–4; Brinkman, 'Charles Darwin's *Beagle* Voyage'.

[65] Darwin, *Origin of Species*, 1. The philosopher in question was John Herschel, noted earlier in this chapter (p. 88). For further discussion of the discoveries of that voyage, see Keynes, *Fossils, Finches and Fuegians*.

During this round-the-world voyage, Darwin made a series of observations about the forms of plant and animal life in various regions, and collected a series of specimens on which he would subsequently reflect on his return to England. The methodology that Darwin describes is that of the accumulation of observations, and attempting to develop a theory which would offer their most satisfactory explanation.

As Darwin reflected on those observations, they seemed to indicate intractable problems and shortcomings with existing understandings of the origins of the biological realm, especially the idea of 'special creation' offered by religious apologists such as William Paley in his *Natural Theology* (1802).[66] While Paley's theory offered explanations of these observations which appealed to Darwin as a young man, he came to see them as increasingly cumbersome and forced.[67]

A more recently proposed alternative to Paley was Charles Lyell's notion of 'natural creation', which offered a naturalist account of the origin of new species in the light of shifting environmental conditions, which led to the extinction of older species and their replacement by newer forms, better adapted to the altered conditions.[68] Yet Lyell was not able to propose a satisfactory mechanism for this process. While a number of explanations could be offered for what Darwin observed during his voyage on the *Beagle*, the question remained which of these explanations was to be considered the best. Many of Darwin's observations did not seem to fit easily within the parameters of existing theories. A better explanation, Darwin believed, had to lie to hand.

The natural philosopher William Whewell used a rich visual image to communicate the capacity of a good theory to make sense of, and weave together, observations. 'The facts are known but they are insulated and unconnected... The pearls are there but they will not hang together until some one provides the string.'[69] The 'pearls' are the observations; the 'string' is a theory

[66] On Paley, see Topham, 'Biology in the Service of Natural Theology'; McGrath, 'Chance and Providence in the Thought of William Paley'; Brunnander, 'Did Darwin Really Answer Paley's Question?' For the cultural influence of Paley's *Natural Theology* at this time, see Fyfe, 'Publishing and the Classics'.

[67] In fairness to Paley, his concern was not (as in the case of Darwin) to explain why there are certain specific complex things rather than others; his concern was to explain why complex things exist in the first place: Brunnander, 'Did Darwin Really Answer Paley's Question?' It should be noted that both Huxley and Darwin used Paley's literary strategies to promote a naturalistic approach to the biosphere: Lightman, *Victorian Popularizers of Science*, 372–7.

[68] Kinch, 'Geographical Distribution and the Origin of Life', 113–17; Rupke, 'Neither Creation nor Evolution'.

[69] Whewell, *Philosophy of the Inductive Sciences*, vol. 2, 36. As has often been pointed out, Whewell's theory of induction is open to criticism: see, for example, Snyder, 'The Mill–Whewell Debate'. Such as Whewell's authority at the time that Darwin included a quotation from Whewell alongside one from

that *connects* and *unifies* the data. Such a theory, Whewell asserted, allows the 'colligation of facts', establishing a new system of relations with each other, unifying what might have otherwise been considered to be disconnected and isolated observations. The 'pearls' were the often puzzling observations that Darwin had accumulated; but what was the best string on which to thread them? Two representative sets of observations may be noted.

First, Darwin noted that many creatures possessed 'rudimentary structures', which have no obvious or predictable biological function—such as the nipples of male mammals, the rudiments of a pelvis and hind limbs in snakes, and wings on many flightless birds. How might these be explained on the basis of Paley's theory, which stressed the importance of the individual design of species? Why should God design redundancies? Second, Darwin's research voyage on the *Beagle* disclosed an uneven geographical distribution of life forms throughout the world.[70] In particular, Darwin was impressed by the peculiarities of remote island populations, such as the finches of the Galápagos islands. While Paley's doctrine of special creation could account for this, it did so in a manner that seemed forced and unpersuasive.

So what could be inferred from these puzzling observations, to which others could easily be added?[71] What was the best explanation of these observations? The best string to connect them? While Darwin knew that his theory of natural selection was not the only possible explanation of the biological data, he believed that it succeeded in making better sense of observations that were otherwise puzzling. The hypothesis of natural selection, he argued, possessed greater explanatory power than its rivals, such as the doctrine of independent acts of special creation, as set out in the writings of William Paley, or Lyell's proposal of 'autochthonous generation'.[72] 'Light has been shown on several facts, which on the belief of independent acts of creation are utterly obscure.'[73]

Yet while Darwin's theory illuminated natural phenomena, Darwin was quite clear that his theory *did* not predict, and *could* not predict. In a letter praising the perspicuity of F. W. Hutton, Darwin singled out this point for special comment.

Francis Bacon at the beginning of *The Origin of Species*. (A third epigram by Joseph Butler was added in the second edition.)

[70] For a good account, see Bowler, 'Geographical Distribution in the *Origin of Species*'.

[71] For example, Darwin was puzzled by the question of fossil vertebrate succession: see Keynes, *Fossils, Finches and Fuegians*, 99–114; Brinkman, 'Charles Darwin's *Beagle* Voyage', 364–9.

[72] Darwin's argumentative analysis parallels what is now known as 'inference to the best explanation': Nola, 'Darwin's Arguments in Favour of Natural Selection and against Special Creationism'.

[73] Darwin, *Origin of Species*, 203. Darwin regularly uses the phrase 'several facts' throughout *The Origin of Species* to refer to 'a body of observations'.

> He is one of the very few who see that the change of species cannot be directly proved, and that the doctrine must sink or swim according as it groups and explains phenomena. It is really curious how few judge it in this way, which is clearly the right way.[74]

Darwin's phrase 'the doctrine must sink or swim according as it groups and explains phenomena' is significant. The nature of the scientific phenomena was such that prediction was not possible for Darwin. This led some philosophers of science, most notably Karl Popper, to suggest unwisely that Darwinism was not really scientific.[75]

Yet more recent studies, especially in the philosophy of biology, have raised interesting questions about whether prediction really is essential to the scientific method. This issue emerged as important in the pre-Darwinian nineteenth-century debate between William Whewell and John Stuart Mill over the role of induction as a scientific method.[76] Whewell emphasized the importance of predictive novelty as a core element of the scientific method; Mill argued that the difference between prediction of novel observations and theoretical accommodation of existing observations was purely psychological, and had no ultimate epistemological significance. Darwin, however, added a comment to the final edition of the *Origin of Species*, making it clear that his theory was to be tested by its capacity to accommodate the observational evidence.

> It can hardly be supposed that a false theory would explain, in so satisfactory a manner as does the theory of natural selection, the several large classes of facts above specified. It has recently been objected that this is an unsafe method of arguing; but it is a method used in judging the common events of life, and has often been used by the greatest natural philosophers.[77]

[74] Darwin, *Life and Letters*, vol. 2, 155. Hutton deserves attention as a perceptive interpreter of Darwin: see, for example, Stenhouse, 'Darwin's Captain'.

[75] Popper, 'Natural Selection and the Emergence of Mind'. This is not, of course, to exempt Darwin from criticism on points of importance. It is surely fair, for example, to note that Darwin's concept of variation within a species seems to be shaped by a very nineteenth-century understanding of fixed social hierarchies: Rose and Rose, 'The Changing Face of Human Nature', 10.

[76] Snyder, 'The Mill–Whewell Debate'. Snyder elsewhere argues that Whewell's views on induction have been misunderstood, and merit closer attention as a distinctive approach: Snyder, 'Discoverers' Induction'.

[77] Darwin, *Origin of Species*, 6th edition, 444.

Darwin fits comfortably into the tradition of natural philosopher, as this evolved from Bacon to Newton. Yet his contribution to its future development arguably lies fundamentally in an insight articulated most clearly in the closing statement of *The Descent of Man* (1871):

> We must acknowledge, as it seems to me, that man with all his noble qualities, with sympathy which feels for the most debased, with benevolence which extends not only to other men but to the humblest living creature, with his god-like intellect which has penetrated into the movements and constitution of the solar system – with all these exalted powers – Man still bears in his bodily frame the indelible stamp of his lowly origin.[78]

The passage clearly strikes a moral tone, locating humanity within the natural order, with imperatives to sympathy and benevolence to other components of that order. Yet it also raises the question of the continuing impact of humanity's evolutionary history on its present condition. Although Darwin confines his reflection to humanity's 'bodily frame', others would suggest that human instincts and patterns of thought might similarly be shaped by the evolutionary process.[79]

Yet where Darwin can be argued to have remained within the broad tradition of natural philosophy, his leading interpreter pressed for the separation of philosophy and natural science, and raised concerns about those who suggested that nature could be a moral inspiration for humanity.

Thomas H. Huxley: The Separation of Philosophy and Natural Science

Although many leading empirical philosophers of the middle of the nineteenth century saw the natural sciences and philosophy as interconnected,[80] Huxley made a bold and decisive distinction between philosophy and science which shaped popular thinking in England during the final quarter of the nineteenth

[78] Darwin, *Descent of Man*, vol. 2, 405.

[79] For this agenda, see Wilson, *Sociobiology*, 5. Critics might point out that this approach seems to rest on the somewhat un-Darwinian belief that human nature, unlike the remainder of the biosphere, was essentially 'fixed' in the Pleistocene epoch, and has not evolved significantly since then: see, for example, Rose and Rose, 'The Changing Face of Human Nature', 12–13. For a fuller statement of these and other concerns, see Rose and Rose, eds., *Alas, Poor Darwin*.

[80] For a good discussion of the 'continued entanglement of science and philosophy even as they were being prized apart', see Cowles, 'The Age of Methods' (quotation at p. 722).

century. Philosophy is the 'answer to the question, What can I know?', and science is the response 'to the inquiry, What do I know?'[81] By the later Victorian period, many saw Huxley as embodying the *persona* of this emerging scientific culture, creating a normative public understanding of science through his public lectures and publications.[82] Huxley's introductory 'science primers' secured him a wide popular readership, and helped shape the new cultural confidence in the natural sciences. 'Natural philosophy' now came to be seen as an outdated way of referring to the natural sciences, which had been left behind as a result of the rapid scientific advances of the nineteenth century.

So what *do* we know? Huxley's answer, though easily stated, is perhaps less easily applied. 'In strictness all accurate knowledge is **Science**; and all exact reasoning is scientific reasoning.'[83] Yet this statement, found in what was intended to be a popular science textbook, is probably to be seen as a pedagogical overstatement, formulated with the need for simplicity of argument in mind. Elsewhere, Huxley is much more guarded and cautious, concerned to emphasize what *can* be known through science, while applying the maxim *ignoramus et ignorabimus* ('we do not know and we will not know') to the shadowy region that lay beyond the application of the scientific method.[84] Yet Huxley's influential formulation of the relation of science and philosophy can be seen to fit into an overall cultural narrative relating 'how science, through the provision of cognitive norms, came to serve as a model for all forms of purposive behaviour.'[85]

While some argued that Huxley's celebrated agnosticism was ultimately devised to evade awkward questions about the grounds and justification of his own ideas,[86] Huxley himself saw it as the inevitable and proper outcome

[81] Huxley, *Hume*, 48–9.

[82] See especially White, *Thomas Huxley*, 32–66, 100–34. For Huxley's dislike of the term 'scientist', which he considered to be a vulgar Americanism, see White, *Thomas Huxley*, 1–2. Huxley's networks were also of significance to his agendas. He was, for example, a member of the X-Club—a private society whose members could 'plot together on how to achieve common goals, such as the advancement of research, the infiltration and control of important scientific institutions and societies, and the bid to undermine the cultural authority of the Anglican clergy': Lightman, 'Huxley and Scientific Agnosticism', 272. In the 1880s, comparisons were often noted between Hume and Huxley on this point: see, for example, the discussion of 'Huxley's Hume' in McCosh, *Agnosticism of Hume and Huxley*, 42–56.

[83] Huxley, *Introductory Science Primer*, 16 (bolding in original).

[84] Huxley, 'The Progress of Science', 104. For a good account of Huxley on the limits of science, see Stanley, *Huxley's Church and Maxwell's Demon*, 80–118.

[85] This is Gaukroger's own summary of his historical investigations of the emergence and triumph of a scientific culture: Gaukroger, *Civilization and the Culture of Science*, 423. See also his more philosophical reflections on this theme in Gaukroger, *The Failures of Philosophy*, 261–82.

[86] Desmond, *Huxley*, 378, 389–90; Lightman, 'Huxley and Scientific Agnosticism', 272–4.

of a scientific approach to knowledge, which was willing to accept both the provisionality and the fallibility of its conclusions. He made this point as follows in a series of lectures to 'working men', delivered at the Museum of Practical Geology in 1862:

> Men of science do not pledge themselves to creeds; they are bound by articles of no sort; there is not a single belief that it is not a bounden duty with them to hold with a light hand and to part with cheerfully, the moment it is really proved to be contrary to any fact, great or small.[87]

Huxley was not, however, entirely consistent on this point, elsewhere pointing out the need to accept certain unprovable assumptions entailed by the scientific method:

> The one act of faith in the convert to science, is the confession of the universality of order and of the absolute validity in all times and under all circumstances, of the law of causation. This confession is an act of faith, because, by the nature of the case, the truth of such propositions is not susceptible of proof. But such faith is not blind, but reasonable.[88]

Although acknowledging the importance of Hume as a philosopher, Huxley's emphasis on the advancement of human civilization through scientific progress led him to detach scientific research, reasoning, and application from its earlier philosophical contextualizations. Whereas earlier natural philosophers saw nature as an important stimulus to ethical reflection, Huxley—following Hume—declined to ground moral reflection in the structures or processes of the natural world.

For Huxley, scientists were 'graduates in the University of Nature',[89] who had learned how to study the order and laws of the natural world. In a lecture of 1868, Huxley declared that:

[87] Huxley, *On Our Knowledge of the Causes of the Phenomena of Organic Nature*, 150. It is possible that Huxley is here responding to the Humean predicament in demonstrating how a purely inductive science, founded solely on the interpretation of experience, can prove its own foundation.

[88] Francis Darwin, ed., *The Life and Letters of Charles Darwin*, vol. 2, 200. Remarks such as this led Paradis to conclude that Huxley's emphasis on a justified belief in natural order trumped his agnosticism, which tended to be deployed primarily in polemical contexts: Paradis, *T. H. Huxley*, 96–7; cf. 100–13.

[89] For this phrase, see Huxley, *Science and Hebrew Tradition*, 8.

> Nature is still continuing her patient education of us in that great university, the universe, of which we are all members. Those who take honours in Nature's university, who learn the laws which govern men and things and obey them, are the really great and successful men in this world.[90]

Yet while Huxley's early essays on the importance of science education reproduce some themes traditionally associated with natural philosophy,[91] particularly achieving harmony with the natural order, his enthusiasm for this older vision appears to have faded over time.

While some later Victorian thinkers tended to see evolution in terms of a progressive ethic of self-improvement,[92] Huxley rejected this interpretation of the evolutionary process. His 1893 Oxford lecture on 'Evolution and Ethics' is perhaps one of the most important expressions of the gap that Huxley had opened up between the study of the natural order and the human quest for civilization and values.[93] Where earlier transformist accounts of evolution were easily assimilated to an ethic of self-improvement, Darwin's approach—especially in its appeal to Malthus's *Essay on Population*—presented evolution in terms of competition between species.[94]

In some prefatory comments to this lecture, Huxley identified his intention to resolve 'the apparent paradox that ethical nature, while born of cosmic nature, is necessarily at enmity with its parent'.[95] For Huxley, human beings are animals who have triumphed in the 'struggle for existence' on account of their capacity for violence and destruction. Yet with the emergence of civilization, the animal characteristics which allowed human beings to triumph in

[90] Huxley, *Science & Education*, 84–5. This lecture was given at South London Working Men's College.

[91] For example, Huxley's 1852 essay 'On the Educational Value of the Natural History Sciences' declares that a student ignorant of physiology is 'blind to the richest sources of beauty in God's creation; and unprovided with that belief in a living law, and an order manifesting itself in and through endless change and variety': Huxley, *Science & Education*, 65.

[92] Lamarck's theory of evolution was clearly open to this interpretation; see, for example, Spenser's reworking of this approach as an extension of a philosophy of self-actualization: Bowler, 'Darwinism and Victorian Values', 138. For late Victorian reflections on the political implications of Darwinism and Lamarckism, see Hale, *Political Descent*, 335–51.

[93] For the best study of this lecture, see Paradis, '*Evolution and Ethics* in Its Victorian Context'. The original text of the lecture can be found in Huxley, *Evolution and Ethics*, 46–116. The important prolegomenon to this lecture, written in June 1894, can be found in Huxley, *Evolution and Ethics*, 1–45. The prestigious Romanes Lectures were established as an annual public lecture at the University of Oxford. The first was delivered in 1892 by W. E. Gladstone, the British Prime Minister; Huxley's was the second such lecture.

[94] For the political impact of Darwin's theory in Victorian England, see Hale, *Political Descent*.

[95] Huxley, *Evolution and Ethics*, viii.

this struggle are no longer seen to be 'reconcilable with sound ethical principles'.[96]

Ethics, for Huxley, is now a principled resistance to precisely those animal qualities that had earlier secured human domination of the living world, and the Darwinian processes that underlie them. We must therefore learn to conquer and subdue the natural animal instincts that linger within us. Our hereditary history continues to shape our present—and it must be resisted, even though it cannot be eradicated. 'The practice of that which is ethically best – what we call goodness or virtue – involves a course of conduct which, in all respects, is opposed to that which leads to success in the cosmic struggle for existence.'[97] While the natural process of evolution may explain the *origins* of human ethics, it cannot now function as the *basis* of ethics. Evolution 'may teach us how the good and the evil tendencies of humanity may have come about; but, in itself, it is incompetent to furnish any better reason why what we call good is preferable to what we call evil than we had before'.[98]

Huxley thus offers a Darwinian basis for the rejection of traditional natural philosophies (though he does not use the term 'natural philosophy' in this lecture).[99] This is expressed most forcefully in his rejection of the 'stoical summary of the whole duty of man, "Live according to nature"', which seems to imply that the cosmic process is 'an exemplar for human conduct'.[100] At the risk of overstating this important point, Huxley comes close to suggesting that the close observation of natural world shows us how *not* to behave. This contrasts sharply with one of the core themes of early modern natural philosophy—that of learning *from* nature, and not simply learning *about* nature. This idea—perhaps most dramatically expressed by Alexander Pope in his *Essay on Man* (1733–4): 'Great Nature spoke; observant Men obey'd'[101]—was now firmly relegated to an outmoded and unusable past.

[96] Huxley, *Evolution and Ethics*, 52–3.

[97] Huxley, *Evolution and Ethics*, 81–2. Note also the 1895 comments of Balfour, *The Foundations of Belief*, 70: 'Not only does Nature take no interest in our general education, not only is she quite indifferent to the growth of enlightenment, unless the enlightenment improve our chances in the struggle for existence, but she positively objects to the very existence of faculties by which these ends might, perhaps, be attained.'

[98] Huxley, *Evolution and Ethics*, 80.

[99] Huxley's target here is the form of eighteenth-century cosmic optimism, found in writers such as Alexander Pope and Joseph Addison, which Basil Willey famously described as 'cosmic Toryism': Willey, *The Eighteenth Century Background*, 43–56.

[100] Huxley, *Evolution and Ethics*, 73–4. Huxley argues that this difficulty might be averted if the term 'nature' is here understood as 'pure reason'.

[101] Pope, *An Essay on Man*, 37 (III.199). The personification of nature, evident in Pope's line, was widely used at this time to enhance its rhetorical authority: Daston, 'Attention and the Values of Nature in the Enlightenment', 123–5.

This chapter has documented the gradual parting of the ways between the natural sciences and philosophy, and the dissolution of the notion of 'natural philosophy' as a viable field of study and research. Yet this need not be the end of the story of natural philosophy. In the second part of this work, we shall consider how this category might be reimagined and repurposed, given the increasingly urgent need of supplementing a scientific approach to the natural world with one that is attentive to wider questions, raised—yet not answered—by the natural sciences.

PART II

A RECONCEIVED NATURAL PHILOSOPHY: EXPLORING A DISCIPLINARY IMAGINARY

6

Reconceiving Natural Philosophy

Laying the Foundations

In the first part of this work, I considered the gradual emergence and subsequent fragmentation of natural philosophy. This 'long, broad, and deep conversation with its history'[1] allows us to understand how earlier generations framed natural philosophy, what its benefits were understood to be, and why its vision faded. Natural philosophy does not emerge from this analysis as a static or fixed, but rather as an *evolving* discipline, developing capacities to engage new questions, while retaining an interest in older ones. While some readers might be surprised at the length of this genealogical analysis, it is essential to any informed attempt to learn from the past, and reflect on how we might renew and retrieve this enterprise. If wrong turns have been taken, it is clearly important to consider how things can be remedied.

In the second part of this study, I will make a case for the retrieval of a natural philosophy, and explore how this might be done. While there are good reasons for suggesting that this could be a productive and timely undertaking, it is not a new or original enterprise, in that such a process of critical reappropriation of earlier forms of natural philosophy can be instanced from the medieval, Renaissance, and early modern periods.[2] Contemporary concerns about the interaction between humanity and the natural world, not least the urgent question of fostering human respect for nature in the light of environmental degradation, suggest the need for a conceptual space which is clearly comparable to that once associated with 'natural philosophy', as this was understood in the early modern period.

My concern in this book is with retrieving this conceptual space, developing an extended disciplinary imaginary which—like its classic predecessors—links attentiveness to the natural world to the quest for understanding, wisdom, and well-being. Yet critics will rightly point out that such a proposal

[1] Antognazza, 'The Benefit to Philosophy of the Study of Its History', 161.

[2] For a good example, see the creative reworkings of Pliny's *enkuklios paideia* at comparable points in western intellectual history: Doody, *Pliny's Encyclopedia*. Comparable strategies of retrieval can be found in other disciplines, including theology: see, for example, Sarisky, ed., *Theologies of Retrieval*.

faces significant challenges. It seems both pointless and impossible. 'Natural philosophy', such critics might reasonably suggest, is simply an obsolete form of natural science, and there would seem to be little to be gained from retrieving such an outdated way of thinking. In any case, the subsequent rise of multiple intellectual disciplines has led to an irreversible fragmentation of intellectual discourse and reflection. Both these criticisms are important, and need to be taken seriously. In the next two sections, I shall engage both these concerns.

Why We Need a Critical Retrieval of Natural Philosophy

In beginning this process, it is helpful to reflect on why engaging nature is so important and challenging. Nature designates something vast, unconstrained by the categories of the human mind and uncontrolled by instrumental reason, which demands we see it as it really is, rather than be satisfied by these imaginatively diminished partial accounts of its greater whole. Where instrumental reason seeks to reduce and control nature, natural philosophy is driven by the instinct to preserve, respect, and appreciate the complexity of nature, allowing itself to be mastered by nature, rather than imposing its controlling categories upon it. Nature is experienced as a conceptual immensity, which resists reduction to propositional language. This book is an exercise in the philosophy of retrieval, which sets out to explore what might be gained through a critical reappropriation of earlier forms of natural philosophy, particularly those that flourished during the remarkably creative period in which a scientific culture gradually emerged during the early modern era.

Although it is often suggested that natural philosophy is simply a transient formative episode in the emergence of the natural sciences, this judgement needs to be challenged. As Walter Lüthy points out, we need to explain why seventeenth-century science was seen as philosophical, and why the philosophy of this period was seen as scientific.[3] It is, of course, entirely possible that this is an insight specific to the social and cultural specifics of that bygone age; yet it is also possible that these writers had grasped something that we need to recover.

Recent work on the emergence of disciplinarity in western academia has highlighted the historical situatedness and contingency of our own

[3] Lüthy, 'What to Do with Seventeenth-Century Natural Philosophy?', 175. This analysis could be extended to take account of the wide range of intellectual territories enfolded within this field.

contemporary 'disciplinary imaginary',[4] and the damage this has inflicted on attempts to create and maintain holistic accounts of the natural world. Yet the institutionalized disciplinary distinctions and divisions that seem permanent and natural are social creations—ways of imagining the intellectual geography and organization of human knowledge production. The ensuing 'disciplinary ordering of knowledge production' is a matter of contingency,[5] and is open to reappraisal and reimagining. If the early modern imaginary—or group of imaginaries—that we call 'natural philosophy' is seen to offer benefits that are otherwise lost, it is possible to construct a new disciplinary imaginary that will retrieve these benefits.

There are good reasons for suggesting that a critical retrieval and reappropriation of a natural philosophy in the twenty-first century is both timely and appropriate. Such a retrieval does not amount to an artificial resuscitation of past habits of thought and imagination, but is rather to be seen as a critical assessment of what was so important about natural philosophy in the past, and how this can be recaptured without trapping us in a bygone age. Five considerations may be noted, to which others might easily be added.

1. There is growing interest in recognizing the significance of early modern natural philosophy in stimulating the art of living well, particularly in terms of fostering 'practical regimens and formative disciplines' that help enact human wisdom.[6] Daniel Defoe's novel *Robinson Crusoe* (1719) describes how a shipwrecked sailor finds himself alone on a desert island, and is transformed by this 'hermeneutic and therapeutic' encounter with his untamed natural environment.[7] At times, Crusoe seems to be portrayed as someone who stands in the tradition of Robert Boyle and Thomas Sprat, and sees the natural order of a desert island, unspoiled by human agency, as a means of enabling a transformation of life and a purification of thought.

There was a clear moral dimension to early modern natural philosophy, seen especially in its 'capacity to serve as a practice that cultivates the moral person'.[8] Classic Chinese natural philosophy similarly echoed the importance

[4] I here draw on Goodstein, *Georg Simmel and the Disciplinary Imaginary*, especially her discussion of how and why the 'modern disciplinary imaginary' came into being (13–134).

[5] Goodstein, *Georg Simmel and the Disciplinary Imaginary*, 4.

[6] Corneanu, *Regimens of the Mind*, 7. Corneanu suggests that early modern writers such as Bacon, Boyle, and Hooke retrieved the ancient view of philosophy as 'fundamentally *padeia* or *askesis* rather than simply *theoria*'.

[7] Corneanu, 'Passions, Providence, and the Cure of the Mind'.

[8] Sprat, *History of the Royal Society of London*, 342–3. The background here is Edward Reynolds's *Treatise of the Passions and Faculties of the Soul of Man* (1640), which suggests that ethical and intellectual advance is predicated on the 'healing of the mind' through appropriate mental regimes.

of living in harmony with the natural world.[9] These perspectives may helpfully be set alongside the classical Stoic view that the ultimate human goal was to 'live in conformity and harmony with nature' (*congruenter naturae convenienterque vivere*).[10] Yet this moral aspect of the engagement with nature is now largely absent from scientific reflection. Steven Weinberg is one of many to insist that the natural sciences have now abandoned any interest in the human quests for goodness, justice, or love, having outgrown such concerns.[11] While this view is influential, it is clear that something of potential importance has been lost. The wisdom of our age may indeed be that natural philosophy is simply an older and outdated version of the natural sciences; an alternative judgement, however, is that the latter present a restricted and impoverished account of the broader vision of natural philosophy.

2. Engaging with the natural world was seen by many early modern natural philosophers as a means of correcting the deficiencies of approaches to philosophy based purely on human reasoning. These, some argued, were limited by human weakness or compromised by original sin, and needed recalibration against an external standard.[12] While such attitudes can be found earlier, as in Bede's meticulous attempt to use an evidence-based approach to the study of nature in eighth-century England,[13] they are particularly evident in Francis Bacon's war against the 'Idols of the Mind'—that is, preconceived or culturally shaped attitudes which distort or impoverish our vision of the natural world.

Bacon thus advocated the practice of natural philosophy—especially the meticulous collection of observations of the natural world and the use of experimental procedures—as the best means of countering inbuilt human propensities to error and self-deception. Natural philosophy, he suggested, might expand our minds 'to the amplitude of the world' rather than reduce 'the world to the narrowness of [our] minds'.[14] This point is found in many recent reflections on the human encounter with an untamed nature, particularly through 'wilderness experiences'. These seem to facilitate a receptive mode of

[9] Chenyang, 'The Confucian Ideal of Harmony'. For a modern reappropriation of this theme in the works of Fang Dongmei (Thomé H. Fang), see Hermann, 'A Critical Evaluation of Fang Dongmei's Philosophy of Comprehensive Harmony'.

[10] The view of the Stoic writer Cato, cited in Cicero, *de finibius bonorum et malorum*, III.vi.26.

[11] Weinberg, *To Explain the World*, 45.

[12] For a thorough exploration of this theme, see Harrison, *The Fall of Man and the Foundations of Science*. For related themes focusing on figures such as Prometheus, see Ziolkowski, *The Sin of Knowledge*.

[13] Ahern, *Bede and the Cosmos*.

[14] Bacon, *Works*, vol. 2, 436. Cf. Giglioni, 'Learning to Read Nature'.

existence, in which we become fully awake to the world around us, creating a heightened state of awareness of its beauty and unfathomed significance.[15]

3. Early modern forms of natural philosophy were not merely attempts to make sense of the natural world, but also means of enabling human beings to flourish within this world. Although the human proclivity to want to make sense of things is now perhaps the most familiar feature of this discursive space,[16] this perception has been shaped by the tendency on the part of popular historians of science to assimilate 'natural philosophy' to 'natural science'. Many early modern natural philosophers saw their task as including the cultivation of respect and reverence for the natural order, appreciating its affective impact on humanity, and understanding better the place of human beings within this realm and the patterns of behaviour that were appropriate to their location within nature.

Robert Boyle, for example, spoke of the natural philosopher as a 'priest in the temple of nature', encouraging an attitude of reverence towards the natural world, which in turn recognizes the beauty (and not merely the intelligibility) of nature.[17] The nineteenth-century American wilderness authors Ralph Waldo Emerson and Henry David Thoreau took a holistic view of humanity's relation with the natural order, seeing this as saturated with moral and spiritual possibilities.[18] It is clear that any attempt to encourage human respectfulness towards the natural order will have to go beyond—and against—the tendency within the natural sciences to think of a scientific instrumentality which is about achieving power and control over the natural world.[19]

4. According to the mathematician Marston Morse, 'the urge to understand is the urge to embrace the world as a unity.'[20] The psychologist Abraham Maslow expressed a similar point in his notion of a 'peak experience'—a 'clear perception (rather than a purely abstract and verbal philosophical acceptance) that the universe is all of a piece and that one has his place in it'.[21] In the

[15] McDonald et al., 'The Nature of Peak Experience in Wilderness'.

[16] Chater and Loewenstein, 'The Under-Appreciated Drive for Sense-Making'.

[17] Boyle, however, only uses this image once, and his idea of treating natural philosophy as a form of devotional practice secured little traction: Gaukroger, *Emergence of a Scientific Culture*, 151–3. For the best discussion of Boyle's notion of a 'priest of nature', see Ben-Chaim, 'Empowering Lay Belief', 52–4.

[18] Aaltola, 'Wilderness Experiences as Ethics', 284.

[19] Cf. Horkheimer, *Eclipse of Reason*, 176: 'The disease of reason is that reason was born from man's urge to dominate nature.' For analysis of the problem, see Leiss, 'Modern Science, Enlightenment, and the Domination of Nature'; Denham, 'The Cunning of Unreason and Nature's Revolt'.

[20] Morse, 'Mathematics and the Arts', 90.

[21] Maslow, *Religion, Values and Peak-Experiences*, 59.

seventeenth century, natural philosophy provided a means of affirming the overall unity of the natural world while respecting its individual elements disclosed through observation—including the human observer of nature. Most forms of natural philosophy of this period were integrative, their fluid and porous boundaries loosely coordinating what we might today think of as the quite distinct disciplines of the natural sciences, philosophy, ethics, music, and theology. Goethe, for example, was often critical of philosophical abstractions, holding that nature had to be grasped as a coherent whole.[22]

> With any given phenomenon in nature – and especially if it is significant or striking – we should not stop and dwell on it, cling to it, and view it as existing in isolation. Instead we should look about in the whole of nature to find where there is something similar, something related. For only when related elements are drawn together will a whole gradually emerge that speaks for itself and requires no further explanation.[23]

In comparing the roles of scientists and poets, the Irish Victorian poet William Allingham contrasted a scientific mentality which wanted to 'break up the whole into little bits, for analysis' with its poetic counterpart, which 'reconstructs the shattered world, and shows it complete and beautiful.'[24] Science thus concerned itself with the dissection of the natural world, breaking down its complexities into its core factual components; poetry brought these back together again, allowing us to discern their deeper significance. Both these elements were enfolded within older visions of natural philosophy; they are now needlessly seen as disconnected.[25]

5. Reflecting on older approaches to natural philosophy might stimulate the exploration of fresh ways of considering the relationship between philosophy and the natural sciences, which has become somewhat strained in recent years through the rise of strong forms of 'scientism', which see the natural sciences as being capable of answering all valid human questions, and thus displacing philosophy as the natural means of exploring issues such as value and meaning.[26] This represents an important challenge to philosophy, and requires careful reflection. One way of responding to this is to return to a period when such dialogues were commonplace, and assumed to be grounded

[22] A point stressed by Bortoft, *The Wholeness of Nature.* [23] Goethe, *Scientific Studies*, 203.

[24] Allingham, 'On Poetry', 525. For an account of the debates between science and poetry in the Victorian age, see Huber, 'Competing for Eternity'.

[25] For a corrective, see McLeish, *The Poetry and Music of Science.*

[26] For comment, see Gaukroger, *The Failures of Philosophy*, 261–76.

in some broader understanding of the natural world—a theme that was articulated in the concept of natural philosophy, particularly those approaches which emerged during the early modern period. There is no question of returning to those viewpoints; the intention is to bring them into critical and constructive conversation with the concerns of the twenty-first century, and see what illumination they might offer.

These concerns clearly provide a motivation for retrieving a viable form of natural philosophy to illuminate and inform some of the dilemmas and concerns of the twenty-first century. If a discipline called 'natural philosophy' had not existed in the past, we would need to invent it today. So can we, like Kepler and Newton, think of our universe as a coherent whole, capable of being represented by a single unified theory? Or is it so complex that the most we can hope for is to accumulate a series of fragmented insights that can only be partially integrated? And what is the role of the imagination in helping us to render this world, both in terms of visualizing its structures, while at the same time developing the 'constitutive linkage of imagery with affect',[27] in order that our accounts of the world enfold the emotions, feelings, and desires which mark a proper human relationship with the world?

To begin to answer this question, we need to explore the consequences of a general trend towards disciplinary *specificity*, which now so often takes the form of disciplinary *isolation*.

The rapid growth of scientific disciplines has necessarily and legitimately led to an irreversible fragmentation of what was once seen as the single undertaking of 'natural philosophy'. Any proposal to retrieve the category of natural philosophy would require a reversal of this historical process, which would seem to be quite unrealistic. This is a serious concern, and we shall explore it further in the next section.

Disciplinary Specificity and the Fragmentation of Knowledge

The classic expressions of early modern natural philosophy—such as Kepler's *Harmonices Mundi* (1619)—often seem to embody the ideal of the *uomo universale*, the Renaissance polymath who was able to master multiple disciplines and integrate them within a coherent worldview.[28] The shape of

[27] Lennon, *Imagination and the Imaginary*, 1.

[28] A good example, not referenced earlier in this study, lies in the Baroque polymath Georg Philipp Harsdörffer (1607–58), whose contributions to the development of a natural philosophy were complemented by significant contributions to the worlds of art, music, and mathematics: Heinecke,

professional academic culture of the twenty-first century, however, is characterized more by the phenomenon of the 'disciplinary silo' or 'intellectual ghetto' which is disconnected from other fields of academic endeavour. It is increasingly difficult to master the literature of even an academic subdiscipline, let alone make connections with others, particularly in different disciplinary domains. Harvey Graff rightly points out that we must avoid thinking of disciplinarity as a problem, or idealize the intellectual past as if it were a lost interdisciplinary paradise.

> Consider claims made for the classical era in Greece, the Renaissance, or the Enlightenment in relationship to disciplines and interdisciplines. Plato cannot be claimed as the first interdisciplinary thinker if there were no disciplines. Nor can Aristotle. This assertion is anachronistic, indeed ahistorical. Although equating major intellectual movements before the nineteenth century with interdisciplinarity can be a powerful force rhetorically and a rhetorical reach for precedent and genealogy, it distorts more than it clarifies.[29]

Disciplinarity evolved for good reasons. It is not an intellectual aberration, but a legitimate and arguably necessary development, which nevertheless has clear disadvantages—above all, a tendency towards an impoverishing compartmentalization of intellectual discussion, and a fragmentation of our vision of reality. While the practical challenges of mastering a substantial body of information across disciplines must be given due weight,[30] there are other disciplinary factors that underlie the potential fragmentation of human knowledge. Perhaps the most significant of these relates to methodological diversity: each intellectual discipline offers its own angle of approach to the study of the natural world, developing research methods which are adapted to its specific objects of study. 'It is the nature of the object that determines the form of its possible science.'[31]

'Naturphilosophie bei Georg Philipp Harsdörffer'. On the phenomenon of the Renaissance polymath, see Burke, *The Polymath*, 26–46.

[29] Graff, 'The "Problem" of Interdisciplinarity in Theory, Practice, and History', 790. Cf. Lloyd, *Disciplines in the Making*.

[30] For the problems of information overload in the natural philosophies of the later Renaissance, and the strategies devised to deal with this, see Ogilvie, 'The Many Books of Nature'; Blair, *Too Much to Know*, 62–116.

[31] Bhaskar, *The Possibility of Naturalism*, 3. For an extensive discussion of this point, see McGrath, *The Territories of Human Reason*, 19–89.

Although I have used the language of perspectives or angles of approach thus far to point to the need for multiple investigative and representational approaches to the natural world, others prefer to speak of 'levels'. For example, in their influential 1958 paper 'The Unity of Science as a Working Hypothesis',[32] Peter Oppenheim and Hilary Putnam argued for a stratified 'layer-cake' model of reality, consisting of a series of layers or levels within the natural world.[33] For each such layer or level in nature, there is a corresponding form of science, adapted to the specific identity of that particular level or stratum. The methods and concepts of that form of science are determined by the object under study in that specific stratum, rather than by methods of investigation which are applicable to other levels. The nature of the object under consideration thus determines the form which its study should take. Although Oppenheim and Putnam suggest that laws of the sciences are in principle formally derivable from lower-level laws, which many argue are ultimately those of physics, others (such as John Dupré) argue this model needs to be reframed to retain its stratified approach without apparently entailing such a reductionist metaphysic.[34]

Yet there is no good reason why such multiple insights, arising from different disciplines and their associated methodologies, should not be held together in some broader vision of reality. While the term 'natural philosopher' has now fallen out of general use, many would point to Albert Einstein as the twentieth century's nearest equivalent. Widely acclaimed, like Newton before him, as a scientific 'genius', Einstein was noted for his commitment to the quest for intellectual unification.[35] Many were impressed by Einstein's willingness, as a scientist, to articulate an ethical and political vision, particularly in the aftermath of the Second World War. Einstein often spoke of the importance of 'striving after the rational unification of the manifold'.[36] Where some would argue that, since science, ethics, and religion use different

[32] Oppenheim and Putnam, 'The Unity of Science as a Working Hypothesis'.

[33] The term 'level' fulfils many functions in science. We can speak of levels of abstraction, being, causation, description, explanation, function, and generality. For a good discussion of this point and its significance, see Craver and Bechtel, 'Top-Down Causation without Top-Down Causes', especially 548–51. See also Wimsatt, 'The Ontology of Complex Systems'.

[34] Dupré, *The Disorder of Things*. For a criticism of this approach, see Davies, 'Explanatory Disunities and the Unity of Science'.

[35] For an excellent account of Einstein's successive attempts to craft unified field theories, see van Dongen, *Einstein's Unification*. For the evolution of Einstein's thought, see Pais, *'Subtle Is the Lord…'*.

[36] Einstein, *Ideas and Opinions*, 44–9. Van Dongen points out that ten of Einstein's articles include the German term 'Einheitliche' in their title: van Dongen, *Einstein's Unification*, 58. See further Sauer, 'Einstein's Unified Theory Program'.

intellectual methods, they are for that reason *incompatible*, Einstein simply treated them as *different*.[37]

Yet although Einstein presented himself as aiming for unity, harmony, and simplicity in both scientific and non-scientific matters, the basis of that unity and harmony is ultimately somewhat elusive. Einstein in effect offers a *bricolage* of unintegrated insights and perceptions,[38] all of which are treated with respect, even if it remains unclear what methods he used to bring them together as a 'big picture' or 'unified theory'. Richard Crockatt points out there is 'a radical disjunction between the mental processes [Einstein] used to address scientific problems and those he employed to deal with moral, social, and political issues'.[39] It seemed as if Einstein was a 'compartmentalized individual whose separate selves operated almost independently of each other'.[40]

One possible explanation of this observation lies in Einstein's emphasis on the importance of intuition (*Anschauung*) in scientific thinking, which first became prominent in his 1909 paper 'On the Development of Our Intuition of the Existence and Constitution of Radiation'.[41] This idea, of course, is evident in earlier natural philosophers such as Goethe, who had a strong sense of the role of intuitive understanding in the human engagement with nature. 'The mind may perceive the seed, so to speak, of a relation which would have a harmony beyond the mind's power to comprehend or experience once the relation is fully developed.'[42]

More recently, the palaeontologist and evolutionary biologist Stephen Jay Gould, noted for his informed immersion in the humanities, set out an intellectual vision of 'quilting a diverse collection of separate patches into a beautiful and integrated coat of many colors'.[43] For Gould, the multiple disciplines enfolded within the natural sciences and humanities were all to be

[37] For Einstein's views on religion, see Jammer, *Einstein and Religion*. Einstein rejected the notion of a 'personal God', defining God as 'a superior mind that reveals itself in the world of experience': Einstein, *Ideas and Opinions*, 255. For the origins of the cultural myth that science and religion are permanently and necessarily at war with each other, see Hardin et al., eds., *The Warfare between Science and Religion*. For scholarly analysis of its failings, see Brooke, *Science and Religion*; Harrison, *The Territories of Science and Religion*.

[38] For the importance of this metaphor, see Johnson, 'Bricoleur and Bricolage'.

[39] Crockatt, *Einstein and Twentieth-Century Politics*, 34.

[40] Crockatt, *Einstein and Twentieth-Century Politics*, 35.

[41] Crockatt, *Einstein and Twentieth-Century Politics*, 36. Cf. Einstein, 'Entwicklung unserer Anschauungen über das Wesen und die Konstitution der Strahlung'. There are some interesting parallels here with Henri Poincaré's 1905 essay 'Du rôle de l'intuition et de la logique en mathématiques'. For a comparison of Poincaré and Einstein on this point, see Miller, 'Cultures of Creativity'. For the role of intuition in mathematics at this time, see Stekeler-Weithofer, *Formen der Anschauung*.

[42] Goethe, *Scientific Studies*, 9. For further discussion, see Mensch, 'Intuition and Nature in Kant and Goethe'.

[43] Gould, *The Hedgehog, the Fox, and the Magister's Pox*, 15. For Gould's relation to the humanities, see York and Clark, 'The Science and Humanism of Stephen Jay Gould'.

respected and valued for what they were—yet also on account of the greater vision of reality that they yielded when woven together. Gould thus resisted E. O. Wilson's account of the unity of human knowledge, which privileges the natural sciences,[44] arguing for

> a consilience, a "jumping together" of science and the humanities into far greater and more fruitful contact and coherence – but a *consilience of equal regard* that respects the inherent differences, acknowledges the comparable but distinct worthiness, understands the absolute necessity of both domains to any life deemed intellectually and spiritually "full," and seeks to emphasize and nurture the numerous regions of actual overlap and common concern.[45]

Gould does not offer a rigorous theoretical account of this process of 'jumping together'; his approach is rather to affirm that this can and should be done, and to point to an imaginative framework—such as 'quilting'—which enables his readers to visualize how this process of bringing and holding together multiple insights might work.

Gould's approach can be located within the growing interest within the philosophy of science at that time in conceptualizing unity as 'interconnection' rather than as reduction.[46] This approach was developed during the nineteenth century by the American philosopher Charles S. Peirce. In his discussion of the nature of theoretical explanation, Peirce points out that a good theory 'adds something to [observations]...because the addition serves to render intelligible what without it, is unintelligible.'[47] The philosophical enterprise is that of creating connections between seemingly disconnected observations, giving rise to a larger vision of the world. Where earlier rationalist accounts of nature often displayed intolerance of complexity and a fixation on reductive approaches to explanation,[48] there is growing sympathy for celebrating complexity, and finding ways of enfolding this within a more capacious vision of the world.

[44] Segerstrale, 'Wilson and the Unification of Science.'

[45] Gould, *The Hedgehog, the Fox, and the Magister's Pox*, 259. For a critical account of Gould's approach, see McGrath, 'A Consilience of Equal Regard.'

[46] See, for example, Grantham, 'Conceptualizing the (Dis)Unity of Science,' 141–5. Note also Giere's account of how 'overlapping theoretical perspectives' might be brought together: Giere, *Scientific Perspectivism*, 92–3. Giere here offers a 'perspectival realism' which affirms a pluralism of models about one and the same system without drawing anti-realist consequences. A similar view is found in Rueger, 'Perspectival Models and Theory Unification.'

[47] McKaughen, 'From Ugly Duckling to Swan,' 466.

[48] An approach found in Wilson, *Consilience*, 291: 'all tangible phenomena, from the birth of stars to the workings of social institutions, are based on material processes that are ultimately reducible, however long and tortuous the sequences, to the laws of physics.'

This point is integral to Mary Midgley's philosophical critique of reductionism, which is helpful in recapturing something of the ethos of what was once known as 'natural philosophy'. She argues that, while there is only one world, it is 'a big one'—and thus demands developing multiple methods of investigation and representation that are appropriate to its complexity.[49] 'For most important questions in human life, a number of different conceptual toolboxes always have to be used together.'[50] Science is part of this 'one world', and cannot cut itself off from other forms of knowledge within that world—many of which are not 'scientific' in nature. Science is thus interconnected with, and to some extent dependent upon, other forms of human knowledge production.[51]

This basic principle of intellectual respect for the complexity of the world leads Midgley to argue for multiple ways of engaging it, in order to grasp and capture that complexity.

> On the one hand, I want to emphasize that there really *is* only one world, but also – on the other – that this world is so complex, so various that we need dozens of distinct thought-patterns to understand it. We can't reduce all these ways of thinking to any single model. Instead, we have to use all our philosophical tools to bring these distinct kinds of thought together.[52]

Midgley's approach, taken alongside the other points made in this section, allows us to respond to this important concern about disciplinary proliferation and the ensuing intellectual fragmentation. On the one hand, these multiple disciples are to be seen as *necessary and proper*; yet on the other, each disciplinary approach (or ensemble of approaches) has to be recognized as *incomplete*, offering only a partial account of a complex world. Each such approach is to be respected; it is, however, to be seen as illuminating only certain aspects of nature, not its totality. We need, as Gould points out, to recognize the 'different light that each [discipline] can shine upon a common quest for deeper understanding of our lives and surroundings in all their complexity and variety',[53] and thus work towards a 'consilience of equal regard'.

Midgley championed philosophical resistance to the constriction of our mental visions of reality, stressing that the quality of our thinking ought to be

[49] Midgley, 'One World but a Big One'. Cf. Marmodora, 'Whole, but Not One'.
[50] Midgley, 'Dover Beach', 219.
[51] Midgley, *Science as Salvation*, 108.
[52] Midgley, *What is Philosophy for?*, 193. Cf. 201: There is 'one world but many windows'.
[53] Gould, *The Hedgehog, the Fox, and the Magister's Pox*, 251.

as deep and complex as the world that we hope to understand. To retrieve a natural philosophy is thus to map an intellectual domain enfolding many local territories of rationality, each with its own distinct identity and approach, yet which contributes towards a greater overall account of what Midgley terms our 'big world'. Each intellectual microcosm needs to be affirmed and respected for what it is, yet also seen as part of a greater vision that can properly be described as 'natural philosophy'.

Mary Midgley on Mapping the Natural World

In recent years, there has been growing interest in the use of 'spatialization metaphors' in making sense of our world,[54] in that these allow the visualization of complex domains of knowledge, enabling their distinct identities to be respected while at the same time disclosing ways in which they may be correlated.[55] Aware of the importance of such metaphors, Midgley developed the idea of using multiple maps of the world as a means of simultaneously affirming its fundamental unity, while being attentive to its complexity. For Midgley, the natural sciences have 'shown us the enormous complexity of the real world', and in doing so have made 'the idea of a single, convergent, reductive explanation out of date'. In trying to do justice to this complex world, we need to develop not a single map, but 'a collection of maps, all of them incomplete, which together gradually shape our understanding of a new piece of country'.[56] Philosophy, Midgley suggests, is 'always looking for helpful connections';[57] the metaphor of mapping allows the interconnectedness yet distinctiveness of individual disciplines to be affirmed, while insisting that they are to be seen as compatible and complementary accounts of a greater reality.

For Midgley, we thus have to use multiple maps, each of which is incomplete in itself, if we are to do intellectual and imaginative justice to our vast and complex natural world.[58] For example, a physical map of Europe shows us the features of its landscapes. A political map shows the borders of its nation states. The first might be useful to a tourist interested in mountain scenes, the second to a refugee seeking asylum. Different maps provide

[54] Lakoff and Johnson, *Metaphors We Live By*, 17.

[55] For illustration of such forms of intellectual spatialization, see Shiffrin and Börner, 'Mapping Knowledge Domains'; Skupin and Fabrikant, 'Spatialization Methods'.

[56] Midgley, 'Mapping Science', 195–6.

[57] Midgley, *What Is Philosophy For?*, 72.

[58] Midgley, *Science and Poetry*, 170–213.

specific information about the same reality. We need multiple mappings of reality because each such mapping is incomplete.

> No map shows everything. Each map concentrates on answering a particular set of questions. Each map "explains" the whole only in the sense of answering certain given questions about it – not others. Each set of questions arises out of its own particular background in life – out of its own specific set of problems, and needs answers relevant to those problems.[59]

Midgley's mapping metaphor has an obvious application in retrieving the domain of a natural philosophy. Each individual discipline uses its own specific methods to develop its own map of the world. While each of these is adequate for its own purposes, it is incomplete. Natural philosophy could be conceived as the coordinating intellectual domain, a disciplinary imaginary which allows these maps to be superimposed, and their insights and information to be woven together into a grander vision of the natural world than any single discipline can achieve individually. Midgley's visual image thus invites us to see or imagine natural philosophy as a grand map of the natural world, an imagined spatial domain resulting from the overlay of multiple disciplinary maps, bringing together separate disciplines such as philosophy and physics.[60] They may be distinct in practice; they can, however, be brought together imaginatively, by inviting us to decide to see them as interconnected, and experience them accordingly.

Yet this form of integration is not limited to the correlation of academic disciplines—such as philosophy, physics, biology, theology, aesthetics, or ethics. It also extends to broader domains of knowledge, as we shall see in the next section.

Karl Popper's Three Worlds

How can we map our complex world, ensuring we are attentive to its cognitive and affective significance? One of the more important aspects of early modern natural philosophy is its ability to connect with the affective and contemplative aspects of life, enfolding both the subjective and the objective aspects of the human encounter with the natural world. This suggests that

[59] Midgley, 'Pluralism: The Many Maps Model', 11.
[60] For a good example of this approach, see Maxwell, *In Praise of Natural Philosophy*.

Karl Popper's concept of the 'three worlds'—objective, subjective, and theoretical—might offer a suitable conceptual framework within which a retrieved vision of natural philosophy might hold together. Popper set out this framework in a paper entitled 'Epistemology without a Knowing Subject':

> Without taking the words "world" or "universe" too seriously, we may distinguish the following three worlds or universes: first, that world of physical objects or of physical states; secondly, the world of states of consciousness or of mental states, or perhaps of behavioral dispositions to act; and thirdly, the world of objective contents of thought, especially of scientific and poetic thoughts and of works of art.[61]

These three worlds, Popper argues, are not isolated and disconnected realms, in that they interact with each other. Popper holds that the third of these worlds 'is a natural product of the human animal, comparable to a spider's web',[62] thus distinguishing himself from Plato's belief that such ideas have an independent existence.

Popper's approach was not well received,[63] with particular criticism being directed against the third of these proposed worlds, which was seen as being imprecise and incoherent.[64] Popper's ontology seemed to his critics to be an unstable hybrid, drawing both on Plato and Frege on the one hand, and the natural sciences on the other. Yet while Popper's derivation of his three worlds is not entirely satisfactory, the framework he offers has a certain intuitive plausibility, particularly in relation to cultural issues.

This intuitive plausibility may help explain the impact of the 'three worlds' and why it has been so widely used, especially as a framework of cultural analysis. For example, the Peruvian novelist Mario Vargas Llosa used Popper's framework in a literary context,[65] arguing that this categorization of the 'three worlds' offered a helpful framework for coordinating the wide range of human responses to the world.

61 Popper, 'Epistemology without a Knowing Subject', 333. For the emergence of this idea, see Boyd, 'Popper's World 3', 226–37.

62 Popper, 'Epistemology without a Knowing Subject', 338.

63 For example, Popper's discussion of language in World Three has been criticized as 'woefully inadequate and ignorant of developments in the philosophy of language': Sceski, *Popper, Objectivity and the Growth of Knowledge*, 121. For more detailed discussion, see Bunge, *Scientific Materialism*, 137–57.

64 For the main lines of criticism, see Cambier, 'The Evolutionary Meaning of World 3', 292–6.

65 Urroz, 'Karl Popper y Mario Vargas Llosa'.

> In the almost infinite series of nomenclatures and classifications that the wise and the mad have proposed to describe reality, Karl Popper's categories are the most transparent: the *first world* comprises material things or objects, the *second* the subjective and private precincts of the mind, and the *third* the products of the spirit.[66]

In clarifying the difference between the second and the third of Popper's worlds, Vargas Llosa suggested that the second world consisted of 'the entire private subjectivity of each individual, the non-transferable ideas, images, sensations, and feelings of each person'. The human constructions which make up the third world, 'although born in the individual subjectivity, have become public'. These include scientific theories, ethical principles, literary works, religious ideas, and other cultural notions.

The cultural importance of Popper's approach has been explored by Louise DeSalvo, who notes that literary works exist as physical objects (World One, in terms of Popper's categories); in subjective, conscious experience (World Two); and, by virtue of their contents, as conceptual objects that can interact with worlds One and Two (World Three).[67] This framework allows her to identify some problems with contemporary literary criticism, as well as suggest some solutions. The problem, she suggests, is that literary criticism has veered between taking 'an extreme objective stance toward a literary text' that aims to determine its essential qualities, and 'an extreme subjective stance' which focuses 'exclusively on the experience of the reader'.[68]

DeSalvo points to ways of resolving this tension within a Popperian framework, seeing literature as 'an interacting set of events or processes' through which we try to make sense of our environment,[69] recognizing these three different worlds and forging connections between them. More importantly, she highlights the importance of subjecting 'individual views of the world to the critical scrutiny of other people', and 'testing one's vision of the world against others' visions of the world and readjusting one's own view in light of what one learns'.[70] What way of visualizing, of imagining, can be developed, which offers the most satisfactory rendering of objective and subjective worlds? And what can be learned from the history of such imaginings of the

[66] Vargas Llosa, 'Updating Karl Popper', 1020.

[67] DeSalvo, 'Popper in the Realm of Literary Criticism'. Cf. Creed, 'René Wellek and Karl Popper on the Mode of Existence of Ideas in Literature and Science'.

[68] DeSalvo, 'Popper in the Realm of Literary Criticism', 177.

[69] DeSalvo, 'Popper in the Realm of Literary Criticism', 187.

[70] DeSalvo, 'Popper in the Realm of Literary Criticism', 189–90.

natural world, which we considered in some detail in the first section of this study?

In my view, Popper's broad-brush depiction of three 'worlds' can be understood and used in an essentially heuristic manner, seeing them as mapping three distinct yet interconnected aspects of the human engagement with the natural world without making ontological assumptions concerning their nature, or the precise nature of their interconnections. Like DeSalvo, I am dissatisfied with the extreme objectivism of some scientific accounts of nature, and the extreme subjectivism of other approaches. Just as DeSalvo finds Popper's account of theoretical constructions helpful in retrieving more integrated approaches to literary engagement, his framework of the 'three worlds' also provides a framework for the reconstruction of natural philosophy as a way of imagining the natural world that both preserves and integrates its objective and subjective aspects.

Such a framework, it must be stressed, is not *observed*; like a scientific theory, it is *inferred* and *constructed* by the human interpretative agent in order to make sense of what is observed and experienced—to allow us to see or imagine our world so that its complexity and coherence can be grasped.[71] This naturally leads us to reflect on whether natural philosophy could now be helpfully seen as an 'imaginary'—'an enabling but not fully explicable symbolic matrix within which a people imagine and act as world-making collective Agents'.[72] We shall discuss this point further in the following chapter.

We are now in a position to consider how a retrieved natural philosophy can connect with and hold together each of Popper's 'three worlds'. The issues are best discussed if we begin by considering his theoretical world and then move on to deal with his objective and subjective worlds. Although this involves looking at Popper's 'World Three' before his two other worlds, it allows a smoother presentation of the overall argument for the retrieval of a viable natural philosophy. We begin by considering a way of conceiving and retrieving a way of envisaging natural philosophy (World Three), and then proceed to explore how this connects up with the objective natural realm (World One) on the one hand, and our subjective experience of nature (World Two) on the other.

[71] To use Popper's terms, what is being proposed is a *conception* of natural philosophy which exists in World Three, and is capable of interpreting *and holding together* both our observations of the natural world (World One), and our affective responses to nature (World Two).

[72] Gaonkar, 'Toward New Imaginaries', 1.

7
Theory
The Contemplation of Nature

As we have seen, Renaissance and early modern natural philosophy located what would now be described as 'natural science' within a broader theological, ethical, and affective context, which offered a loose conceptual or imaginative unification of the many facets of the natural world, while encouraging attentiveness to each of its aspects. This was not seen as a contrived unnatural alliance of divergent disciplines, but simply as an appropriate imaginative correlation of multiple facets of a complex world. Some such form of interdisciplinary engagement, whether known by that term or not, can be seen throughout intellectual history.[1]

We begin our reflections on how we might retrieve a viable concept of natural philosophy by stepping into the third of Popper's worlds, which is made up of human intellectual constructions and artefacts—products of the human mind such as scientific theories, moral values, languages, mathematical constructions, and socially shared ideas about certain 'things'. Although Popper's 'World Three' was initially regarded with suspicion by his critics, it is increasingly being recognized as the 'keystone of his thought, holding its arches together, unifying and extending his ideas, allowing a comprehensive vision of our place in the universe'.[2]

Theōria: A Way of Beholding the World

The Greek term *theōria*, which underlies our modern term 'theory', has the deeper sense of 'beholding' or 'contemplating', a term which enfolds both gaining a deeper understanding of the world and developing appropriate practices within it. There is a direct continuity here with the use of the term to

[1] Note the comment of Graff, *Undisciplining Knowledge*, 5: 'Interdisciplinarity is part of the historical making and ongoing reshaping of modern disciplines. It is inseparable from them, not oppositional to them.'

[2] Boyd, 'Popper's World 3', 222.

refer to contemplation in religious contexts, as the widespread use of *theōria* in early Greek-speaking Christianity makes clear.[3] Yet as we noted earlier, Aristotle tended to consider *theōria* as sterile and pointless (pp. 21–2), in that it seemed to serve no useful purpose in advancing human happiness. This judgement can now be seen as having been superseded in the light of the growing recognition of the importance of a sense of coherence and perception of wholeness for human mental well-being.[4]

The act of discerning order within the world and being able to make connections often relieves metaphysical anxiety, reassuring us of the possibility of responsible and rational inhabitation of the world. The point of a good scientific theory is to identify patterns and establish connections within nature. In his first novel, C. P. Snow—a physical chemist who later gained fame as a novelist—spoke of the sense of coherence and beauty that resulted from reflecting on Dmitri Mendeleev's Periodic Table of the Elements (1869). The chemical world could be seen in a new way, in which the surface experience of disorder gave way to a deeper sense of ordered elegance.

> For the first time I saw a medley of haphazard facts fall into line and order. All the jumbles and recipes and hotchpotch of the inorganic chemistry of my boyhood seemed to fit into the scheme before my eyes – as though one were standing beside a jungle and it suddenly transformed itself into a Dutch garden. "But it's true," I said to myself. "It's very beautiful. And it's true."[5]

Some philosophers have noted the relationship between discerning coherence and experiencing resilience. Iris Murdoch, for example, wrote of 'the calming whole-making tendencies of human thought',[6] evident in our attempts to discern patterns within our world, and impose a coherent structure upon it. Although Murdoch does not use this phrase, she recognizes the importance of an 'inner consilience'—an imagined unification or correlation of what initially appears to be discordant and disconnected. Murdoch is, of course, aware that this can easily become the act of imposing an order of our own

[3] For a good account of the role of *theōria* in engaging nature and worship, see Foltz, *The Noetics of Nature*, especially 158–74. See also Blowers, 'Beauty, Tragedy and New Creation', 10–13.

[4] See, for example, Eriksson and Lindström, 'Antonovsky's Sense of Coherence Scale and Its Relation with Quality of Life'; Mayer et al., 'Enhancing Sense of Coherence and Mindfulness'.

[5] Snow, *The Search*, 33. A 'Dutch garden' refers to a formalized garden introduced into England by William of Orange in the 1690s.

[6] Murdoch, *Metaphysics as a Guide to Morals*, 7. For further discussion, see Browning, *Why Iris Murdoch Matters*, 1–54; Badhwar, *Well-Being*, 52–80.

desiring and making, and consequently stresses the importance of a truthful attentiveness to the natural world, which helps us break free from self-serving myths and habits of thought. Nature can serve as a decentring corrective to human self-referentiality. Yet the most important discussions of the impact of theory on human well-being originate from within psychology, in response to the empirical evidence that suggests that finding coherence within the world or meaning in events promotes human well-being, and enables us to cope with trauma in life.[7]

Seeing Nature through a Theoretical Lens

Nature is not seen directly, but seen through an informing *theōria*. We do not simply 'see' nature; we *see it as something*. William Whewell argued that we do not see nature as it is, but rather approach it through a theoretical gateway: there is 'a mask of theory over the whole face of nature'.[8] The act of observation incorporates an assumed and often unacknowledged commitment to a *theōria*, which—like Bacon's 'Idols of the Mind'—might prevent us from seeing nature as it really is.

The remarkable transition from a Ptolemaic to a Copernican worldview over the course of the seventeenth century highlighted a theme which is essential to any proper understanding of a natural philosophy. Observation is not a neutral process, but is shaped by assumptions, explicit or implicit, about *what* is being observed. The recognition of the theory-laden nature of observation has highlighted the point that an apparently 'natural' or 'unbiased' beholding of the natural world is actually informed and to some extent determined by an assumed theory.[9] Part of the 'paradigm shift' described by Thomas Kuhn involves seeing the world through a new *theōria* and appreciating the difference this makes. As Kuhn points out, a new theory allows us to observe things in a new way, displacing an older way of interpreting and envisaging the world—with significant cognitive and affective consequences. 'Before [Copernicanism], the sun and moon were planets, the earth was not.

[7] There is a large literature: see, for example, McKnight and Kashdan, 'Purpose in Life as a System That Creates and Sustains Health and Well-Being'; George and Park, 'Existential Mattering'; Costin and Vignoles, 'Meaning Is About Mattering'; for the increasing recognition of religion in creating meaning, see Park, 'Religion as a Meaning-Making Framework in Coping with Life Stress'.

[8] Whewell, *Philosophy of the Inductive Sciences*, vol. 1, 1.

[9] Adam, *Theoriebeladenheit und Objektivität*, 25–49.

After it, the earth was a planet, like Mars and Jupiter; the sun was a star; and the moon was a new sort of body, a satellite.'[10]

Kuhn's point is that the phenomena were unchanged; they were, however, *seen* in a new way. One observer might 'see' the sun rise and set; another might 'see' the earth turning on its axis, leading to the apparent motion of the sun across the heavens. The history of science offers many examples of the same object being seen in different ways as the result of using different theoretical lenses. For example, on 3 December 1714, John Flamsteed, the British Astronomer Royal observed a faint star in the constellation of Taurus, only just visible to the naked eye, which he catalogued as '34 Tauri'; on 13–14 March 1781, the British astronomer William Herschel observed *the same object*, and realized it was a planet, now known as Uranus.[11]

The point to appreciate is that seeing is not a mechanical act of visual capture, comparable to a photographic image; rather, it presupposes a 'hermeneutical consciousness' (Hans-Georg Gadamer), in that seeing is an act of interpretation, which elicits the kind of responses that Popper located in World Three—such as languages, theories, concepts, and symbolic systems of representation and meaning.[12] And, as Aristotle rightly pointed out, *theōria* allows us to envisage a particular way of understanding ourselves and our place in the world, and thus links *theōria* with its appropriate *ēthos* and *askēsis*. For example, a Christian *theōria* leads to nature being seen as divine creation, thus pointing to the need to live according to this view of reality, and developing practices or disciplines that consolidate this way of seeing things.[13] There is thus an implicit connection between *theōria* and *praxis*—a connection that was made explicit by Gadamer.

Hans-Georg Gadamer on Theory and Practice

We have already noted the importance of *theōria* to classic Greek philosophers. For many today, the word 'theory' has overtones of detachment and abstraction, allowing 'the subject to stand over against the object'.[14] If this is the case, it would seem that an appeal to *theōria* is of little use in dealing with this

[10] Kuhn, *The Road since Structure*, 15.

[11] Rawlins, 'A Long Lost Observation of Uranus'; Steinicke, 'William Herschel, Flamsteed Numbers and Harris's Star Maps'. Herschel realized it was a planet both on account of its faint disk, and its motion against the fixed stars in the interval between his two observations.

[12] Kiefer, 'Hermeneutical Understanding as the Disclosure of Truth'.

[13] See further Wirzba, 'Christian *Theoria Physike*'.

[14] Nightingale, *Spectacles of Truth in Classical Greek Philosophy*, 9.

bifurcation between the divergent approaches to nature now offered by science and the humanities, which contrasts with the more holistic vision of natural philosophy.

As an example of this bifurcation, we may consider the tensions in mid-nineteenth-century England between those who saw nature in essentially objective scientific terms, and those who stressed its subjective impact on the beholder, especially in terms of arousing a sense of beauty and wonder. The former assumed that 'the world of natural objects, of bare, clear, downright facts, is unproblematically given';[15] the latter regarded this as unacceptably reductionist, failing to give due weight to the aesthetic and affective impact of nature on humanity. The human observer is *affected* by the act of observation. The point to appreciate is that theory does more than link observations in the mind of the beholder, enfolding and connecting them with epistemic filaments, like those of Whitman's 'Noiseless Patient Spider';[16] theory also *connects the human observer with nature*, rather than merely establishing connections between one observed aspect of nature and another.

Gadamer points out that, in its original classic Greek context, *theōria* was not concerned with an abstract and detached spectatorial account of the world; it rather gives rise to a transformative contemplative journey with specific moral outcomes.[17] 'Theorizing' was not understood as an impartial and detached apprehension of an object at a distance, but was ultimately an act of self-transformation, in that it created a redefined sense of the individual self.[18] This point is echoed by Martha Nussbaum, who observes that there are limits to a constant critical detachment towards texts or the world; at points, we have to allow or force ourselves to be receptive and 'porous' to what is being encountered.[19]

In retrieving these aspects of the classical notion of *theōria* in his essay 'In Praise of Theory', Gadamer argues that, while the prevailing sense of 'theory' is now that of detached observation,[20] this is a significantly diminished account of the classic notion.

> We must ask a double question: is there perhaps more to theory than what the modern institution of science represents to us? And, is practice too

[15] Belsey, *Critical Practice*, 9. Cf. Tate, 'Poetry and Science'.
[16] Whitman, *Leaves of Grass*, 343.
[17] Nightingale, *Spectacles of Truth in Classical Greek Philosophy*, 128. For the political enactment of *theōria*, see Lefèvre, 'Der Tithonus Aristons von Chios und Ciceros Cato'.
[18] Nightingale, *Spectacles of Truth in Classical Greek Philosophy*, 13–14.
[19] Nussbaum, *Love's Knowledge*, 281–2.
[20] I here follow the analysis in Brogan, 'Gadamer's Praise of Theory'; Konchak, 'Gadamer's "Practice" of *Theoria*'.

> perhaps more than the application of science? Are theory and practice correctly distinguished at all when they are seen only in opposition to each other?[21]

Theory is 'seeing what is'; yet this seeing is an *act of interpretation*. It is a 'hermeneutic concept which means that it is always referred back to a context of supposition and expectation.'[22]

Gadamer notes Aristotle's emphasis on the human desire for knowledge, partly in order to survive, but more importantly to flourish and achieve happiness. Yet while Aristotle held that our 'greatest joy' lay in such knowledge, Gadamer stresses that he also knew the importance of human practices arising from this way of 'seeing' (*Schauen*), thus holding theory and practice together.[23] Gadamer stresses that the term 'theory' has its origins in the observation of Greek religious festivals, in which the observer was also a participant in these events. 'The type of vision that Gadamer associates with *theōria* is not that of a neutral observer, whose emphasis is on the control of an object or to turn it to their own purposes by explaining it, but rather involves participation and more relational experience.'[24] *Theōria* does not denote a singular detached act of observation 'that establishes what is present', but rather involves attentive participation in what is *found* to be present.[25]

Gadamer argues that the bifurcation between the 'theoretical' and the 'practical' is in fact a modern development, which does not adequately capture Aristotle's assumption of the fundamental interconnection of theory and practice.[26] 'Both Aristotle and Gadamer equally acknowledge the inseparability of the practice of understanding and theoretic awareness. This hermeneutic circle that mutually defines theory and practice transforms our understanding of both theory and practice.'[27]

Gadamer's retrieval of *theōria* offers both an intellectual and a pragmatic stimulus to holding together theory and practice, as well as the intellectual and affective dimensions of the human engagement with the natural world. Emphasizing that *theōria* simultaneously enfolds observation and participation, Gadamer expands the category of theory to go beyond scientific and

[21] Gadamer, 'Praise of Theory', 24. For the political and social importance of 'reducing the gap between our thoughts and praxis', see Morello, 'Charles Taylor's "Imaginary" and "Best Account" in Latin America', 332.

[22] Gadamer, 'Praise of Theory', 31.

[23] Gadamer, 'Praise of Theory', 31.

[24] Konchak, 'Gadamer's "Practice" of *Theoria*', 457.

[25] Gadamer, 'Praise of Theory', 20–1.

[26] On this point, consider the relation of what Aristotle termed *ta phusikēs theōrias* ('theoretical natural science') with *ta ēthikēs theōrias*: cf. Reeve, *Action, Contemplation, and Happiness*, 160–1.

[27] Brogan, 'Gadamer's Praise of Theory', 152.

utilitarian instrumentality, enabling it to engage areas of life that lie beyond the scope of objective science. 'The possibilities of science are limited in a far more fundamental way. There will always be areas that fundamentally cannot be approached by objectification and treated as methodical objects. Many of the things in life are of this kind.'[28] In his revisionist account of the notion, Gadamer presents *theōria* as bridging theory and practice on the one hand, and objective and subjective aspects of the human engagement with the world on the other. This can easily be mapped onto Popper's 'World One' and 'World Two'.

Gadamer's reflections also allow an easy alignment with Pierre Hadot's signature claim that the philosophy of the Hellenic age was fundamentally understood and enacted as 'a way of life' (*une manière de vivre*) based on a set of practices intended to foster both discourse and behaviour that accorded with a right understanding of nature and the good life.[29] A disciplined contemplation of the world enabled human self-transformation, leading to forms of practices and disciplines which served to foster wisdom.

> Ancient philosophy, and the "science" it made possible, was first and foremost about the advocacy for a way of life and the disciplines that enabled its practitioners to live well (however that was conceived). *Theoria*, the way of seeing being recommended by a philosophical school, was inextricably connected to an *ethos* or way of being in the world.[30]

This leads us to reflect further on how we are to understand natural philosophy as such a 'way of seeing', which is an imaginative undertaking that is clearly located within Popper's 'World Three'.

Natural Philosophy as a Disciplinary Imaginary

During the 1990s, the notion of 'imaginaries' began to gain traction as a way of visualizing the ways in which individuals hold together multiple aspects of a complex reality. This idea emerged in the later writings of Jacques Lacan, who used the term *imaginaire* to designate a human subject's capacity to achieve integration of seemingly disconnected experiences.[31] Where many

[28] Gadamer, 'Praise of Theory', 29. [29] Hadot, 'La philosophie antique'.
[30] Wirzba, 'Christian *Theoria Physike*', 213.
[31] For the development of Lacan's views, see Licitra Rosa et al., 'From the Imaginary to Theory of the Gaze in Lacan'.

philosophers offered objective theoretical accounts of this process, which can be justified on rational grounds,[32] others have argued that this is really about understanding how individuals *imagine* their social and personal worlds.[33] This turn from third person objectivities to first person subjectivities places a new emphasis on the role of the imagination in organizing perceptions of the world.[34] Perhaps the most familiar of this category of approaches is the 'social imaginaries', which is particularly associated with the philosopher Charles Taylor.

Reacting against the pre-eminence of epistemology in modern philosophy, Taylor points to the importance of 'social imaginaries' in shaping how we see the world and visualize its multiple aspects. Drawing on Heidegger's notion of *Verworfenheit*, Charles Taylor argues that we find ourselves 'thrown into' an historically given way of thinking and imagining. Taylor's point is that we inherit or embrace a way of seeing things, and subsequently rationalize this, providing a theoretical justification for connections that are already present in this 'social imaginary'.[35] While the intellectual visions of Kepler, Boyle, and Newton appear to have been shaped by prevailing 'social imaginaries' that regarded such a richer outcome as culturally natural and intuitively plausible, many today would suggest that this attitude is no longer 'given', but now has to be reconstructed through an act of reimagination.

Yet we are not trapped by or within existing or inherited imaginaries; these are products of the human mind that can be recreated and reinvented. For Cornelius Castoriades, one of the earlier advocates of the idea of a 'social imaginary', such imaginaries are autonomous self-constructions, and thus are not to be seen as 'right' or 'wrong'. A new social imaginary can thus be constructed and inhabited, through a process of interrogating our own constructions and thus creating a new social world through imagining a preferable alternative.[36] This work suggests that we might retrieve an older vision of natural philosophy through challenging prevailing disciplinary imaginaries, and constructing alternative ways of imagining the natural world. Natural philosophy can be retrieved as a 'disciplinary imaginary', whose intellectual coherence is created within the mind of the observer, rather than imposed by the force of prevailing academic conventions.

As we shall see, there are some important outcomes of seeing a retrieved natural philosophy in this way. In her important study of the role of

[32] For example, Giere, *Scientific Perspectivism*.
[33] Gaonkar, 'Toward New Imaginaries', 4.
[34] For the turn from the 'Reasonable to the Imaginal', see Bottici, 'Imagining Human Rights', 112–15.
[35] Taylor, *Modern Social Imaginaries*, 23–5.
[36] Canceran, 'Social Imaginary in Social Change', 32.

'imaginaries' in constituting our experience of the world, Kathleen Lennon makes the point that such 'imaginaries' offer us frameworks for organizing our experience of the world, and allowing us to engage both its cognitive and affective senses. An 'imaginary is the shape or form in terms of which we experience the world and ourselves; a gestalt which carries significance, affect, and normative force'.[37] It allows us to connect up aspects of our experience of the world, in that they are not merely the way we think about that world; they are also the means by which we 'feel our way around' that world.[38]

Lennon's account of an imaginary thus allows us to see ways of connecting Popper's 'World One' and 'World Two' through something that belongs to his 'World Three'. The objective and subjective aspects of the natural world are thus engaged through an imaginative lens that is not observed or experienced, but is *constructed*—not in isolation from observation and experience, but as a means of enfolding and affirming them.

> The imaginary texture involves a gestalt, a schema or organizing form, which we find in the world as experienced by us. This gestalt we have suggested, following Kant and Merleau-Ponty, is neither imposed nor simply discovered, but emerges from a creative interplay between corporeal subjects and the world within which they are placed, and to which they are sensible.[39]

Lennon does not consider the realm of the imaginary to be 'a domain of illusion posited in opposition to a "real"; it is rather that by which the real is made available to us'.[40] An imaginary is not something that we whimsically and arbitrarily impose upon the natural world; rather, it emerges plausibly from our interaction with that world. Lennon's position is open to criticism;[41] it does, however, open a way of holding together themes that are intuitively seen to be interconnected, with a view to exploring their relationships, and the outcome of this more expansive way of engaging the world.

This observation helps us understand how we can imagine a domain which extends across present disciplinary boundaries—boundaries that arose for specific historical reasons, yet which can be transcended imaginatively, even

[37] Lennon, *Imagination and the Imaginary*, 73.

[38] Lennon, *Imagination and the Imaginary*, 1.

[39] Lennon, *Imagination and the Imaginary*, 53. Note that Lennon thus considers that an imaginary is 'both disclosed and created': Lennon, *Imagination and the Imaginary*, 138.

[40] Lennon, 'Imaginary Bodies and Worlds', 2.

[41] For example, her ambiguity about whether an imaginary is 'a shape we project onto the world' or 'the shape of the world as it is': Kind, 'Imagination and the Imaginary, by Kathleen Lennon', 1247–8.

if they remain an important means of obtaining and structuring both our knowledge of the natural world, and the academic structures within which this is researched and taught. The increasing tendency to speak of 'disciplinary imaginaries' highlights concerns about potentially arbitrary boundaries between different areas of knowledge production,[42] which leads to the isolation and fragmentation of discussions that properly transgress such boundaries—including many that were once enfolded in the generous conceptual space of 'natural philosophy'. It emphasizes that there are no 'right' articulations of disciplinary boundaries, and invites the development of individually satisfactory reimaginations of intellectual territories. Historically, disciplines tend to fragment and diversify, often for highly pragmatic reasons. Yet there is no good reason why a wider disciplinary imaginary should not be conceived, countering this tendency towards increasing specialization, which would be capable of encouraging dialogue across multiple domains of interest and concern.

The concept of a 'disciplinary imaginary' can be used as an interpretative lens to understand how seventeenth-century writers—such as Kepler, Galileo, and Boyle—envisaged the intellectual domain of natural philosophy, seeing it as enfolding philosophy, theology, physics, biology, astronomy, mathematics, and—in Kepler's case—music. These three writers offered, not so much a personal *synthesis* of these fields, but a personal *imagined* vision of their interconnectedness—a 'disciplinary imaginary', which they and others in this cultural location found both intuitively plausible and intellectually satisfying.

The case of Newton is particularly instructive. While some older accounts of Newton suggested that his synthesis of theology, philosophy, and science reflected a rigorous intellectual framework, more recent studies suggest that there is 'no support for the notion that there is some simple conceptual or methodological coherence to his work'.[43] Newton's concept of natural philosophy appears to represent a personal assimilation of its multiple components, in effect representing an individual disciplinary imaginary that resonated with others during this formative period.

One of the chief virtues of retrieving this disciplinary imaginary is its receptiveness towards multiple disciplinary conversations and correlations, opening the way to potentially enriching approaches to fields such as environmental ethics, natural law, science and religion, and scientifically engaged

[42] A point stressed by Goodstein, *Georg Simmel and the Disciplinary Imaginary.*
[43] Iliffe, *Priest of Nature*, 14.

philosophies and theologies.[44] Natural law, for example, can be understood as resting on the belief that the cosmos has a moral, and not merely a physical, order,[45] thus extending the scope of its investigation beyond the natural sciences. This idea is encountered in many forms, including the Confucian notion of seeking 'harmony' with heaven and earth.[46] This would enable reconnection across such artificial disciplinary divides, and recapture something of the original spirit of natural philosophy, which links attentiveness to the natural world with the acquisition of both understanding and wisdom. It does not require abandoning present-day understandings of disciplinary identity and integrity, but rather to imagine how they might be brought into productive conversation and correlation.

In this chapter, we have sketched some aspects of Popper's World Three—the world of human mental creations, which enable us to make sense of what we observe around us, and experience within us. Yet these theories and imaginaries do more than help us understand our world; they encourage and facilitate a deeper level of engagement with nature, which is conducive to human wisdom, flourishing, and well-being. In the next chapter, we shall turn to consider Popper's 'World One', the objective world of objects that exist around us.

[44] Like Josh Reeves, I would certainly see natural philosophy as *enfolding* the field now known as 'science and religion', although I do not agree with his suggestion that this field 'should be seen as an updated version of natural philosophy'. Cf. Reeves, 'The Field of Science and Religion as Natural Philosophy'.

[45] For this point, see Brown, *The Ethos of the Cosmos*; Crowe, 'Metaphysical Foundations of Natural Law Theories'.

[46] For the interpretative issues associated with this Confucian idea of correlating human existence with a cosmic moral order, see Slater, 'Two Rival Interpretations of Xunzi's Views on the Basis of Morality'.

8

Objectivity

Understanding the External World

We now turn to consider Popper's 'World One'—the 'world of physical objects or of physical states' that played such an important role in the emergence of natural philosophy in the seventeenth century in the thought of Kepler, Galileo, Boyle, and Newton. For the philosopher Mary Midgley, 'the natural sciences are wholly dedicated to talking about objects. That is their job.'[1] Scientific theories (located in World Three) represent human attempts to explain what is observed in the external world of objects (located in World One).

Science and the Quest for Objectivity

For many, the quest for objectivity of judgement, avoiding subjectivity and bias, is one of the most distinctive and important features of today's natural sciences.[2] Science is universally applicable, irrespective of its geographical or sociological location of its practitioners, or their religious or political beliefs. This quest for an objectivity which transcended partisan commitments is deeply rooted in the history of early modern natural philosophy. This was especially the case in seventeenth-century England, when natural philosophy was seen as a discipline capable of bridging the religious and political divides of that period that had been exacerbated by the Civil War.[3]

The concept of 'objectivity' was considered to be an epistemic virtue in the late seventeenth century, in that it pointed to the possibility of a universal, incorrigible knowledge, which was not contaminated or tainted by partisan

[1] Midgley, *Science and Poetry*, 12.

[2] See, for example, Daston, 'Objectivity and the Escape from Perspective'; Kuukkanen, 'Autonomy and Objectivity of Science'.

[3] For a good account of the issues, see Hunter, 'Latitudinarianism and the "Ideology" of the Early Royal Society'.

concerns or cultural precommitments.[4] Francis Bacon's notion of scientific self-distancing, enacted through the neutral instrumentality of experimentation, was mirrored in an increasing emphasis on intellectual accountability.[5]

Bacon saw experimentation as a means of ensuring objectivity of judgement, countering the potential for self-deception in all human understanding which he expressed in his 'Doctrine of Idols'. Any such distortions in human judgement could be corrected by the unquestionable neutrality of the experimental instrument.

Yet others questioned this judgement, pointing out the inevitable subjectivity of judgement that accompanied the interpretation of instrumental observation. The British natural philosopher Margaret Cavendish rejected 'the validity of the subject-object boundary', while highlighting the significance of the restricted knowledge of nature that was possible for humanity on account of the location of the human observer within—not above—nature.

> Cavendish's assertion that nature cannot be wholly known arises not merely out of a confrontation with the enormity of nature's infinite, but also out of a belief that man is inextricably a part of the nature he seeks to know: there simply exists no outside vantage point from which to view and thereby to control some object called nature.[6]

For Cavendish, the Baconian claim to 'objectivity' could not be detached from the human tendency to *create*, and not merely *communicate*, knowledge.[7] It is an important point, even though it ran counter to the dominant trends in the natural philosophy of the age.

Virtuosi, according to this consensus, were those whose ideas were informed by neutral and detached observation, experiment, and critical reflection on the natural order on the one hand, and who expressed their evidence-based opinions in an undogmatic, cautious, and gracious manner on the other, aware of their own corrigibility and fallibility.[8] Boyle and Newton, for example, both stressed the provisional nature of experimental findings, and believed that the only way to discover what kind of world God

[4] As argued by Daston, 'Objectivity and the Escape from Perspective'; Murphy and Traninger, eds., *The Emergence of Impartiality*.

[5] Zagorin, 'Francis Bacon's Objectivity and the Idols of the Mind'.

[6] Keller, 'Producing Petty Gods', 457.

[7] For the broader context, see de Rycker, 'A World of One's Own'.

[8] The Italian term *virtuosi* was widely adopted within the Royal Society to designate what are in effect amateur gentlemen precursors of today's professionalized scientists. Cf. Yeo, *Notebooks, English Virtuosi, and Early Modern Science*, 1–36.

had created was to investigate that world empirically, using neutral tools of investigation that ensured objectivity of judgement.[9]

In modern natural sciences, the distinction between objectivity and subjectivity primarily concerns the extent to which the various implicit and explicit biases of the knower can be neutralized or eliminated from the process of knowledge production. This point is widely accepted, even though it clearly needs nuancing. Nicholas Rescher, for example, argues that the essence of an objective judgement is that it possesses 'a cogency compelling for everyone alike (or at least all normal and sensible people), independently of idiosyncratic tendencies and inclinations'.[10] This, he suggests, means trying to identify and eliminate 'biases, idiosyncrasies, predilections, personal allegiances, and the like'[11] from our attempts to understand the world and ourselves.

There is, however, an obvious problem with this approach, in that it appears to treat pre-understandings simply as 'bias'. To observe the world is not simply to record what is there; it is to impose or employ some implicit descriptive framework in interpreting what is observed. Observation is an interpretative process, laden with implicit theoretical precommitments.[12] Scientific progress does not take place by *eliminating* the biasing effect of such pre-understandings, but by *revising and refining* what come to be recognized as inadequate pre-understandings, and replacing them with more developed understandings that are better placed to accommodate those data and experiences. Daniel Dennett develops this point further, noting that natural scientists might incorporate unacknowledged and unverified philosophical ideas in their theories. 'There is no such thing as philosophy-free science; there is only science whose philosophical baggage is taken on board without examination.'[13]

Yet some would argue that science is objective in a second sense, holding that the natural sciences engage only with the objects of the physical world, with the intention of achieving a rational understanding of the world. This approach, which became particularly important in the later nineteenth century, regards any form of emotional, affective, or aesthetic engagement with nature as inappropriate. The natural sciences *report* on nature, avoiding any affective response to nature, or any sense that the observer is participating in what is observed. Engagement with the objective world is thus held to lead to impersonal quantified accounts of nature, deliberately setting to one side any

[9] Wojcik, 'Pursuing Knowledge', 184. [10] Rescher, *Objectivity*, 7.
[11] Rescher, *Objectivity*, 16.
[12] Ochs, 'Transcriptions as Theory'; Adam, *Theoriebeladenheit und Objektivität*.
[13] Dennett, *Darwin's Dangerous Idea*, 21.

subjective engagement with the natural world—such as an experience of wonder at its beauty.[14]

Yet this emphasis upon objectivity simply reified subjectivity as the 'other', in that objectivity cannot be articulated without an opposing subjectivity.[15] This artificial and unsustainable separation of the objective and subjective is out of place within a natural philosophy. The two may certainly be *distinguished*, as in Popper's useful framework of the 'three worlds'; yet this distinction is intended to allow each of them to be considered in greater individual depth, to explore the way in which they can relate to each other, and to appreciate the outcomes of this process of correlation, especially in the multifaceted human engagement with the natural world.

The natural sciences aim to make rational sense of our world, placing particular emphasis on its *intelligibility*.[16] Especially in popular scientific writing, the scientific enterprise is articulated primarily in ways that privilege the domains of the rational or cognitive, and play down the emotional, imaginative, or affective domains.[17] It is certainly true that human beings seem to have an innate desire and capacity to make sense of our world, perhaps arising from some deeper evolutionary instinct.[18] Yet this 'understanding' cannot be framed simply in terms of logical analysis and quantification. As the history of science makes clear, this 'logic of scientific discovery' often involves an appeal to the imagination in generating hypotheses, and an appeal to the human sense of beauty in evaluating the scientific theories that result.[19] The journey of scientific discovery does not end in a cold rationalism; indeed, this journey might not even *begin* with a desire to make sense of things. As many, including the poet Hermann Hesse, have pointed out, an experience of wonder is often both the emotional trigger and imaginative gateway to a pilgrimage of intellectual discovery.[20]

[14] See Daston and Galison's reflections on the 'scientific objectivity' which emerged during the nineteenth century: Daston and Galison, *Objectivity*, 197–205.

[15] For Steven Shapin, the notions of objectivity and subjectivity can be likened to 'Doppelgänger, the good child and its evil twin, where the one is the positive image, the other the negative'. Shapin, 'The Sciences of Subjectivity', 171.

[16] Dear, *The Intelligibility of Nature*, 173.

[17] For a good account of this problem, see McLeish, *The Poetry and Music of Science*, 262–300.

[18] For comments on this tendency, see Chater and Loewenstein, 'The Under-Appreciated Drive for Sense-Making'.

[19] McAllister, 'Is Beauty a Sign of Truth in Scientific Theories?'; McLeish, *The Poetry and Music of Science*, 72–127.

[20] Hesse, *Mit dem Erstaunen fängt es an*, 7–10.

Thomas Nagel on the Limits of Objectivity

For Thomas Nagel, 'there are things about the world and life and ourselves that cannot be adequately understood from a maximally objective standpoint.'[21] Nagel clearly thinks of a spectrum of possibilities existing between the objective and subjective, and is particularly critical of those who consider it possible for an observer to adopt a purely detached objective account of every aspect of existence. 'We may think of reality as a set of concentric spheres, progressively revealed as we detach gradually from the contingencies of the self.'[22] Nagel's point is not that there is a distinct 'subjective' metaphysical realm, but that the single reality of our world requires engagement and apprehension from both objective and subjective perspectives or points of view, while noting the difficulties for an external observer in understanding another's 'point of view'.

Nagel thus argues that there are limits to a purely objective account of the world which render it incomplete as a basis for humane life and thought. In his influential essay 'What Is It Like to Be a Bat?', Nagel maps out some significant lines of concern about such an excessive objectivity, even if he does not quite manage to articulate a clear alternative.[23] For Nagel, the nature of human consciousness is such that a subjective perspective on the world is proper and inevitable. The physical sciences take the form of 'a domain of objective facts...that can be observed and understood from many points of view'. Yet a subjective perspective expresses what it feels like to be a given subject. And for that subject, this constitutes the 'subjective facts' that not only 'embody a particular point of view', but which are 'accessible only from one point of view'.[24] Nagel illustrates this point by considering how someone might enter into another person's point of view: 'The subjective character of the experience of a person deaf and blind from birth is not accessible to me, for example, nor presumably is mine to him. This does not prevent us each from believing that the other's experience has such a subjective character.'[25]

Nagel's analysis is important to the natural philosophical project developed in this volume for two reasons. First, he highlights the importance of our subjective perceptions of the world for us. How the world appears to me is not the same as how it appears to you. Second, it highlights a difficulty noted

[21] Nagel, *The View from Nowhere*, 7. [22] Nagel, *The View from Nowhere*, 5.

[23] Nagel, *Mortal Questions*, 165–80. For helpful engagements with Nagel on these points, see Foss, 'Subjectivity, Objectivity, and Nagel on Consciousness'; Ratcliffe, 'Husserl and Nagel on Subjectivity and the Limits of Physical Objectivity', 355–9.

[24] Nagel, *Mortal Questions*, 171–2. [25] Nagel, *Mortal Questions*, 170.

earlier in exploring the specific accounts of natural philosophy offered by Kepler, Boyle, and others—namely, that these tend to take the form of personal syntheses or viewpoints whose internal rationalities are not entirely accessible to us today, in that they are individual constructions which were clearly considered plausible and reasonable to those who constructed them, as well as others who engaged them. There is clearly a cultural element to these ways of thinking, which is not self-evident or intrinsically plausible to those in a different cultural context.

Nagel's analysis confirms both that an objective scientific knowledge of our physical world is important, and that there is more that needs to be said about that world, not least its impact on human subjective experiential worlds, and its role in catalysing and informing a quest for meaning in life, and the pursuit of the good life. Natural philosophy sought to capture this wider vision of scientific knowledge, and use it to develop practices and values which enabled meaningful inhabitation of the natural world.

Yet this broader vision of the human engagement with the natural order faced a significant challenge in the twentieth century, through the intellectual movement now known as 'scientism'. In the next section, we shall explore this development, and how it relates to the retrieval of a natural philosophy.

Scientism: The Redundancy of Philosophy?

Most philosophers take the view that each academic discipline has the ability to illuminate part of a larger picture, and thus exists in a collaborative relationship with other forms of knowledge. This view has, however, met with resistance from a group of writers who want to grant epistemic privilege to the natural sciences. While such views are found mainly among scientific popularizers, often accompanied by unsubstantiated rhetoric about the 'death of philosophy', they can also be found in more sophisticated forms which merit serious attention. For example, Edward O. Wilson acknowledges that many now regard the natural sciences, social sciences, and humanities as separated 'by an epistemological discontinuity, in particular by possession of different categories of truth, autonomous ways of knowing, and languages largely untranslatable into those of the natural sciences'.[26] Yet for Wilson, the unifying agenda of the Enlightenment holds that everything is 'ultimately reducible . . . to the laws of physics'.[27]

[26] Wilson, 'Resuming the Enlightenment Quest', 17.
[27] Wilson, *Consilience*, 291.

The term 'scientism' is now widely used to refer to reductive views such as that of Wilson,[28] which allocate the fundamental right to discursive justification exclusively to the natural sciences. The Italian philosopher Evandro Agassi suggests that the defining feature of scientism lies in its endemic reductionism, which insists that 'the whole is identical with a horizon which is actually partial, no matter how spacious it might seem.'[29] Scientism is sometimes described using unhelpful value-laden judgements in referring to the extent to which the natural sciences can inform or determine philosophical, ethical, or religious issues. For example, Susan Haack frames scientism in terms of an 'exaggerated kind of deference to science', or an 'excessive readiness' to accept a scientific pronouncement as authoritative.[30]

But who decides whether such attitudes are 'exaggerated' or 'excessive'? There is no extra-disciplinary 'view from nowhere'—comparable perhaps to Henry Sidgwick's 'point of view of the universe'[31]—that allows us to determine objectively the extent to which the natural sciences *ought* to be involved in the deliberations and decisions of other disciplines—or, indeed, the extent to which other disciplines, particularly history and philosophy, ought to be involved in scientific reflection.

The philosopher Alex Rosenberg offers a definition of scientism, which is a helpful starting point for reflection on interdisciplinary knowledge production. For Rosenberg, scientism

> is the conviction that the methods of science are the only reliable ways to secure knowledge of anything; that science's description of the world is complete in its fundamentals; and that when "complete," what science tells us will not be surprisingly different from what it tells us today.[32]

If Rosenberg's view is accepted, it follows that the category of natural philosophy becomes redundant, in that the natural sciences are able to answer any

[28] There is a growing literature in this field, particularly in relation to the questions the rise of scientism raises for philosophy: Kidd, 'Reawakening to Wonder'; Kitcher, 'The Trouble with Scientism'; Williams and Robinson, eds., *Scientism*; Stoljar, *Physicalism*; Boudry and Pigliucci, eds., *Science Unlimited?*; Göcke, *After Physicalism*; Ridder et al., eds., *Scientism*. The concept of 'scientific naturalism' can be stated in various forms, which Lynne Rudder Baker characterizes as 'eliminative naturalism', 'reductive naturalism', and 'nonreductive naturalism': Baker, 'Naturalism and the Idea of Nature'.

[29] Agazzi, 'Fede, ragione e scienza', 95.

[30] Haack, *Defending Science*, 17–18. For concerns about the use of such pejorative language, see Clarke and Walsh, 'Scientific Imperialism and the Proper Relations between the Sciences'.

[31] For a recent defence of this view, set out in Sidgwick's *Methods of Ethics* (1874), see Lazari-Radek and Singer, *The Point of View of the Universe*, especially 94–114.

[32] Rosenberg, *Atheist's Guide to Reality*, 6–7.

valid philosophical questions relating to the natural world, and the place of humanity within it.[33]

Three significant criticisms can be made of scientism, when it is framed in this way. First, its legitimacy rests on what appear to be circular forms of argument. If the natural sciences are proposed as the only reliable source and criterion of knowledge, how can this claim be shown to be 'knowledge', rather than merely an 'opinion', other than by an appeal to the natural sciences themselves? The validity of an epistemic rule *R* is thus confirmed by the application of that same epistemic rule *R*. This charge of circularity could be avoided by justifying the reliability of science by an appeal beyond the sciences, drawing on the support of other disciplines in order to make a cumulative case for this belief. Yet in making such an appeal, the reliability of non-scientific disciplines and authorities is implicitly being conceded.

So how might the methods of the natural sciences be used to prove that those methods are the *only* reliable ways to secure knowledge of anything? Mikael Stenmark points out that it is very difficult to devise a credible scientific experiment which would show that science determines the limits of reality, or demonstrates that other methods of investigating of the world are inadequate or invalid.[34] In any case, there appears to be a category error here, in that what are being assessed are actually second-order *philosophical claims about science*, which cannot be verified experimentally. This suggests that the confirmation or refutation of such an approach to science would therefore have to be conducted on a *philosophical*, rather than a *scientific*, basis.[35]

Second, scientism, as defined by Rosenberg, fails to give due weight to the provisional nature of science. Rosenberg suggests that when it is 'complete, what science tells us will not be surprisingly different from what it tells us today'. I am not aware of any compelling empirical evidence for this bold assertion. One obvious problem is the phenomenon of radical theory change in scientific history. The scientific consensus over time has shifted dramatically and often unpredictably.[36] The scientific enterprise is not yet terminated or completed, and its present findings are incomplete, and in some cases possibly erroneous.[37] In his inaugural lecture at Cambridge University in 1879,

[33] Other definitions might be noted, such as that offered by Pigliucci, 'New Atheism and the Scientistic Turn in the Atheism Movement'.

[34] Stenmark, *Scientism*, 22–3.

[35] For such a critique, see Kidd, 'Reawakening to Wonder'.

[36] For a good account of the philosophical complexity of this question, see Rescher, 'Some Issues Regarding the Completeness of Science and the Limits of Scientific Knowledge'.

[37] For the contrary view that our future scientific understanding is not likely to be significantly deeper than at present, see Wigner, *Symmetries and Reflections*, 215.

the physicist James Clerk Maxwell warned against any complacency on this matter.

> This characteristic of modern experiments – that they consist principally of measurements – is so prominent, that the opinion seems to have got abroad, that in a few years all the great physical constants will have been approximately estimated, and that the only occupation which will then be left to men of science will be to carry on these measurements to another place of decimals.[38]

Today's theories might turn out to be transient staging-posts, rather than final resting places; they might be *better* than those they had displaced, but they are not *right* for that reason. The 'received wisdom' of the natural sciences is not static or fixed, but is rather subject to constant review, so that 'the price of scientific progress is the obsolescence of scientific knowledge.'[39] It is important to recall that the scientific consensus of the first decade of the twentieth century was that the universe was more or less the same today as it always had been; yet this once fashionable and seemingly reliable view had been eclipsed by the seemingly unstoppable rise of the theory of cosmic origins generally known as the 'Big Bang'.[40]

James Ladyman, in defending a more cautious concept of 'scientism', points out that 'fallibilism is integral to the scientistic spirit, and it is entirely in accord with it to say that science is the worst source of knowledge about the world apart from all the rest. The history of science teaches us that even cherished laws may be subject to revision.'[41] As Arthur Koestler remarked, during his exploration of changing scientific ideas about the universe: 'The progress of science is strewn, like an ancient desert trail, with the bleached skeletons of discarded theories which once seemed to possess eternal life.'[42] The *history* of science may thus help us reflect on what its potential *futures* might be.[43]

A third concern arises from the influential 'critical realist' analysis of Roy Bhaskar. For Bhaskar, ontology determines epistemology; it is the nature of an

[38] Maxwell, *Scientific Papers*, vol. 2, 244.
[39] Daston, 'Scientific Error and the Ethos of Belief', 1.
[40] Kragh, *Conceptions of Cosmos*; Kragh, *Higher Speculations.*
[41] Ladyman, 'Scientism with a Humane Face', 115.
[42] Koestler, *The Ghost in the Machine*, 178.
[43] This point, it should be added, applies to philosophy as well: Hatfield, 'The History of Philosophy as Philosophy'; Antognazza, 'The Benefit to Philosophy of the Study of Its History'. A fuller discussion of these concerns would include the consequences of the observation that the cognitive processes that lead to the production of scientific knowledge are social in character: see the issues raised in Longino, *The Fate of Knowledge*, 77–144.

object that determines both what can be known about it, and how it is to be known. Bhaskar's critical realism offers a helpful way of critiquing scientism as the improper imposition of a research method appropriate for one level of reality onto every aspect of the natural and social world.[44] For Bhaskar, 'the nature of the object' determines 'the form of its possible science'. Scientism, Bhaskar notes, denies that there are 'any significant differences in the methods appropriate to studying social and natural objects'.[45]

If Bhaskar is right, scientism—whether intentionally or not—reduces reality to what can be known through the application of one specific research method. Epistemology is thus allowed to determine ontology, in that the use of one specific research method determines what is 'seen'—and hence judged to be real. Scientism is, on this approach, blind to the existence of levels of reality that cannot be engaged by the methods of the natural sciences—methods, it must be added, which were actually developed with other research tasks in mind. Examples of such levels which are inaccessible using these scientific methods include questions of existential meaning and moral value.[46]

These three points, when taken together, point to the continuing importance of the disciplines of the history and philosophy of science in helping us to understand both the nature and limits of scientific knowledge, and the manner in which a revived natural philosophy might offer a fuller and more satisfying account of the place of the natural sciences in the human quest for wisdom. To live meaningfully in this world, human beings need more than a purely scientific analysis of our universe is able to provide. As Eleonore Stump caustically (yet perceptively) remarked, there is no good reason to suppose that 'left-brain skills alone will reveal to us all that is philosophically interesting about the world'.[47]

Neo-Confucianism: Scientism and Natural Philosophy in Twentieth-Century China

Although the capacity of the natural sciences to provide authoritative and comprehensive answer to all of life's questions has been discussed extensively

[44] Bhaskar, *The Possibility of Naturalism*, 2–3; Bhaskar, *A Realist Theory of Science*, 36–40.

[45] Bhaskar, *The Possibility of Naturalism*, 2.

[46] Bhaskar himself is open to 'ethical naturalism'—the view that values can be derived from facts: see, for example, his discussion in Bhaskar, *Scientific Realism and Human Emancipation*, 177. Others within the critical realist movement have, however, argued that Bhaskar is inconsistent here, in that his supposedly factual premises already include assumptions about values, and he fails to explain how we can obtain access to these objective moral values: see, for example, Hostettler and Norrie, 'Are Critical Realist Ethics Foundationalist?'; Elder-Vass, 'Realist Critique without Ethical Naturalism and Moral Realism'.

[47] Stump, *Wandering in Darkness*, 24.

in western academia, it is too easy to overlook developments in twentieth-century China, which have important consequences for any informed reflection about the scope and future of natural philosophy. Several forms of natural philosophy can be discerned within China's rich cultural history, with roots that can be traced back to the classical period. Daoism, Confucianism, and Buddhism all developed forms of natural philosophy, often focusing on the issue of achieving 'harmony' between heaven and earth.[48] Yet these did not lead to the emergence of the natural sciences in their western form,[49] characterized by the use of the experimental method, and a Baconian agenda of transforming nature through an enhanced understanding of its functionalities.

In 1919, however, things changed. Following the collapse of the of the Qing dynasty in 1912 and the end of the First World War in 1918, Chinese nationalists called for a rejection of traditional values—which were widely associated with Confucianism—and the adoption of the western ideals of 'Mr. Science'.[50] The 'May Fourth Movement' increasingly saw western science as a means of modernizing China, and enabling it to achieve its potential, particularly in the face of western cultural and political encroachment.[51] Traditional Confucian values came to be seen as reactionary, inhibiting the social change and technological advances that were essential if China was to become a modern nation state. Modernization of China came to become associated with an idealist belief in 'the power of science to modify society by using the methods, values, and ideas underlying science'.[52] Gradually science came to be seen not simply as an instrument or a technique for achieving cultural and social change, but as an *ideology*—a way of determining values and meaning.[53]

The form of scientism that emerged in China in the early decades of the twentieth century had two main elements: a 'conception of science as an all-inclusive system of nature which not only informs us of objective reality concerning the physical universe but prescribes an outlook on human life and society as well'; and the core conviction that science represented a 'mode of

[48] For the various approaches found in these traditions, see Qiu, 'Onitsura's Makoto and the Daoist Concept of the Natural'; Berthrong, 'Confucian Views of Nature'.

[49] This is the view of Fung Yu-Lan, set out in his influential 1922 article 'Why China Has No Science'. For the ensuing scholarly debate over the historical emergence of science in a Chinese context, see Elman, *On Their Own Terms*. The legacy of Joseph Needham in engaging the question of the development of science in China continues to be controversial: see, for example, Finlay, 'China, the West, and World History in Joseph Needham's *Science and Civilisation in China*'.

[50] Mitter, *A Bitter Revolution*, 3–68, 102–49.

[51] See the important analysis in Hui, 'The Fate of "Mr. Science" in China'.

[52] Ouyang, 'Scientism, Technocracy, and Morality in China', 178.

[53] For a good discussion of the issues, see Chiu, *Thomé H. Fang, Tang Junyi and Huayan Thought*, 180–97. For the history of Chinese science at this time, see Elman, 'Toward a History of Modern Science in Republican China'.

thinking, a methodology which promised to be the only valid way of understanding life and the world'.[54] A good example of this approach in the 1920s is found in the works of the geologist Ding Wenjiang, who advocated a scientific rationalism which restricted itself to categorization of observations of the physical world. 'The scientific method is nothing but the division into classes of facts in the world and the search for their order.'[55]

In a lecture given at Qinghua University on 14 February 1923, Zhang Junmai argued that this growing trend towards scientism was destructive of some traditional cultural values, and drew on the writings of European philosophers such as Henri Bergson (1859–1941) who had established themselves as critics of scientism and positivism in opposing the overextension of science.[56] Zhang Junmai's criticism focused on the impoverishment of the subjective, which was the inevitable outcome of science's attempt to objectify reality, and reduce this to a unified worldview. Yet these points were not made in dialogue with traditional Chinese intellectual traditions, but rather their European counterparts.

In the late 1950s, however, a traditional alternative began to emerge. 'Neo-Confucianism' presented itself as maintaining continuity with older and distinctively Chinese understandings of the natural world, and offering answers to important questions that could not be answered on the basis of science alone.[57] The potential tensions between 'science' and 'philosophy'—two categories that do not easily fit into traditional Chinese taxonomies of knowledge—were of no small importance in the genesis of Neo-Confucianism.[58] In the past, Chinese intellectuals had located the intelligibility of the world in the metaphysical worldview offered by traditional Chinese religion and philosophy. While recognizing that the natural sciences offered a new way of making the external world intelligible, there were increasing concerns that this form of scientism proved incapable of engaging deeper questions of value and meaning.

Neo-Confucian intellectuals did not deny the usefulness of scientific discourse as such, but rather pointed out the need to complement its 'objective logical causal mode of thinking with an approach which promised to yield a subjective, direct, and empathic comprehension of the world'.[59]

[54] Chang, 'New Confucianism and the Intellectual Crisis of Contemporary China', 283. Cf. Oldstone-Moore, 'Scientism and Modern Confucianism'.

[55] Cited in Stock, *The Horizon of Modernity*, 206–7.

[56] For the importance of this lecture, see Stock, *The Horizon of Modernity*, 204–6.

[57] Chang, 'New Confucianism and the Intellectual Crisis of Contemporary China', especially 282–5.

[58] Stock, *The Horizon of Modernity*, 197–8.

[59] Chang, 'New Confucianism and the Intellectual Crisis of Contemporary China', 285.

This development, particularly when seen against the widespread revival of Confucianism in contemporary China,[60] points to the kind of synergy between science and other forms of human knowledge that is characteristic of western natural philosophy, and the need to offer an account of the world which was attentive to both its objective and subjective dimensions.

Although the natural sciences can indeed be considered to be universal in terms of their methods of investigation and criteria of theory choice, there is a growing recognition that science is culturally embedded, interacting with other cultural stakeholders. The emergence of 'natural philosophy' in the West, especially during the early modern period, points to the intersections and interactions of what might now be called 'science' with wider social, political, and philosophical issues. The rise of Neo-Confucianism can be seen as enabling Chinese scientists to retain their cultural identity, offering a framework which both encourages the development of the sciences and recognizes the need to engage deeper questions of meaning and value that are not themselves adequately or reliably engaged by the scientific method.[61]

This naturally leads us to move on to consider Popper's 'World Two' in the next chapter, and reflect on its importance for a retrieved natural philosophy.

[60] Kang, 'A Study of the Renaissance of Traditional Confucian Culture in Contemporary China'.
[61] See the discussion in Cha, 'Modern Chinese Confucianism', especially 486–8.

9

Subjectivity

An Affective Engagement with Nature

Popper's 'World Two' enfolds our subjective personal perceptions and experiences. In common with Popper's other two worlds, this world is better seen as broadly descriptive, rather than precisely analytic. Yet it is a world that many intuitively recognize as plausible and meaningful, mapping broad aspects of human existence and reflection that are not adequately rendered by a purely objective account of reality. For many, this aspect of human existence is of defining importance.[1] Yet it seems to be excluded by a purely objective account of the natural sciences. David Hume, for example, contrasted the objectivity of human reason with the risk of a subjectivity based on taste:

> The distinct boundaries and offices of *reason* and of *taste* are easily ascertained. The former conveys the knowledge of truth and falsehood: the latter gives the sentiment of beauty and deformity, vice and virtue. The one discovers objects as they really stand in nature, without addition or diminution: the other has a productive faculty, and gilding or staining all natural objects with the colours, borrowed from internal sentiment, raises in a manner a new creation.[2]

Yet it is one thing to make knowledge dependent on taste, and quite another to acknowledge the existential importance of the subjective world of human experience, and attempt to weave this into an overall account of the human encounter with nature. In what follows, we shall consider the importance of some subjective aspects of human existence.

[1] For example, see Agamben, *The Open*, 13–16.

[2] Hume, *Enquiries Concerning Human Understanding and Concerning the Principles of Morals*, 294. For a detailed discussion of this aspect of Hume's thought, see Kail, *Projection and Realism in Hume's Philosophy*, 147–203.

Human Subjectivity: Space and Time versus Place and History

As human beings, we are both observers of nature, and part of nature. We are participants in the historical process, who try to make sense of it as it unfolds.[3] Yet this is only one way of framing our inhabitation of the world of space and time. One of the most striking features of Newton's *Principia Mathematica* is its mathematization of the notions of absolute time and absolute space.[4] Although Newton regarded space and time as distinct entities, this view was successfully challenged by Hermann Minkowski in a lecture of 1908, in which he reconceptualized Einstein's new theory of special relativity, formulated in 1905, in terms of a four-dimensional 'spacetime', which unified the notions of space and time.[5] This development raised some difficult philosophical questions, as a unified 'spacetime' seemed to bear little relationship to how humans subjectively experience living in space and time.[6]

In his 1928 classic *The Nature of the Physical World*, Arthur Eddington noted how there was a significant disparity between the pure objectivity of physics and the subjective experiential world of individuals. While stressing the physical adequacy of Minkowski's representation of the world, Eddington noted that 'something must be added to the geometrical conceptions comprised in Minkowski's world before it becomes a complete picture of the world as we know it.'[7] For Eddington, who played a critical role in confirming Einstein's theory of General Relativity in 1919, this 'picture as it stands is entirely adequate to represent those primary laws of Nature'; it is not, however, adequate to engage our inner perceptions of the passage of time. Something important seems to be missing from such a purely objective account of time.

From the perspective of physics, Einstein held that the 'distinction between past, present and future has only the meaning of a persistent illusion'.[8] A purely physical account of the present moment, as distinguished from the past and future, can easily be offered; yet this fails to account for why human

[3] For reflections on the subjective human perception of the passage of time, see Phillips, ed., *The Routledge Handbook of Philosophy of Temporal Experience*.

[4] Janiak, *Newton as Philosopher*, 130–62.

[5] Minkowski, 'Raum und Zeit'.

[6] For the response of twentieth-century philosophers to this development, see Slowik, 'The Fate of Mathematical Place'.

[7] Eddington, *The Nature of the Physical World*, 68. For a philosopher's assessment of Eddington, see Merleau-Ponty, *Philosophie et théorie physique chez Eddington*.

[8] Speziali, ed., *Albert Einstein—Michele Besso Correspondence*, 537–8: 'Für uns gläubige Physiker hat die Scheidung zwischen Vergangenheit, Gegenwart und Zukunft nur die Bedeutung einer wenn auch hartnäckigen Illusion.'

beings both consider the present to be distinct from the past and the future, and regard it as having special significance. Past, present, and future can be represented *chronologically and spatially* using a world line—yet their significance cannot be represented *existentially* in this way. Most people find it difficult to think neutrally and dispassionately about the transition from a past in which we did not exist, through a present in which we live and think, to a future in which we will no longer exist. Rudolf Carnap, reflecting on his discussions with Einstein about the 'present moment' at Princeton during the late 1940s, suggested that Einstein had concluded that scientific descriptions cannot satisfy our human needs, in that there is something essential about the 'present moment' that lies beyond the reach of science.[9]

The difficulty here is that the objective scientific language of 'time' and 'space' is inadequate to account for these aspects of human existence. For example, an objective account of the physical geography of Australia is not capable of capturing the subjective importance of features of that landscape to Australian aboriginal populations, in terms of the memories which are linked with certain places, or to the meaning that individuals and communities attach to them.[10] A purely objective account of physical geography is incapable of identifying and preserving these aspects of the cultural landscape. Alongside the spatial realities of physical geography, there exist important and identity-giving networks of memory and attachment to place.[11] A holistic account of the significance of such landscapes needs to bring both these objective and subjective aspects together. Each demands quite different methods of investigation and representation in order to do them justice.

For Aristotle, a 'place' (*topos*) was essentially an inert container, an abstract point that was not fundamentally different from any other point.[12] Yet Aristotle fails to take account of history, memory, and the human tendency to attach significance to places. We have seen how this is significant in relation to Australian aboriginal culture; it is also a highly significant issue in relation to the land of Israel, in which these factors play a decisive role in Jewish tradition.[13] Walter Brueggemann captured this entanglement of history, memory, belonging, and place in ancient Israel in his landmark work *The Land*, in which he argues that to make sense of the preoccupations of ancient Israel, a

[9] Schilpp, ed., *The Philosophy of Rudolf Carnap*, 37–8.

[10] Paton, 'The Mutability of Time and Space as a Means of Healing History in an Australian Aboriginal Community'.

[11] Krichauff, *Memory, Place and Aboriginal-Settler History*, 93–109.

[12] Morison, *On Location*, 133–73.

[13] Vanderhooft, 'Dwelling beneath the Sacred Place'. For the politics of naming places in modern Palestine, which reflects issues relating to 'place' and 'history', see Peteet, 'Words as Interventions', 156–65.

fundamental distinction had to be made between 'space' and 'place': 'Place is space which has historical meanings, where some things have happened which are now remembered and which provide continuity and identity across generations.'[14] Brueggemann's analysis of the history of Israel rightly emphasized the manner in which specific *places* play a critically important place in human life, not least in that they function as anchor points for memory, identity, and aspiration. They continue to be important for many Jews to this day.[15]

To be clear, there is nothing wrong with the objectivity of physical geography, or any other science. My point is simply that there is more that needs to be said. The case of natural landscapes is particularly significant, in the light of their affective impact on people—not merely in evoking a sense of beauty or awe, but in terms of their memories, associations, and attachments, which are so important to some individuals and people groups in shaping personal identity.[16] The problems arise when we are asked to believe that there is nothing more to reality than what the physical realm discloses. This foreclosure and abbreviation of a complex reality may be appropriate for certain analytical enterprises, but it fails to deal with the complexities of the human interaction with the natural world.

The conceptual space once known as 'natural philosophy' allowed engagement with both the objective and subjective aspects of time and space, holding them together in a creative tension. It is not that one is right and the other wrong; it is that each represents a different, yet important, aspect of the complex dynamic of human existence within the natural order. They offer different modes of envisaging and representing our place in the universe, shaping our understanding of who we are, and how we should inhabit the natural world. They both need to be integrated within a richer and deeper understanding of our world—such as that which might be offered by a reconceived natural philosophy.

Beauty and Wonder: An Affective Engagement with Nature

Many would argue that one of the gateways to the study of the natural sciences is a subjective experience of wonder or awe at the immensity or beauty

[14] Brueggemann, *The Land*, 5.
[15] See, for example, Kaplan, 'Time, History, Space, and Place'.
[16] Tilley and Cameron-Daum, *Anthropology of Landscape*, 1–22.

of the natural world.[17] These experiences both compel attentiveness towards nature, and reward it. People often experience a sense of wonder at the beauty of some aspects of nature—such as sunsets, night skies, and expansive landscapes.[18] Yet beauty elicits desire, creating a sense of longing for participation that goes beyond mere observation. This, as C. S. Lewis observed, is an *immersive* desire.

> We do not want merely to see beauty, though, God knows, even that is bounty enough. We want something else which can hardly be put into words – to be united with the beauty we see, to pass into it, to receive it into ourselves, to bathe in it, to become part of it.[19]

Lewis's point is that we are not merely drawn *towards* beauty; we are drawn *into* it, in a process that demands what John Cottingham has helpfully styled an 'epistemology of involvement' rather than an 'epistemology of detachment'.[20] There are clear parallels with Gadamer's account of the relation of theory and participation, noted earlier (pp. 133–5). The observer of nature is affected by this encounter, and drawn into a deeper relationship with the natural world.

This point is particularly significant in relationship to landscapes and environmental issues. The environmental philosopher Holmes Rolston draws attention to the implications of this observation for attempts to construct a coherent account of the human response to natural landscapes that is appropriate to the environmental challenges that we face.

> We do not always need science to teach us what happens on landscapes, though science enriches that story.... Science brings insight into continuing organic, ecological, and evolutionary unity, dynamic genesis; but such unity may also have already been realized by pre-scientific peoples in their inhabiting of a landscape. Science can engage us with landscapes too objectively,

[17] For good accounts of this, see Haralambous and Nielsen, 'Wonder as a Gateway Experience'; Tallis, *In Defence of Wonder*, 1–22.

[18] See especially Fisher, *Wonder, the Rainbow, and the Aesthetics of Rare Experiences*; Mayne, *This Sunrise of Wonder*.

[19] Lewis, *Essay Collection*, 104.

[20] Cottingham, *Philosophy of Religion*, 23–4. Cottingham distinguishes his 'epistemology of involvement' from an 'epistemology of submission', in which subsequent questioning is excluded. For an assessment of this approach, see Kanterian, 'Naturalism, Involved Philosophy, and the Human Predicament'.

> academically, disinterestedly; landscapes are also known in participant encounter, by being embodied in them.[21]

As we noted in the previous section, these multiple responses to natural landscapes and vistas highlight the dangers of limiting the human engagement with nature to what can be measured or weighed.[22]

For Rolston, one of the issues concerns the sensitization of the human observer to what is being observed—a process which is certainly informed by science, yet which transcends scientific or utilitarian considerations. Engagement with a natural landscape—as with nature in general—has two focal points: first, that 'aesthetic experience must be participatory, relating an actual beholder to a landscape'; and second, 'that nature is objective to such beholders, actually known in the physical and biological sciences'.[23] These can be correlated; they are not, however, identical. 'We humans carry the lamp that lights up value, although we require the fuel that nature provides.'[24] There has to be a way of holding together objective and participatory engagement with such landscapes.

In observing and valuing nature, we bring a plurality of interpretative frameworks to bear, which are often unconnected in terms of the manner in which these are derived and acquired, the manner in which they are integrated within an individual's mind, and the manner in which they are applied. These theories may originate from rational reflection, or from the radical epistemic transformations that are often associated with religious conversion. William James noted that this often led to 'a transfiguration of the face of nature', in which a 'new heaven seems to shine on a new earth'.[25]

As Rolston stresses, human beings develop their own distinct outlooks on nature, which are shaped by many factors—such as personal experience, religious beliefs, scientific knowledge, and the desire for happiness—which are woven together to create a way of seeing—and hence valuing—the natural world.[26] The observer of a beautiful landscape might find herself overwhelmed with its beauty, intrigued as to what scientific explanations might be offered for its shape and structures, and impelled to preserve its beauty

[21] Rolston, 'Does Aesthetic Appreciation of Landscapes Need to Be Science-Based?', 381.
[22] The complaint of Romantic poets such as Keats: see pp. 95–6.
[23] Rolston, 'Does Aesthetic Appreciation of Landscapes Need to Be Science-Based?', 377.
[24] Rolston, 'Value in Nature and the Nature of Value', 15.
[25] James, *The Varieties of Religious Experience*, 151.
[26] There are important parallels here with the construction of 'social imaginaries', which inform the way we understand and engage the natural, social, and personal worlds. For such imaginaries, see Gaonkar, 'Toward New Imaginaries'; Taylor, *Modern Social Imaginaries*.

through appropriate ecological strategies. These responses are connected primarily by the order-creating mind of the individual observer.[27]

Yet a question might arise here: if beauty is a subjective matter, how are we to account for the widespread interest in aesthetic evaluations of supposedly objective scientific theories?[28] Not only are such theories often described in subjective terms as 'elegant', 'beautiful', or 'ugly'; many of the core terms historically associated with 'natural philosophy' are clearly dependent on human aesthetic judgements—such as symmetry, simplicity, coherence, elegance, and harmony.[29] In 1954, the brilliant quantum theorist Paul Dirac declared that 'physicists generally have come to believe in the need for physical theory to be beautiful, as an overriding law of nature. It is a matter of faith rather than of logic.'[30] This resonance between aesthetic and scientific theorizing is clearly significant, even if it is not fully understood.

> The hallmarks of scientific understanding are similar to an aesthetic feature associated with literature, music, and the visual arts. It is the feature described as coherence, harmony, and inevitability of fit. Aesthetics thus plays an epistemic role in science as an indication of understanding.[31]

This puzzling resonance between theoretical beauty and scientific truth can be accounted for in a number of ways—for example, through an appeal to the traditional Thomist idea that beauty is a transcendental reality, or the suggestion that supposedly aesthetic judgements are actually disguised epistemic assertions.[32] Yet whatever its explanation may be, this correlation is an important indication of the need to interconnect subjective human perceptions and objective scientific accounts of the world.[33]

[27] For a good example of this complexity of observation, see Williams, 'Representations of Nature on the Mongolian Steppe'. Williams weaves together the complex scientific and political dimensions of these landscapes, noting the divergent perspectives of different observers (e.g., urban intellectuals, and Mongolian herders).

[28] See, for example, Kosso, 'The Omniscienter'; McAllister, 'Is Beauty a Sign of Truth in Scientific Theories?'

[29] For a highly original study of the significant correlations and interaction between musical harmony and scientific understanding, see Chua, *Absolute Music and the Construction of Meaning*.

[30] Dirac, 'Logic or Beauty?', 268.

[31] Kosso, 'The Omniscienter', 39.

[32] On the former, see Ramos, *Dynamic Transcendentals*, 72–9; on the latter, Todd, 'Unmasking the Truth beneath the Beauty'.

[33] There are clearly limits to this parallel between truth and beauty; while it is persuasive in the domains of physics and mathematics, it is less so in the biological sciences.

Responding to Nature: Science, Poetry, and the Imagination

What language should we use to try and represent nature? In the case of the physical world, the most appropriate language is that of mathematics, which proves remarkably capable of rendering its contours and interconnections. We have already noted Galileo's firm conviction that mathematics was the natural language of the cosmos.[34] This broad picture, which emerged within early modern natural philosophy, arguably reached its peak in Newton's *Principia*. The 'mathematization' of nature was admirably suited to objects in Popper's 'World One'—for example, in describing and then explaining the orbits of the planets. But what of Popper's 'World Two'?

One of the reasons for offering an extended conversation with the history of natural philosophy in the first part of this work is to make it clear that what we now know as 'science' evolved slowly in constant and mutually enriching dialogue with other disciplines that would now be seen as constituting a different intellectual domain—the humanities. Tom McLeish makes the important and entirely fair point that 'claiming science as an exclusive property of the modern world removes the deep and slow cultural development of an imaginative and creative engagement with nature that develops, at least chronologically, alongside the story of art in its own multitude of forms.'[35] Each has an important place in the spectrum of human knowledge and cultural engagement.

It seems as if a different language has to be used in order to express the affective impact of the natural world upon us—a language that is capable of *evoking* something of this impact, and not merely describing it. It is now widely accepted that different disciplines develop their own distinct research methods, conceptualities, and languages, adapted to their specific areas of concern and exploration.[36] Poetic language aims to describe the impact of nature on human beings, not through the precise analytical language of mathematics, but using language that both displays and *elicits* desire, amazement, wonder, or awe.

Some might suggest that this linguistic dichotomy could be mapped onto the neurological lateralization of the left and right human brain hemispheres, perhaps in the manner suggested by Iain McGilchrist in *The Master and His*

[34] Palmerino, 'The Mathematical Characters of Galileo's Book of Nature'.
[35] McLeish, *The Poetry and Music of Science*, 12.
[36] McGrath, *The Territories of Human Reason*, 76–9.

Emissary (2009). While concerns might be raised about aspects of his approach,[37] McGilchrist avoids any simplistic equation of the left hemisphere with the cold rationalities of scientific logic and the right with the intuitive capacities of artistic creativity, yet notes the complex interaction between our inclination to dissect and our desire to make whole: 'Our talent for division, for seeing the parts, is of staggering importance – second only to our capacity to transcend it, in order to see the whole.'[38]

In its original forms, natural philosophy was as much about the celebration of the totality of nature, as an analysis of its individual components. The personification of nature—as in 'Mother Nature'—was one of several devices used to affirm the fundamental unity of the natural world, while encouraging exploration of its multiple aspects.[39] Yet the fragmentation of natural philosophy in England during the nineteenth century led to tensions between those who saw nature in essentially objective scientific terms, and those who stressed its subjective impact on the beholder, especially in terms of arousing a sense of beauty and wonder. The former assumed that 'the world of natural objects, of bare, clear, downright facts, is unproblematically given';[40] the latter regarded this as unacceptably reductionist, failing to give due weight to the aesthetic and affective impact of nature on humanity.

Robert Hunt, a Victorian poet who was sympathetic to the sciences, nevertheless expressed concern about the suppression of the imagination in an age that seemed to be increasingly shaped by concerns about matters of fact and questions of utility.[41] In his *Poetry of Science* (1848), Hunt offered what is in effect a disciplinary imaginary, in which the natural sciences are concerned with an 'inductive search' for basic principles, philosophy with the deduction of 'large generalities from the fragmentary discoveries of severe induction', and poetry unifies the 'facts of the one and the theories of the other', allowing observation of the world to be connected with 'exalted ideas'.[42] It is not necessary to accept Hunt's specific understanding of individual disciplines to appreciate the point he is making; in effect, his plea is for a unified vision of reality that transcends the specific individual forms of human knowledge

[37] See, for example, the perceptive comments in de Haan, 'McGilchrist's Hemispheric Homunculi'.

[38] McGilchrist, *The Master and His Emissary*, 93.

[39] For the important consequences of this way of envisaging nature, see Liu et al., '"Mother Nature" Enhances Connectedness to Nature and Pro-Environmental Behavior'.

[40] Belsey, *Critical Practice*, 9. Cf. Tate, 'Poetry and Science'.

[41] Hunt develops this point in his novel *Panthea* (1849). For a detailed analysis, see Tait, 'The "True Philosophy" of Robert Hunt'.

[42] Hunt, *Poetry of Science*, 17–18.

production, and a language that is capable of doing justice to the beauty and wonder of nature.[43]

Hunt's point is developed by the environmental activist Wendel Berry, who argues that the language of science fails to excite and engage those who could—and should—be concerned about the destruction of natural habitats. How, he asks, can these people be plausibly invited to 'save' a world which their technological language has reduced to an assemblage of utterly featureless abstractions? A new language has to be found that is capable of evoking compassion and concern for its protection and renewal on the part of human agents.

> The problem, as it appears to me, is that we are using the wrong language. The language we use to speak of the world and its creatures, including ourselves, has gained a certain analytical power... but has lost much of its power to designate what is being analyzed or to convey and respect or care or affection or devotion toward it.... It is impossible to prefigure the salvation of the world in the same language by which the world has been dismembered and defaced.[44]

Hunt's unificationist aspirations are echoed in the poetry of Rebecca Elson, an observational astronomer at the University of Cambridge until her early death in 1999, at the age of 39. Elson's poems—published posthumously as *A Responsibility to Awe*—are remarkable attempts to explore the 'antithesis between matters of the mind, such as mathematical and logical reasoning, and matters of the heart, which connotes her more creative, intuitive, and emotional faculties'.[45] How are we to visualize dark matter, when it cannot be seen? And what is its deeper significance?

> Above a pond,
> An unseen filament
> Of spider's floss
> Suspends a slowly
> Spinning leaf.[46]

[43] There are important parallels here with the religious 'beholding' of the natural world: see Blowers, 'Beauty, Tragedy and New Creation'.

[44] Berry, *Life is a Miracle*, 8.

[45] Heuschling, 'Don't Ask the Questions You've Been Taught by Science', 46.

[46] Elson, *A Responsibility to Awe*, 15.

In this remarkable haiku-like poem, Elson 'transformed the "dark matter" of her scientific study into poetic mystery, even into the mystery of her own impending death'.[47] Elson's poetry allowed her to explore the feelings and existential anxieties that contemporary cosmology evoked within her—feelings that the objective language of science could not adequately or appropriately express, but which were integral to Elson's perceptions of the cosmos, and her place within it.

Elson and Hunt can both be seen as what Mary Midgley describes as 'bridge builders' between science and poetry, who were convinced that the 'antithesis between thought and feeling' was false, and needed to be overcome or subverted.[48] Other examples could easily be given—think, for example, of Wittgenstein's positive reflections on the mutual engagement of poetry and philosophy.[49] The musical notion of harmony can be expressed objectively at the mathematical level, but is experienced subjectively by the listener; as Kepler showed, it could also be connected with a wider frame of discourse within natural philosophy.[50] Midgley herself cites an entire untitled early poem of 1864 by Gerard Manley Hopkins to illuminate the need for an informed 'reshaping of our whole conceptual background' in response to the emergence of a scientific culture:[51]

> It was a hard thing to undo this knot.
> The rainbow shines, but only in the thought
> Of him that looks. Yet not in that alone,
> For who makes rainbows by invention?
> And many standing round a waterfall
> See one bow each, yet not the same to all,
> But each a hand's breadth further than the next.
> The sun on falling waters writes the text
> Which yet is in the eye or in the thought.
> It was a hard thing to undo this knot.

Midgley's choice is judicious and appropriate, although she does not offer a commentary on the poem. Hopkins makes the point that the rainbow is not a

[47] Fiske, 'The Poetic Mystery of Dark Matter', 845.
[48] Midgley, *Science and Poetry*, 74–7.
[49] Perloff, *Wittgenstein's Ladder*, 181–218.
[50] Stephenson, *The Music of the Heavens*. For an excellent account of how music could be reconnected with the discourses of theology, biology, philosophy, chemistry, and physics, see Chua, *Absolute Music and the Construction of Meaning*, especially 12–22.
[51] Midgley, *Science and Poetry*, 78–9. For the poem, see Hopkins, *The Major Works*, 29.

fabrication, residing solely in the observer's imagination. Yet while two observers might indeed see a rainbow, and agree it is not an invention, they each see it *in different ways*. It is 'not the same to all'. While Hopkins holds that the rainbow is something that is objectively present and external to an observer, it cannot be isolated from the subjective perceptions of each individual observer.[52] Their response to these aspects of nature will mingle the cognitive and affective, the objective and subjective. For example, Christian theologians have traditionally acknowledged a range of subjective responses to the natural world,[53] including an 'affective' appreciation in response to the beauty of nature; an 'affective–cognitive' appreciation triggered by the beauty which is disclosed through detailed study of the natural order; and a 'cognitive' appreciation arising from contemplation of the harmonious functioning of the natural order as a whole.[54] While some of these are informed by the natural sciences, others are clearly not.

Yet—and this is the critical point—these subjective and objective accounts of the natural world can be woven together, in a comprehensive account of what any individual makes of nature, and how this is perceived to be significant to that individual. Popper's three worlds can helpfully be *distinguished*, but they cannot be *separated* in the lives of individuals. The issue is not how we divide up our experiences of this complex world—for example, into the objective, subjective, and theoretical. It is rather that, however we might choose to organize and distinguish these experiences, the process of distinguishing them cannot be detached from the task of bringing them back together again. As McGilchrist rightly observed, we might begin with the division of nature, allowing us to see its parts; yet what is of greater importance is the human capacity to 'see the whole'.[55] And that, as we shall see in the final chapter of this work, is what a reimagined natural philosophy might be able to do.

[52] For discussion of this poem, see Hatch, 'Gerard Manley Hopkins and Victorian Approaches to the Problems of Perception', 170–3.

[53] I here follow Schaefer, 'Appreciating the Beauty of Earth', 25–36.

[54] Moulin, 'Beauty as Natural Order'.

[55] McGilchrist, *The Master and His Emissary*, 93.

10

Natural Philosophy

Recasting a Vision

In his elegant essay in praise of natural philosophy, Nicholas Maxwell identifies two great problems of learning that confront humanity as it faces an uncertain future: 'learning about the nature of the universe and our place in it, and learning how to create as good, as wise, as civilized a world as possible'.[1] Significant progress in achieving the first objective, he argues, was made during the seventeenth century; the second, however, seems further away than ever, as an increased ability to exploit the natural world leads to the destruction of natural habitats, the extinction of species, and the crisis of climate change. This demands new ways of thinking, he suggests, that require a redrawing of the intellectual landscape, leading to a transformed science and transformed philosophy becoming a single domain of thought—natural philosophy.[2]

As earlier chapters will have made clear, I am in sympathy with many of Maxwell's concerns and aspects of his analysis of the problems we now face, including the question of how we cultivate wisdom in a culture dominated by purely explanatory notions of truth.[3] In this final chapter of this study, I shall set out a more modest way of retrieving natural philosophy which I believe can avoid some of the potential difficulties that might stand in the way of Maxwell's ambitious project of reframing the relation of science and philosophy in a new style of natural philosophy, while at the same time expanding its scope. Those difficulties relate primarily to the institutional embeddedness of disciplines in western academia, which so easily can become an obstacle to interdisciplinary undertakings, or to radical disciplinary realignments. The challenge is to *reimagine* natural philosophy.

[1] Maxwell, *In Praise of Natural Philosophy*, 218.

[2] Maxwell, *In Praise of Natural Philosophy*, xi.

[3] A theme singled out for particular discussion in the assessments of Maxwell found in McHenry, ed., *Science and the Pursuit of Wisdom*. For a full statement of Maxwell's own views on wisdom, see Maxwell, *From Knowledge to Wisdom*.

On Bridging Popper's Three Worlds

Popper's three worlds can be seen as distinct aspects of the human engagement with the natural world, possessing a coherence and unity within the mind of the individual beholder. Distinguishing these three worlds does not amount to disconnecting them; it is rather a mapping exercise which facilitates an understanding of their mutual relationship, and helps to identify the intellectual territory that a natural philosophy was once understood to occupy. The question is whether (and how) we can weave these components together in a coherent whole.

I prefaced this book with a short passage from John Banville's novel *Kepler*, describing how Kepler's theoretical unification of the solar system allowed him to grasp the fundamental interconnection of things. Before, he had 'voyaged into the unknown' and created 'fragmentary and enigmatic charts apparently disconnected with each other'. Yet following his theoretical breakthrough, he came to see that these were not 'maps of the islands of an Indies, but of different stretches of the shore of one great world'.[4] The point that Banville makes in this passage is that a set of seemingly disconnected fragmentary charts were brought together in Kepler's fertile imagination to disclose a map of a single greater reality. The individual islands remained; yet they could now be seen in a new way—as interconnected features of a larger and more complex landscape.

But how might such unification proceed? For Isaiah Berlin, the 'ideal of a unified system of all the sciences, natural and humane, has been the programme of the modern Enlightenment'.[5] This lingering agenda of the Enlightenment for the unification of the multiple domains of human knowledge underlies the biologist Edward O. Wilson's influential manifesto *Consilience: The Unity of Human Knowledge* (1998), which is the culmination of the unificationist programme he initiated twenty years earlier with the publication of his *Sociobiology*.[6] Though ingenuously framed as a responsible quest for a foundation of the unity of human knowledge, Wilson is actually proposing the subordination of the humanities to the norms and methods of

[4] Banville, *Kepler*, 179. See also Banville, 'Beauty, Charm, and Strangeness'. For Banville's narrative of early modern science, see Booker, 'Cultural Crisis Then and Now'; Fiorato, *The Relationship between Literature and Science in John Banville's Scientific Tetralogy*.

[5] Berlin, *The Proper Study of Mankind*, 327–8.

[6] Segerstrale, 'Wilson and the Unification of Science'; Carbonell, 'Wilson and Gould', 346–47.

the natural sciences, by which they are to be judged. 'We have the common goal of turning as much philosophy as possible into science.'[7]

The Enlightenment's 'dream of intellectual unity' having foundered through an accelerating process of disciplinary fragmentation,[8] Wilson suggests that this dream can now only be realized by permitting the natural sciences to re-establish this unity through judicious extension of its methods to other disciplines. For Wilson, consilience is ultimately a 'metaphysical world view', which is 'allegiant to the habits of thought that have worked so well in exploring the material universe'.[9]

Wilson's attempt to unify human knowledge by reducing it to physics has been subject to considerable criticism. The most effective response came from the palaeontologist Stephen J. Gould, who disputed Wilson's interpretation of the term 'consilience' (borrowed from the philosopher William Whewell). For Gould, Whewell does not treat consilience as a method of *reduction*. Where Wilson interprets consilience as the 'unification of all knowledge along a single chain of rising complexity', Gould insisted that it was about recognizing 'irreducible different ways of knowing'.[10] Where Whewell 'regarded the humanities (particularly moral and religious reasoning) as a set of logically and inherently separate ways of knowing', Wilson wants to subsume them within a 'single reductionist chain' offered by the natural sciences.[11] Where Wilson tends to think of science as a superpower exercising influence over smaller intellectual nations, Gould tends to think of them as autonomous nations, engaged in dialogue and collaboration for their greater good.[12]

Consilience thus 'arises from a patchwork of independent affirmations, not by subsumption under an imposed ensign of false union'.[13] In making this incisive criticism of Wilson's unificatory imperialism, Gould offers an alternative intellectual vision of 'quilting a diverse collection of separate patches into a beautiful and integrated coat of many colours'.[14] Though Gould's 'patches' are linked together by theoretical threads, they nevertheless retain their distinct disciplinary identities. Why, he asked, should we not 'enjoy the

[7] Wilson, *Consilience*, 11. This goal is perhaps most evident in Wilson's stated goal of reducing the humanities to the laws of physics: Wilson, *Consilience*, 291.

[8] Wilson, *Consilience*, 14–44. [9] Wilson, *Consilience*, 9.

[10] Gould, *The Structure of Evolutionary Theory*, 254.

[11] Gould, *The Hedgehog, the Fox, and the Magister's Pox*, 255. See the analysis in McGrath, 'A *Consilience of Equal Regard*'.

[12] Gould, *The Hedgehog, the Fox, and the Magister's Pox*, 155–7.

[13] Gould, *The Hedgehog, the Fox, and the Magister's Pox*, 20.

[14] Gould, *The Hedgehog, the Fox, and the Magister's Pox*, 15.

differences' between these disciplines, while at the same time 'find some meaningful order in the totality?'[15]

A similar point was made by Mary Midgley in her critique of aggressive scientific reductionism, when she pointed out that science is not 'an isolated monarchy', but is rather 'a republic, doing constant business with the other republics around it'.[16] Midgley is inviting us to *imagine* the breadth and depth of a broad disciplinary domain which, like the Swiss Confederation, brings and holds together a set of distinct cantons, each with their own history, languages, and boundaries. Midgley's critique of the *monarchia* of the natural sciences is a helpful way of challenging the colonial outlook of scientism, making the critically important point that the natural sciences are part of a wider federal framework of human knowledge production.

This, however, does not resolve the question of the disciplinary location of a retrieved and renewed natural philosophy. Where others might propose a new discipline of natural philosophy, I offer something more modest: an understanding of natural philosophy as a voluntary confederation of disciplinary territories, none of which are required to give up their distinct identities and interests, but which can join together in exploring the greater question of understanding how humans interact with the natural world cognitively, affectively, aesthetically, and morally. I shall explore this in what follows.

Seeing though Many Eyes: A Reimagined Natural Philosophy

John Cottingham makes the point that the human response to a complex reality is multi-layered, carrying 'a rich charge of symbolic significance that resonates with us on many different levels of understanding, not all of them, perhaps, fully grasped by the reflective, analytic mind'.[17] The notion of a 'disciplinary imaginary', discussed in Chapter 7, clearly has potential for retrieving the original vision of natural philosophy, holding together these multiple levels of insight. What the Italian philosopher Chiara Bottici describes as the recent transition from 'the Reasonable to the Imaginal'[18] offers us what seems to me to be the best solution to the issue of disciplinary fragmentation—to reimagine their relationships. As we saw earlier (pp. 164–5), this kind of 'disciplinary imaginary' allows us to visualize bringing Cottingham's 'levels of

[15] Gould, *The Hedgehog, the Fox, and the Magister's Pox*, 190.

[16] Midgley, 'Mapping Science', 195.

[17] Cottingham, *Philosophy of Religion*, 8.

[18] Bottici, 'Imagining Human Rights', 112–15. For Bottici's development and application of this approach, see Bottici, *Imaginal Politics*, 54–71.

understandings' and Midgley's 'multiple maps' together, without becoming trapped in technocratic discussions of the precise conceptual formulation and justification of this enterprise.

The natural philosopher may indeed be an individual observer of nature; yet she has at her disposal multiple windows through which nature can be seen, each disclosing its own distinct insights.[19] This goes beyond Aristotle's use of other observers or agents—such as farmers and fishermen—in order to expand his factual knowledge about the natural world on the island of Lesbos;[20] it is about adopting the *theoretical lenses* used by others, in the expectation that these will allow us a deeper quality of engagement with the natural world than any single way of seeing it. This idea is expressed in Proust's declaration that the 'only true voyage of discovery' is 'to possess other eyes, to behold the universe through the eyes of another, of a hundred others.'[21]

The natural human capacity to incorporate other perspectives or visualizations of reality into their own way of thinking reflects this ability to construct and hold together such complex renderings of the world.[22] A natural philosophy might 'behold the universe' through a number of different disciplinary eyes, each distinct yet offering perspectives that can be woven together. In what follow, I shall note four representative windows or lenses on the natural world, each of which is represented in early modern natural philosophy.

1. *The Scientific Eye*. Here, I use the term 'scientific' in its present-day sense, to refer to the study of what Steven Weinberg describes as an impersonal world governed by laws of nature that are established through observation and experiment, and have to be stated mathematically. 'The working philosophy of most scientists is that there is an objective reality and that, despite many social influences, the dominant influence in the history of science is the approach to that objective reality.'[23] Yet as Weinberg rightly points out, scientists know and appreciate the beauty and wonder of nature. 'As we understand more and

[19] For Midgley's use of 'multiple windows' on reality, see Midgley, *The Myths We Live By*, 27. Midgley's point is that the use of any single window offered by a research method produces an impoverished and inadequate account of the world.

[20] On which see Leroi, *The Lagoon*.

[21] Proust, *La prisonnière*, 69: 'Le seul véritable voyage, le seul bain de Jouvence, ce ne serait pas d'aller vers de nouveaux paysages, mais d'avoir d'autres yeux, de voir l'univers avec les yeux d'un autre, de cent autres.'

[22] See, for example, Becchio et al., 'In your Place'; Ward et al., 'Spontaneous Vicarious Perception of the Content of Another's Visual Perspective.'

[23] Weinberg, *Facing Up*, 91.

more about nature, the scientist's sense of wonder has not diminished but has rather become sharper, more narrowly focused on the mysteries that still remain.'[24] This scientific eye, as Weinberg rightly points out, is incapable of dealing with questions of meaning or purpose. Its ultimate quest is a 'final theory'—a scientific principle which explains gravity and all the other forces of nature.

2. *The Aesthetic Eye*. The beauty of nature, and the sense of wonder that it evokes, inspires many to enter scientific research (including, I may add, myself). Yet beauty is more than a gateway to understanding. For Paul Dirac, the beauty of a scientific theory was itself an indicator of its truth. Yet for others, the beauty of nature demands attention in itself, not as a portal to something else. Contemplating the beauty of the night sky or an Alpine landscape focuses on their perceptual qualities, rather than a scientific account of their origins or forms. In one sense, therefore, scientific knowledge can distract from an attentiveness to the experience of natural beauty and its capacity to trigger our 'submerged sunrise of wonder'.[25] The artistic engagement with nature, particularly natural landscapes, requires 'the cultivation of an ability to see beauty'[26]—in other words, an *askēsis* of respectful and loving attentiveness towards nature, which so often demands poetic rather than scientific expression, perhaps facilitated by exploratory, projective, ampliative, and revelatory modes of imagination.[27]
3. *The Ethical Eye*. The ethical aspect of nature concerns both a natural way of behaving, and an appropriate way of behaving towards nature. Appreciating the beauty of nature often evokes a sense of moral responsibility towards preserving its fragile elegance. While a Darwinian account of nature is often thought to emphasize the violence and wastefulness of the evolutionary process, it also highlights the interconnectedness of ecosystems, and the vulnerability of such systems to environmental changes brought about by human agency. There are clear tensions between contemporary awareness of the need to respect nature with the approaches of some early modern philosophers—such as Francis Bacon—who considered natural philosophy as an agent enabling

[24] Weinberg, *Facing Up*, 71. [25] Chesterton, *Autobiography*, 99.
[26] Gussow, 'Beauty in the Landscape', 231.
[27] For these imaginative approaches to natural beauty, see Brady, 'Imagination and the Aesthetic Appreciation of Nature'.

human dominion over nature.[28] While David Hume may have raised questions about whether nature could mandate a specific set of ethical values, today's natural philosophers will clearly see the cultivation of and appropriate respectfulness towards nature as an essential element of an environmental ethic.

4. *The Spiritual Eye.* While natural philosophy can be practised within an agnostic or atheist framework, much early modern natural theology was conspicuously attentive to the spiritual or religious aspects of an engagement with nature—Kepler offering a Lutheran, Galileo a Catholic, and Boyle an Anglican perspective. The enterprise of 'physico-theology' recognized the religious dimensions of natural philosophy, while resisting any encroachment on its territories from theological or ecclesiastical authorities. Natural philosophy was not seen primarily as a gateway to religious belief, but rather as an enterprise that could be correlated with *existing* religious beliefs, adding further depth and texture to them. Although some, such as Richard Bentley,[29] saw natural philosophy as offering an apologetic for religion, many considered that it offered a moderating framework of discussion, discouraging dogmatism.[30] Although dominated by Christian writers in western Europe, natural philosophy has been ecumenical in its religious outlooks, embracing and illuminating Jewish, Islamic, Daoist, and Confucian approaches. It continues to be hospitable to a wide range of religious and spiritual perspectives, while *requiring* none of these.

A Retrieved Natural Philosophy: Two Strategies

So how do these reflections on seeing a complex world through many eyes help us reflect on how natural philosophy might be retrieved? The intellectual and cultural history of the European Renaissance offers us two models of understanding how what are now seen as interdisciplinary sights can be integrated. One is the notion of the *uomo universale*, the idealized polymath of the Renaissance, who was able to master multiple disciplines and integrate

[28] Lancaster, 'Natural Knowledge as a Propaedeutic to Self-Betterment'; Rodríguez-García, 'Scientia potestas est'.

[29] For Bentley's apologetic application of Newtonian natural philosophy, see Calloway, *Natural Theology in the Scientific Revolution*, 117–37; Connolly, 'Metaphysics in Richard Bentley's Boyle Lectures'.

[30] Hunter, 'Latitudinarianism and the "Ideology" of the Early Royal Society'.

them within a coherent worldview.[31] The other is the vision of communal interdisciplinarity, in which scholars, writers, and artists from diverse disciplines would gather in the spacious and secluded gardens of their wealthy patrons to discuss the great questions of life, sharing wisdom across what have now become disciplinary boundaries.[32]

The first approach sees natural philosophy as an individual synthesis, a personal way of imagining how the multiple aspects and levels of the human engagement with the natural world can be held together. This way of conceiving natural philosophy as a disciplinary imaginary enables an individual to see the world through multiple windows, and hold these insights together in a coherent and plausible manner. It respects the privileges of individuality, while responding to its limitations by offering a wider imaginative horizon. Natural philosophy thus becomes an exercise in self-transcendence, a willingness to acknowledge individual deficits and heal these by immersion in the wisdom of other disciplines.

One of the most insightful accounts of this process is found in a late work by C. S. Lewis, which highlighted how reading literature 'heals the wound, without undermining the privilege, of individuality'. Lewis's point is that such a literary immersion enables us to see the world through other people's eyes, opening up new possibilities which transcend the individual's limitations.

> My own eyes are not enough for me, I will see through those of others.... In reading great literature, I become a thousand men and yet remain myself. Like the night sky in the Greek poem, I see with a myriad eyes, but it is still I who see.[33]

If nature is imagined as a theatre, immersion in writers from other disciplines allows us 'to occupy, for a while, their seat in the great theatre, to use their spectacles' in order to augment and expand our own limited vision.[34] However, there are clear limits placed on the capacity of individuals to do this, not least because of the scholarly burdens demanded by responsible

[31] Burke, *The Polymath*, 26–46. For the informative case of Georg Philipp Harsdörffer (1607–58), see Keppler-Tasaki and Kocher, eds., *Georg Philipp Harsdörffers Universalität.*

[32] For a snapshot of this intellectual culture during the Renaissance, see Celenza, *The Lost Italian Renaissance.* For the more recent appeal of this vision of shared wisdom, see Bennett, 'Renaissance Man', especially 72.

[33] Lewis, *An Experiment in Criticism*, 140–1.

[34] Lewis, *An Experiment in Criticism*, 139. Note how Lewis here uses the imagery of the *theōros* (cf. p. 20).

approaches to interdisciplinary projects, and the risk of superficial engagements with its constituent fields.

The second approach addresses some of these concerns about the limits of individual imaginings of natural philosophy. This sees natural philosophy as a form of shared knowledge and reflection involving a coalition or network of individuals, each steeped in the wisdom of their own disciplines, who are willing and able to work collaboratively to offer a grander overall vision of the natural world than any single discipline can offer. Such an enterprise is perhaps more difficult than might be imagined. The issue of disciplinary hegemony remains problematic; some will insist that their viewing windows on nature are at least *privileged*, and some might suggest that they are the only valid means of securing reliable insights into nature. Equally problematic is the tendency of such collaborations to yield an accumulation of unintegrated insights, in effect placing perspectives in parallel columns. Each creates a map of its own territory, which needs to be correlated with other such maps.

This approach allows us to look through different disciplinary windows, and see the world from their perspectives. Happily, each specialist can explain what we are seeing through their window, ensuring that we understand this distinct perspective. Yet this process yields a plurality of insights, a set of different patches that can be quilted together to yield a natural philosophy. Identifying the components is one thing; putting them together to yield an integrated whole is another. Yet this is not a new problem. It is familiar to most scientists, particularly those who resist Nancy Cartwright's view of scientific knowledge as a patchwork of essentially unrelated ideas, and prefer to see scientific knowledge as a web of distinct yet interconnected ideas. 'In the modern state of science, no discovery lives in a cocoon; rather it is built within and upon the entire interconnected structure of what we already know.'[35]

On Integrating Multiple Perspectives

So how might we try to integrate different disciplinary perspectives? How can such *rationalismes régionaux* (Bachelard) be mapped onto a larger intellectual landscape?[36] Some might rightly respond that it is a simple fact that people do weave together ideas which some might consider to be incompatible, in that

[35] Anderson, 'Science: A "Dappled World" or a "Seamless Web"?', 490–1.

[36] Bachelard, *Le rationalisme appliqué*, 119–37. For the related geographical metaphor of intellectual 'territories' governed by different methods and norms, see McGrath, *Territories of Human Reason*.

they result from the application of quite different intellectual methods. The 'weaving' in question here may be psychological, as much as philosophical—namely, accounting for an actual practice which is theoretically questionable, yet which is widely encountered and regarded as significant and productive.[37] Einstein's correlation of science, religion, and ethics is ultimately a matter of personal intuition on his part; yet he clearly saw this integration as satisfying. Einstein in effect offers a *bricolage* of unintegrated insights and perceptions,[38] without clarifying what methods and assumptions govern his bringing them together in this way.

There is a clear parallel here with the common practice of speaking of 'science' in the singular as a way of enfolding the individual natural sciences.[39] As we noted earlier (p. 90), the establishment of the British Association for the Advancement of Science in 1831 was partly motivated by a desire to recapture a sense of scientific unity, which was seen to be threatened by disciplinary fragmentation. While a number of unificatory projects were proposed from the seventeenth to the nineteenth century,[40] none of these have proved entirely satisfying in capturing a common essence of the individual disciplines usually referred to as 'science'.[41] Yet the singular term 'science' has functioned satisfactorily for many as a disciplinary imaginary for two centuries.

It is a matter of observation that human beings regularly bring together multiple perspectives or insights—whether scientific, ethical, political, or religious—without feeling the need to offer a rigorous intellectual justification for doing so.[42] One important approach was developed by the natural philosopher John Wilkins in the seventeenth century—the development of a language that integrated every aspect of natural philosophy, including religion, science, and education.[43] Where some early modern thinkers tried to develop intellectually rigorous conceptual frameworks for such an integration, poets and literary writers of this period argued that the answer lay in what we might designate an *ars combinatoria*, an act of imagination in which nature was

[37] Bromme, 'Beyond One's Own Perspective'. Bromme suggests that 'a stable disciplinary identity and flexibility' appears to be an indicator of interdisciplinary resilience.

[38] For the importance of this metaphor, see Johnson, 'Bricoleur and Bricolage'.

[39] Golinski, 'Is It Time to Forget Science?'

[40] Gaukroger, 'The Unity of Science and the Search for a Unity of Understanding in the Modern Era'.

[41] For Kant's failed attempt at a philosophical synthesis of rationalism and empiricism, see Sperber, 'There's No Success Like Failure'.

[42] The sociologist Christian Smith, for example, has documented how we use multiple narratives to make sense of different aspects of our world, despite the fact that these narratives are often competitive and at points mutually exclusive. See Smith, *Moral, Believing Animals*, 63–94. For the general philosophical problem of integrating multiple perspectives, see Rueger, 'Perspectival Models and Theory Unification'.

[43] Subbiondo, 'Preliminary Reflections on a Philosophical Language'.

conceived as a whole, despite its complexity and diversity.[44] The unity lay in the subjective imagination of the beholder of nature, not in an objective account of its aspects.

A classic example from the field of early modern natural philosophy may help us visualize a possible way ahead. In 1672, Isaac Newton published a paper reporting on some experiments involving passing beams of sunlight through prisms which, he declared, proved that sunlight consists of a heterogeneous mixture of variously coloured light rays.[45] Newton declared that a beam of sunlight could be split up into a continuous spectrum of seven colours: red, orange, yellow, green, blue, indigo, and violet. Medieval descriptions of the rainbow recognized five colours: red, yellow, green, blue, and violet; Newton added orange and indigo to increase this number to seven, possibly because this allowed correlation with the doctrine of harmonious relationships set out in Kepler's *Harmonices Mundi*.[46] Newton's experiment showed that these seven colours were originally present in the sunbeams, rather than being added to them by his experimental equipment; and that they could subsequently be recombined into white light. Newton's optical experiment might serve as a parable for reflecting on how we might reconfigure natural philosophy. Any reliable account of the natural world must enfold analytic, imaginative, symbolic, and poetic forms of understanding, in that we are trying to understand both the world we inhabit, and its impact upon us.

Earlier, we noted Ian McGilchrist's remark that 'our talent for division, for seeing the parts, is of staggering importance – second only to our capacity to transcend it, in order to see the whole.'[47] The range and depth of the human engagement with the natural world is such that it now constitutes a spectrum of disciplines, each arising in response to a perceived research need. Each allows a concentration on a specific range of topics, and for that reason is necessarily incomplete. Natural philosophy now enfolds a spectrum of institutionally disconnected parts, each of which can be isolated for close inspection; yet, like the colours which can be distinguished within Newton's spectrum, they can be brought together again to allow us to see the whole.

[44] See the important discussion in Bauer, 'Naturverständnis und Subjektkonstitution aus der Perspektive der frühneuzeitlichen Rhetorik und Poetik.'

[45] For the experiments and their analysis, see Schaffer, 'Glass Works.' Newton demonstrated that colour is not an inherent property of objects but rather arises within the human eye in response to emissions of light. For Goethe's criticism of Newton's interpretation of these experiments, see Mueller, 'Prismatic Equivalence.'

[46] For this and other possible reasons for Newton's expansion of the spectral range, see Finlay, 'Weaving the Rainbow,' 387.

[47] McGilchrist, *The Master and His Emissary*, 93.

A natural philosophy is the grander vision of nature that arises from transcending disciplinary divisions, affirming the value of all its disciplinary components while actively seeking to discern the larger picture of the natural world that they disclose.

From Theory to Practice: Shaping our Engagement with Nature

Aristotle's programmatic move from theory to practice included the development of appropriate practices or disciplines as an embodied enactment of such a theory, particularly as this was framed by Hans-Georg Gadamer (pp. 133–6). Gadamer's analysis allows us to distinguish the two distinct processes of *looking at* and *seeing* the natural order. For Gadamer, seeing involves a 'hermeneutical consciousness', a process of interpretation which enables us to grasp the meaning of what we look at. This process of seeing is informed by a theory, and leads to certain ways of behaving within and towards the natural world. For early modern natural philosophers, seeing nature properly led both to self-knowledge and self-reformation.[48] To learn *about* nature leads to learning *from* nature.

So how might this expanded disciplinary imaginary of natural philosophy help shape informed attitudes towards the natural order? To use Aristotelian terms, what appropriate practice or discipline (*askēsis*) might arise from, and represent an appropriate expression of, its associated *theōria*? Many themes might be considered here, including the appreciation of the complexity and beauty of nature, the development of gratitude for nature, and reflection on the paradox that the human observer of nature is also part of that same nature. Given limits on space, however, I shall focus on two important outcomes: the cultivation of attentiveness towards nature on the one hand, and of a proper respect for nature on the other.

Human Attentiveness towards Nature

One of the most interesting aspects of early modern natural philosophy is its patient, respectful, and disciplined attentiveness towards nature. Rather than impose predetermined conceptual schemes upon the natural order, natural

[48] Corneanu, *Regimens of the Mind*, 46–78.

philosophers aimed to allow nature to disclose its own intrinsic rationality, through careful observation or judicious experimentation on their part.

The American novelist Henry Miller, who cultivated this *askēsis* of attentiveness to improve the quality of his writing, captured its essence in an often-cited maxim: 'The moment one gives close attention to any thing, even a blade of grass, it becomes a mysterious, awesome, indescribably magnificent world in itself.'[49] Miller here hints at Scotus's notion of *haecceitas*, a theoretical device which accentuates the importance of observing and *appreciating* the distinct identity of any particular aspect of the natural order.[50] Artistically, this line of thought leads to a deeper appreciation of the beauty and diversity of the natural world, which demands to be seen, not merely reduced to manageable categories. John Ruskin expressed this point in his declaration that 'to see clearly is poetry, prophecy and religion – all in one.'[51]

This heightened attentiveness is evident in the English poet Thomas Traherne (1636–74) who exemplifies the outcomes of such a theologically attentive reading of the natural world.

> You never Enjoy the World aright, till you see how a Sand exhibiteth the Wisdom and Power of God: And Prize in evry thing the Service which they do you, by Manifesting His Glory and Goodness to your Soul, far more than the Visible Beauty on their Surface, or the Material Services, they can do your Body.[52]

The poet who is most noted for an attentiveness to nature is Gerard Manley Hopkins, particularly in his poem 'As Kingfishers Catch Fire.'[53] While the extent of Hopkins's debt to Scotus's notion of *haecceitas* remains under scholarly consideration,[54] he has clearly grasped the importance of the respectful celebration of the complexity of nature, and the individual significance of its elements. Yet Hopkins's attentiveness towards nature reflects more than an awareness of its beauty and fragility; it rests on a *theōria* grounded in a theology of creation, which invites us to see the natural order as rich in signs and

[49] Miller, *On Writing*, 37.

[50] Morejón, 'Differentiation and Distinction'. For Edith Stein's use of this Scotist notion, see Alfieri, *La presenza di Duns Scoto nel pensiero di Edith Stein*, 127–212.

[51] Ruskin, *Complete Works*, vol. 6, 333. See further Hersey, 'Ruskin as an Optical Thinker'.

[52] Traherne, *Centuries* I, 27, lines 1–5; in *Thomas Traherne: Poetry and Prose*, 3. Cf. Lane, 'Thomas Traherne and the Awakening of Want'.

[53] Boggs, 'Poetic Genesis, the Self, and Nature's Things in Hopkins'.

[54] Llewelyn, *Gerard Manley Hopkins and the Spell of John Duns Scotus*, 30–42.

symbols pointing to its origins and final goal in God.[55] This *theōria*, which represents 'the church's "sanctified intuition" of the meaning of the world',[56] heightens our attentiveness, creates expectation and anticipation, and invites participation. To study nature is, in some way and to some extent, to deepen both cognitive and affective appreciation of God as its creator.

This same attentiveness to nature underlies the scientific enterprise, which is dependent upon the precise and meticulous observation of nature,[57] prior to its theoretical interpretation. While there are several scientific explanations of what is to be understood by 'attention'—such as the active filtering of perceptual information[58]—its intuitive sense remains helpful in expressing an attitude of focalization and concentration, evident in a willingness to absorb and appreciate the fine detail of the vast world we inhabit, while at the same time exploring how this might inform a deeper understanding of this world.

Iris Murdoch saw such an attentiveness as a moral issue, in that it required a truthful engagement with the world, and an attempt to 'decentre' this process of beholding from the vested interests and personal agendas of the observer. At points, Murdoch suggests that this process can be seen as amounting to the virtue of obedience.[59] For Murdoch, this process is enabled by art.

> Great art teaches us how real things can be looked at and loved without being seized and used, without being appropriated into the greedy organism of the self. This exercise of detachment is difficult and valuable whether the thing contemplated is a human being or the root of a tree or the vibration of a colour or a sound. Unsentimental contemplation of nature exhibits the same quality of detachment: selfish concerns vanish, nothing exists except the things which are seen.[60]

Through its correlation of Popper's 'three worlds', a natural philosophy enables this same 'unsentimental contemplation of nature', while at the same time allowing this to be connected with—but not determined by—the personal subjective world of the individual.

[55] A point developed in detail in Wirzba, 'Christian *Theoria Physike*'.

[56] Blowers, 'Beauty, Tragedy and New Creation', 7; cf. 13.

[57] Daston, 'Attention and the Values of Nature in the Enlightenment'.

[58] For these and other approaches, see Watzl, 'The Nature of Attention'.

[59] Murdoch, *The Sovereignty of Good*, 39.

[60] Murdoch, *The Sovereignty of Good*, 64. See further Browning, *Why Iris Murdoch Matters*, 85–114. Compare this with Tom McLeish's more scientific approach: McLeish, 'The Re-discovery of Contemplation through Science'.

Human Respectfulness towards Nature

It is perhaps a small, but important, transition from cultivating attentiveness towards nature to respecting the natural world. As Murdoch makes clear, there is a moral dimension to both our intellectual reflections about nature, and our ethical responsibilities towards nature. In both cases, there is an underlying moral imperative—to do justice to what is being seen, both intellectually and practically.

Classic natural philosophy regarded human beings as an intellectually privileged part of the natural order, with the ability both to make sense of the natural world, and to live appropriately within it. Yet the biological advances of the nineteenth century which ultimately led to the fragmentation of natural philosophy also confirmed the importance of reflection more closely on the place of humanity within nature disclosed by evolutionary theory, and appreciating the entanglement of human well-being with that of the wider biological world. An integral element of the philosophical agenda of the philosopher Mary Midgley is the recognition that human beings emerged from within the natural order, and are dependent upon the sustainability of that order for their future survival.[61]

Yet Murdoch's emphasis on the importance of morality raises an important question. In the early modern period, many would have agreed with the classical view that nature disclosed a moral way of life, and that morality lay in discerning and enacting this. Yet this view was called into question by David Hume's argument that there was no simple logical connection between the observation of nature and the construction of moral values.[62] Hume was not suggesting that it was impossible to derive moral values from the natural world; rather, he highlighted the need for bridging theories to establish a connection between the 'is' of observation and the 'ought' of moral obligation.

One such 'bridging theory' which was of major importance to early modern natural philosophy was the Christian notion that the world was God's creation.[63] This theoretical reading of the natural world was echoed in early modern physico-theology, enabling nature to be interpreted as God's

[61] Midgley, *Animals and Why They Matter*. For analysis, see Cooper, 'Animals, Attitudes and Moral Theories'.

[62] See Pigden, ed., *Hume on 'Is' and 'Ought'*. Putnam and others have argued that the dichotomy of facts versus values has collapsed following the widespread rejection of positivism: Putnam, *The Collapse of the Fact/Value Dichotomy and Other Essays*, 7–45. Putnam here argues both that certain epistemic values are presupposed in the natural sciences, and that they are entangled with certain 'thick' normative concepts.

[63] See the important study of Wirzba, 'Christian *Theoria Physike*'.

creation,[64] thus pointing to the idea of disclosure of divine beauty and harmony within nature, and to the ethically important notion of respecting the natural world as God's creative work of art.[65] For example, both James Thomson's *The Seasons* (1730) and Alexander Pope's *Essay on Man* (1733–4) can be seen as 'eco-theological' poems,[66] advocating a theologically grounded environmental ethos and ethic, even though they adopt quite different perspectives in developing this theme.[67]

Such a Christian theological approach to nature illuminates what appears to be a general principle: developing an environmental ethic appears to be dependent on articulating a persuasive *interpretation* of the natural world—an interpretation that is neither given within nature nor can be considered to represent an overwhelmingly persuasive public reading of nature.[68] Neo-Confucianism, for example, offers a coherent environmental ethic;[69] this goes far beyond what can be understood of nature on the basis of mere observation. An additional *theōria* is required to develop an appropriate *askēsis* of respect, which is in conflict with interpretations of science which emphasize the role of instrumental reason in enabling the human domination and exploitation of nature.[70]

More pragmatic reasons for respecting nature can, of course, be articulated, without requiring an additional *theōria*, such as that noted in this discussion. The empirical demonstration of the vulnerability of ecosystems, and the potential implication of their failure or collapse for human well-being and survival, offers an important rationale for respecting the natural world. Yet the motivation here is anthropocentric, aiming to elicit action which indirectly preserves the future for humans. While some strands of early modern natural philosophy saw the judicious use of resources within the natural world to improve the human condition as being acceptable, other strands—such that represented by Thomson's *Seasons*—emphasized the human responsibility to preserve and improve nature as a thing of beauty and harmony, to be

[64] Ogilive, 'Natural History, Ethics, and Physico-Theology'; Mandelbrote, 'What Was Physico-Theology For?'

[65] Wirzba, 'Christian *Theoria Physike*'. This point had earlier been challenged in White, 'The Historical Roots of Our Ecological Crisis'. For a scholarly critique of White's unsatisfactory paper, see Harrison, 'Subduing the Earth'.

[66] Sitter, 'Eighteenth-Century Ecological Poetry and Ecotheology'.

[67] Willan, 'The Proper Study of Mankind in Pope and Thomson'.

[68] See Van Buren, 'Environmental Hermeneutics'.

[69] Blakeley, 'Neo-Confucian Cosmology, Virtue Ethics, and Environmental Philosophy'.

[70] Leiss, 'Modern Science, Enlightenment, and the Domination of Nature'.

respected and valued for what it is, rather than as something that serves human interests.[71]

Conclusion

Our story began with Aristotle's observations of the natural world on the Aegean island of Lesbos around the years 346–343 BCE, and his quest for recurring patterns that might help explain how this rich and complex world could be understood. Although others had opened up similar questions in ancient India and China, Aristotle would prove to be the catalyst for an increasingly sophisticated engagement with the natural order in the West, which gave rise to the evolution of what we now know as 'natural philosophy'. Perhaps western natural philosophy is seen at its best in the seventeenth century, when natural philosophers such as Kepler and Newton were able to demonstrate the coherence of the external world, and the remarkable capacity of the human mind to capture this intellectually, and represent it mathematically. Far from being an unsophisticated and primitive ancestor of today's natural sciences, natural philosophy sought to cultivate human wisdom and understanding through an intellectual and affective engagement with the world. It possessed a depth and range that modern scholarship is in the process of uncovering, and that speaks to many of our concerns about the abuse of instrumental reason and the technocratic languages and practices that now distance us from nature.

We clearly need this wisdom and this intellectual discipline today. In 2017, the Museum of Natural History in Paris issued a manifesto, arguing that the environmental crisis demanded a new attentiveness towards the place of humanity within nature, and the responsibilities that this entailed. 'Grounding human beings in nature', it was argued, allowed us to see ourselves as 'active agents within, and also as victims of, the transformation of nature that we ourselves have brought about'.[72] Those themes are not new, nor can they be detached from deeper and broader questions once engaged by a natural philosophy. We need a holistic view of nature if we are to engage such questions about the natural world in their full complexity, and develop responsible answers and strategies in the light of the challenges we face.

[71] Sitter, 'Eighteenth-Century Ecological Poetry and Ecotheology', 21.
[72] Abbadie et al., *Manifeste du Muséum*, 28–9.

This book has argued for an expanded imaginative vision of our encounter with nature, enriching but not contradicting the natural sciences. Yes, human beings long for understanding; yet they also want to know how they fit into a larger scheme of things, what gives meaning to their lives, and what their place is in this strange world that we call 'nature'. William Blake protested that the rationalists of his day had imprisoned the natural world within the narrow confines of their ideologies, and were unable to break free from this self-imposed, constricted, and impoverished account of nature. 'Man has closed himself up, till he sees all things thro' narrow chinks of his cavern.'[73] To be truly human, we need an extension of our vision of our world, expanding our minds and imaginations to embrace the vast natural world for what it is, rather than reducing that world to what we consider manageable and palatable.

The natural sciences are rightly celebrated for their precision and objectivity of judgement—yet can only offer us a partial account of the human engagement with nature, failing to do justice to the passion, emotion, delight, and joy that it evokes within us, or to the 'ultimate questions' that seem to be embedded and embodied in our lives. There is nothing wrong with science, save that there is more that needs to be said and seen. It is a role that the disciplinary imaginary of natural philosophy played in the past. Perhaps it might help us reflect on these same questions today.

[73] Blake, *The Marriage of Heaven and Hell*, 26. Blake here evokes the image of Plato's underground cave.

Bibliography

What follows is a list of works that were engaged in the preparation of this study, not all of which are explicitly discussed or referenced in the text itself.

Aaltola, Elisa. 'Wilderness Experiences as Ethics: From Elevation to Attentiveness.' *Ethics, Policy & Environment* 18, no. 3 (2015): 283–300.

Abbadie, Luc, et al. *Manifeste du Muséum: Quel futur sans nature?* Paris: Muséum d'histoire naturelle, 2017.

Ables, Brent. 'Disagreement and Philosophical Progress'. *Logos & Episteme* 6, no. 1 (2015): 115–27.

Adam, Matthias. *Theoriebeladenheit und Objektivität. Zur Rolle von Beobachtungen in den Naturwissenschafte*. Frankfurt am Main: Ontos Verlag, 2002.

Agamben, Giorgio. *The Open: Man and Animal*. Stanford, CA: Stanford University Press, 2004.

Agazzi, Evandro. 'Analogicità del concetto di scienza: Il problema del rigore e dell'oggettività nelle scienze umane'. *Epistemologia* 2 (1979): 39–66.

Agazzi, Evandro. 'Fede, ragione e scienza'. In *Scienza e fede: Le nuove frontiere*, edited by Paolo dell'Aquila, 83–94. Cesena: Società Editrice Il Ponte Vecchio, 2007.

Ahern, Eoghan. *Bede and the Cosmos: Theology and Nature in the Eighth Century*. London: Routledge, 2020.

Akopyan, Ovanes. *Debating the Stars in the Italian Renaissance: Giovanni Pico Della Mirandola's Disputationes adversus Astrologiam Divinatricem and Its Reception*. Leiden: Brill, 2020.

Alfieri, Francesco. *La presenza di Duns Scoto nel pensiero di Edith Stein: La questione dell'individualità*. Rome: Pontificia Universitas Lateranensis, 2011.

Allingham, William. 'On Poetry'. *Fraser's Magazine for Town and Country* 448, no. 75 (1867): 523–36.

Alwishah, Ahmed, and Josh Hayes, eds. *Aristotle and the Arabic Tradition*. Cambridge: Cambridge University Press, 2015.

Anderson, Douglas. 'The Evolution of Peirce's Concept of Abduction'. *Transactions of the Charles S. Peirce Society* 22, no. 2 (1986): 145–64.

Anderson, Pamela Sue. 'What's Wrong with the God's Eye Point of View: A Constructive Feminist Critique of the Ideal Observer Theory'. In *Faith and Philosophical Analysis: The Impact of Analytical Philosophy on the Philosophy of Religion*, edited by Harriet A. Harris and Christopher J. Insole, 85–99. Burlington, VT: Ashgate, 2005.

Anderson, Philip W. 'Science: A "Dappled World" or a "Seamless Web"?' *Studies in the History and Philosophy of Modern Physics* 32, no. 3 (2001): 487–94.

Andrée, Alexander. 'Peter Comestor's Lectures on the Glossa "Ordinaria" on the Gospel of John: The Bible and Theology in the Twelfth-Century Classroom'. *Traditio* 71 (2016): 203–34.

Angelici, Ruben. Semiotic Theory and Sacramentality in Hugh of Saint Victor. London: Routledge, 2019.

Anstey, Peter R. 'Experimental versus Speculative Natural Philosophy'. In *The Science of Nature in the Seventeenth Century: Patterns of Change in Early Modern Natural

Philosophy, edited by Peter Anstey and John A. Schuster, 215–42. Dordrecht: Springer, 2005.
Anstey, Peter R. 'Francis Bacon and the Classification of Natural History'. *Early Science and Medicine* 17, no. 1/2 (2012): 11–31.
Anstey, Peter R. *John Locke and Natural Philosophy*. Oxford: Oxford University Press, 2013.
Anstey, Peter R. 'Philosophy of Experiment in Early Modern England: The Case of Bacon, Boyle and Hooke'. *Early Science and Medicine* 19, no. 2 (2014): 103–32.
Anstey, Peter R., and Michael Hunter. 'Robert Boyle's "Designe About Natural History"'. *Early Science and Medicine* 13, no. 2 (2008): 83–126.
Anstey, Peter R., and Alberto Vanzo. 'The Origins of Early Modern Experimental Philosophy'. *Intellectual History Review* 22 (2012): 499–518.
Antognazza, Maria Rosa. *Leibniz on the Trinity and the Incarnation: Reason and Revelation in the Seventeenth Century*. New Haven, CT: Yale University Press, 2007.
Antognazza, Maria Rosa. 'The Benefit to Philosophy of the Study of Its History'. *British Journal for the History of Philosophy* 23, no. 1 (2015): 161–84.
Anzulewicz, Henryk. 'Albertus Magnus über die Felicitas Contemplativa als die Erfüllung eines natürlichen Strebens nach Wissen'. *Quaestio* 15 (2015): 457–66.
Ariew, Roger. *Descartes and the First Cartesians*. New York: Oxford University Press, 2014.
Aristotle, *The Nicomachean Ethics*, translated by W. D. Ross. Oxford: Oxford University Press, 2009.
Ashley, Benedict. 'St. Albert and the Nature of Natural Science'. In *Albertus Magnus and the Sciences*, edited by James Weisheipl, 73–102. Toronto: Pontifical Institute of Mediaeval Studies, 1980.
Asúa, Miguel de. 'War and Peace: Medicine and Natural Philosophy in Albert the Great'. In *A Companion to Albert the Great: Theology, Philosophy and the Sciences*, edited by Irven Resnick, 269–97. Leiden: Brill, 2013.
Atkins, Peter. 'The Limitless Power of Science'. In *Nature's Imagination: The Frontiers of Scientific Vision*, edited by John Cornwell, 122–32. Oxford: Oxford University Press, 1995.
Augé, Marc. *Non-Lieux: Introduction à une anthropologie de la surmodernité*. Paris: Éditions du Seuil, 1992.
Ault, Donald D. *Visionary Physics: Blake's Response to Newton*. Chicago: University of Chicago Press, 1974.
Ayala, Francisco J. 'Darwin and the Scientific Method'. *PNAS* 106 (2009): 10033–9.
Ayers, Michael. 'Was Berkeley an Empiricist or a Rationalist?' In *The Cambridge Companion to Berkeley*, edited by Kenneth P. Winkler, 34–62. Cambridge: Cambridge University Press, 2005.
Ayers, Michael, and Maria Rosa Antognazza. *Knowledge and Belief from Plato to Locke*. Oxford: Oxford University Press, 2019.
Bachelard, Gaston. *Le rationalisme appliqué*. Paris: Presses Universitaires de France, 1949.
Bacon, Francis. *Works*. 14 vols. Stuttgart–Bad Cannstatt: Frommann, 1961–3.
Badhwar, Neera K. *Well-Being: Happiness in a Worthwhile Life*. Oxford: Oxford University Press, 2014.
Baehr, Peter. 'The "Iron Cage" and the "Shell as Hard as Steel": Parsons, Weber, and the *Stahlhartes Gehäuse* Metaphor in the *Protestant Ethic and the Spirit of Capitalism*'. *History and Theory* 40, no. 2 (2001): 153–69.
Baghramian, Maria. '"From Realism Back to Realism": Putnam's Long Journey'. *Philosophical Topics* 36, no. 1 (2008): 17–35.

Bagioli, Mario. 'Stress in the Book of Nature: The Supplemental Logic of Galileo's Realism'. *MLN* 118, no. 3 (2003): 557–85.

Baigrie, Brian S. 'The Justification of Kepler's Ellipse'. *Studies in History and Philosophy of Science* 21 (1991): 633–64.

Baigrie, Brian S. 'Catherine Wilson's *The Invisible World*: Early Modern Philosophy and the Invention of the Microscope'. *International Studies in the Philosophy of Science* 12, no. 2 (1998): 165–74.

Bailey, Michael David. 'The Feminization of Magic and the Emerging Idea of the Female Witch in the Late Middle Ages'. *Essays in Medieval Studies* 19, no. 2 (2002): 120–34.

Baker, Lynne Rudder. 'Ontology, Down-to-Earth'. *The Monist* 98 (2015): 145–55.

Baker, Lynne Rudder. 'Naturalism and the Idea of Nature'. *Philosophy* 92, no. 3 (2017): 333–49.

Balfour, Arthur. *The Foundations of Belief*. New York: Longmans, 1895.

Banville, John. 'Beauty, Charm, and Strangeness: Science as Metaphor'. *Science* 281, no. 5373 (1998): 40–1.

Banville, John. *Kepler*. London: Picador, 1999.

Barbour, Julian B. *The Discovery of Dynamics: A Study from a Machian Point of View of the Discovery and the Structure of Dynamical Theories*. Oxford: Oxford University Press, 2001.

Barker, Peter, and Bernard R. Goldstein. 'Realism and Instrumentalism in Sixteenth Century Astronomy: A Reappraisal'. *Perspectives on Science* 6 (1999): 232–58.

Barker, Peter, and Bernard R. Goldstein. 'Theological Foundations of Kepler's Astronomy'. *Osiris* 16 (2001): 88–113.

Barnett, S. J. *The Enlightenment and Religion: The Myths of Modernity*. Manchester: Manchester University Press, 2003.

Barry, Andrew, and Georgina Born, eds. *Interdisciplinarity: Reconfigurations of the Social and Natural Sciences*. London: Routledge, 2013.

Barry, Andrew, Georgina Born, and Gisa Weszkalnys. 'Logics of Interdisciplinarity'. *Economy and Society* 37, no. 1 (2008): 20–49.

Bates, Stanley. 'Refusing Disenchantment: Romanticism, Criticism, Philosophy'. *Philosophy and Literature* 40, no. 2 (2016): 549–57.

Bauer, Barbara. 'Naturverständnis und Subjektkonstitution aus der Perspektive der frühneuzeitlichen Rhetorik und Poetik'. In *Künste und Natur in Diskursen der frühen Neuzeit*, edited by Hartmut Laufhütte, 69–132. Wiesbaden: Harrassowitz, 2000.

Becchio, Cristina, Marco del Giudice, Olga dal Monte, Luca Latini-Corazzini, and Lorenzo Pia. 'In Your Place: Neuropsychological Evidence for Altercentric Remapping in Embodied Perspective Taking'. *Social Cognitive and Affective Neuroscience* 8, no. 2 (2013): 165–70.

Becher, Tony, and Paul Trowler. *Academic Tribes and Territories: Intellectual Enquiry and the Cultures of Disciplines*. 2nd edition. Maidenhead: Open University Press, 2001.

Belliotti, Raymond A. *Is Human Life Absurd? A Philosophical Inquiry into Finitude, Value, and Meaning*. Leiden: Brill, 2019.

Belmonte, Juan Antonio. 'In Search of Cosmic Order: Astronomy and Culture in Ancient Egypt'. *Proceedings of the International Astronomical Union* 5 (2009): 74–86.

Belsey, Catherine. *Critical Practice*. 2nd edition. London: Routledge, 2002.

Bénatouïl, T., and M. Bonazzis. '*Θεωρια* and *Βιοσ Θεωρητικοσ* from the Presocratics to the End of Antiquity: An Overview'. In *Theoria, Praxis, and the Contemplative Life after Plato and Aristotle*, edited by T. Bénatouïl and M. Bonazzis, 1–14. Leiden: Brill, 2012.

Ben-Chaim, Michael. 'The Value of Facts in Boyle's Experimental Philosophy'. *History of Science* 38 (2000): 57–77.

Ben-Chaim, Michael. 'Empowering Lay Belief: Robert Boyle and the Moral Economy of Experiment'. *Science in Context* 15 (2002): 51–77.

Benin, Stephen D. *The Footprints of God: Divine Accommodation in Jewish and Christian Thought*. Albany: State University of New York Press, 1993.

Bennett, J. A. 'Robert Hooke as Mechanic and Natural Philosopher'. *Notes and Records of the Royal Society of London* 35, no. 1 (1980): 33–48.

Bennett, Paul. 'Renaissance Man'. *Landscape Architecture Magazine* 88, no. 8 (1998): 70–5.

Benton, Tim. '"I Am Attracted to the Natural Order of Things": Le Corbusier's Rejection of the Machine'. In *Being Modern: The Cultural Impact of Science in the Early Twentieth Century*, edited by Robert Bud, Paul Greenhalgh, Frank James, and Morag Shiach, 373–85. London: UCL Press, 2018.

Berensmeyer, Ingo. 'Rhetoric, Religion, and Politics in Sir Thomas Browne's *Religio Medici*'. *Studies in English Literature 1500–1900* 46 (2006): 113–32.

Berlin, Isaiah. *The Proper Study of Mankind: An Anthology of Essays*. Edited by Henry Hardy and Roger Hausheer. New York: Farrar, Straus & Giroux, 2000.

Berlin, Isaiah. *The Crooked Timber of Humanity: Chapters in the History of Ideas*. 2nd edition. Edited by Henry Hardy. Princeton, NJ: Princeton University Press, 2013.

Berns, Andrew D. *The Bible and Natural Philosophy in Renaissance Italy: Jewish and Christian Physicians in Search of Truth*. New York: Cambridge University Press, 2015.

Bernstein, Alan E. 'Magisterium and License: Corporate Autonomy against Papal Authority in the Medieval University of Paris'. *Viator* 9 (1978): 291–308.

Bernstein, Richard J. 'The Unresolved Problems of Late Critical Theory'. *History and Theory* 56, no. 3 (2017): 418–32.

Berry, Wendell. *Life Is a Miracle: An Essay against Modern Superstition*. New York: Counterpoint, 2001.

Bersini, Hugues, and Jacques Reisse. *Comment définir la vie? Les réponses de la biologie, de l'intelligence artificielle et de la philosophie des sciences*. Paris: Vuibert, 2007.

Berthrong, John. 'Confucian Views of Nature'. In *Nature across Cultures: Views of Nature and the Environment in Non-Western Cultures*, edited by Helaine Selin, 373–92. Dordrecht: Springer Science, 2003.

Bhaskar, Roy. *Scientific Realism and Human Emancipation*. London: Verso, 1986.

Bhaskar, Roy. *The Possibility of Naturalism: A Philosophical Critique of the Contemporary Human Sciences*. 3rd edition. London: Routledge, 1998.

Bhaskar, Roy. *A Realist Theory of Science*. London: Verso, 2008.

Bialas, Volker. *Johannes Kepler, Astronom und Naturphilosoph*. 2nd edition. Linz: Universitätsverlag Rudolf Trauner, 2013.

Bieri, Hans, and Virgilio Masciadri. *Der Streit um das Kopernikanische Weltsystem im 17. Jahrhundert: Galileo Galileis Akkommodationstheorie und ihre historischen Hintergründe: Quellen, Kommentare, Übersetzungen*. Freiburger Studien zur Frühen Neuzeit. Bern: Lang, 2007.

Binder, P. M. 'Theories of Almost Everything'. *Nature* 455 (2008): 884–5.

Blachowicz, James. 'How Science Textbooks Treat Scientific Method: A Philosopher's Perspective'. *British Journal for Philosophy of Science* 60 (2009): 303–44.

Black, Winston. 'The Quadrivium and Natural Sciences'. In *The Oxford History of Classical Reception in English Literature: Volume 1: 800–1558*, edited by Rita Copeland, 77–89. Oxford: Oxford University Press, 2016.

Blackford, Russell, and Damien Broderick, eds. *Philosophy's Future: The Problem of Philosophical Progress*. Hoboken, NJ: Wiley, 2017.

Blackwell, Richard J. *Galileo, Bellarmine and the Bible*. Notre Dame, IN: University of Notre Dame Press, 1991.

Blair, Ann. 'Tycho Brahe's Critique of Copernicus and the Copernican System'. *Journal of the History of Ideas* 51, no. 3 (1990): 355–77.

Blair, Ann. *The Theater of Nature: Jean Bodin and Renaissance Science*. Princeton, NJ: Princeton University Press, 1997.

Blair, Ann. 'Mosaic Physics and the Search for a Pious Natural Philosophy in the Late Renaissance'. *Isis* 91, no. 1 (2000): 32–58.

Blair, Ann. 'Natural Philosophy'. In *The Cambridge History of Science*, edited by Katharine Park and Lorraine Daston, 363–406. Cambridge: Cambridge University Press, 2006.

Blair, Ann. *Too Much to Know: Managing Scholarly Information before the Modern Age*. New Haven, CT: Yale University Press, 2010.

Blair, Ann, and Kaspar von Greyerz, eds. *Physico-Theology: Religion and Science in Europe, 1650–1750*. Baltimore: Johns Hopkins University Press, 2020.

Blake, Liza. 'Allegorical Causation and Aristotelian Physics in Henry Medwall's Nature'. *Studies in English Literature 1500–1900* 55, no. 2 (2015): 341–63.

Blake, William. *The Marriage of Heaven and Hell*. Boston: John W. Luce, 1906.

Blake, William. *The Complete Poetry and Prose of William Blake*. Edited by David V. Erdman. New York: Anchor Books, 1988.

Blakeley, Donald N. 'Neo-Confucian Cosmology, Virtue Ethics, and Environmental Philosophy'. *Philosophy in the Contemporary World* 8, no. 2 (2001): 37–49.

Blankenhorn, Bernhard. 'How the Early Albertus Magnus Transformed Augustinian Interiority'. *Freiburger Zeitschrift für Philosophie und Theologie* 58 (2011): 351–86.

Blowers, Paul M. 'Beauty, Tragedy and New Creation: Theology and Contemplation in Cappadocian Cosmology'. *International Journal of Systematic Theology* 18, no. 1 (2016): 7–29.

Blum, Paul Richard. *Studies on Early Modern Aristotelianism*. Leiden: Brill, 2012.

Bogen, Jim. '"Saving the Phenomena" and Saving the Phenomena'. *Synthese* 182, no. 1 (2011): 7–22.

Boggs, Rebecca Melora Corinne. 'Poetic Genesis, the Self, and Nature's Things in Hopkins'. *Studies in English Literature, 1500–1900* 37, no. 4 (1997): 831–55.

Bohm, David. *On Dialogue*. London: Routledge, 1996.

Bohm, David. *Wholeness and the Implicate Order*. London: Routledge, 2002.

Bohm, David, and Basil J. Hiley. *The Undivided Universe: An Ontological Interpretation of Quantum Theory*. London: Routledge, 1993.

Bohr, Niels. 'Mathematics and Natural Philosophy'. *Scientific Monthly* 82, no. 2 (1956): 85–8.

Bolton, Robert. 'Science and Scientific Inquiry in Aristotle: A Platonic Provenance'. In *The Oxford Handbook of Aristotle*, edited by Christopher Shields, 46–61. Oxford: Oxford University Press, 2012.

Bolton, Robert. 'Intuition in Aristotle'. In *Rational Intuition: Philosophical Roots, Scientific Investigations*, edited by Lisa M. Osbeck and Barbara S. Held, 39–54. Cambridge: Cambridge University Press, 2014.

Boner, Patrick. Kepler's Cosmological Synthesis: Astrology, Mechanism and the Soul. Leiden: Brill, 2013.

Booker, M. Keith. 'Cultural Crisis Then and Now: Science, Literature, and Religion in John Banville's *Doctor Copernicus* and *Kepler*'. *Critique: Studies in Contemporary Fiction* 39, no. 2 (1998): 176–92.

Boran, Elizabethanne, and Mordechai Feingold, eds. *Reading Newton in Early Modern Europe*. Leiden: Brill, 2017.

Bortoft, Henri. *The Wholeness of Nature: Goethe's Way of Science*. Edinburgh: Floris Books, 1996.

Bottici, Chiara. 'Imagining Human Rights: Utopia or Ideology?' *Law and Critique* 21, no. 2 (2010): 111–30.

Bottici, Chiara. *Imaginal Politics: Images Beyond Imagination and the Imaginary*. New York: Columbia University Press, 2014.

Boudry, Maarten, and Massimo Pigliucci, eds. *Science Unlimited? The Challenges of Scientism*. Chicago: University of Chicago Press, 2017.

Bourg, Julian. 'A Modernist Catholic? Edouard Le Roy's Dual Critique of Scientism and Neo-Scholasticism'. *Modern Schoolman* 78, no. 4 (2001): 317–43.

Bowler, Peter J. 'Darwinism and Victorian Values: Threat or Opportunity?'. *Proceedings of the British Academy* 78 (1992): 129–47.

Bowler, Peter J. 'Geographical Distribution in the *Origin of Species*'. In *The Cambridge Companion to the 'Origin of Species'*, edited by Michael Ruse and Robert J. Richards, 153–72. Cambridge: Cambridge University Press, 2009.

Boyd, Brian. 'Popper's World 3: Origins, Progress, and Import'. *Philosophy of the Social Sciences* 46, no. 3 (2016): 221–41.

Boyle, Robert. *Some Considerations Touching the Vsefulnesse of Experimental Naturall Philosophy*. Oxford: Henry Hall, 1663.

Boyle, Robert. *The Excellency of Theology, Compared with Natural Philosophy*. London: Henry Herringman, 1674.

Boyle, Robert. *The Works of the Honourable Robert Boyle*. Edited by Thomas Birch. 2nd ed. 6 vols. London: Rivingtons, 1772.

Bradley, M. C. 'Hume's Chief Objection to Natural Theology'. *Religious Studies* 43, no. 3 (2007): 249–70.

Brady, Emily. 'Imagination and the Aesthetic Appreciation of Nature'. *Journal of Aesthetics and Art Criticism* 56, no. 2 (1998): 139–47.

Brady, Maura. 'Galileo in Action: The "Telescope" in *Paradise Lost*'. *Milton Studies* 44 (2005): 129–52.

Braillard, Pierre-Alain, and Christophe Malaterre. *Explanation in Biology: An Enquiry into the Diversity of Explanatory Patterns in the Life Sciences*. Dordrecht: Springer, 2015.

Brake, Elizabeth. 'Making Philosophical Progress: The Big Questions, Applied Philosophy, and the Profession'. *Social Philosophy and Policy* 34, no. 2 (2017): 23–45.

Brandt, Reinhart. 'Francis Bacon, Die Idolenlehre'. In *Grundprobleme der großen Philosophen. Philosophie der Neuzeit I*, edited by Josef Speck, 9–34. Göttingen: Vandenhoeck & Ruprecht, 1986.

Brazil, Kevin. 'T.S. Eliot: Modernist Literature, Disciplines and the Systematic Pursuit of Knowledge'. In *Being Modern: The Cultural Impact of Science in the Early Twentieth Century*, edited by Robert Bud, Paul Greenhalgh, Frank James, and Morag Shiach, 77–92. London: UCL Press, 2018.

Bredekamp, Horst. *Galilei der Künstler: Der Mond. Die Sonne. Die Hand*. Berlin: de Gruyter, 2009.

Brennan, Timothy. 'Vico and Modern Scientism'. *Italian Culture* 35, no. 2 (2017): 129–42.

Brigandt, Ingo. 'Beyond Reduction and Pluralism: Toward an Epistemology of Explanatory Integration in Biology'. *Erkenntnis* 73, no. 3 (2010): 195–311.

Brinkman, Paul. 'Charles Darwin's *Beagle* Voyage, Fossil Vertebrate Succession, and "the Gradual Birth & Death of Species"'. *Journal of the History of Biology* 43, no. 2 (2010): 363–99.

Broadie, Sarah. 'Nature and Craft in Aristotelian Teleology'. In *Biologie, logique et métaphysique chez Aristote*, edited by D. Devereux and P. Pellegrin, 389–403. Paris: CNRS, 2010.

Brogan, Walter. 'Gadamer's Praise of Theory: Aristotle's Friend and the Reciprocity between Theory and Practice'. *Research in Phenomenology* 32, no. 1 (2002): 141–55.

Bromme, Rainer. 'Beyond One's Own Perspective: The Psychology of Cognitive Interdisciplinarity'. In *Practising Interdisciplinarity*, edited by Nico Stehr and Peter Weingart, 115–33. Toronto: University of Toronto Press, 2000.

Brooke, John Hedley. *Science and Religion: Some Historical Perspectives*. Cambridge: Cambridge University Press, 1991.

Brooke, John Hedley. 'Wise Men Nowadays Think Otherwise: John Ray, Natural Theology and the Meanings of Anthropocentrism'. *Notes and Records of the Royal Society* 54, no. 2 (2000): 199–213.

Brown, Eric. 'Contemplative Withdrawal in the Hellenistic Age'. *Philosophical Studies* 137 (2008): 79–99.

Brown, Matthew J. *Science and Moral Imagination: A New Ideal for Values in Science*. Pittsburgh: University of Pittsburgh Press, 2020.

Brown, Stuart. 'On Why Philosophers Redefine Their Subject'. *Royal Institute of Philosophy* Supplement 33 (1992): 41–57.

Brown, William P. *The Ethos of the Cosmos: The Genesis of Moral Imagination in the Bible*. Grand Rapids, MI: Eerdmans, 1999.

Browne, Thomas. *Religio Medici*. Canterbury: Moregate, 1894.

Browning, Gary K. *Why Iris Murdoch Matters: Making Sense of Experience in Modern Times*. London: Bloomsbury Academic, 2018.

Brueggemann, Walter. *The Land: Place as Gift, Promise, and Challenge in Biblical Faith. Overtures to Biblical Theology*. 2nd edition. Philadelphia: Fortress Press, 2002.

Brunnander, Björn. 'Did Darwin Really Answer Paley's Question?' *Studies in History and Philosophy of Science. Part C* 44, no. 3 (2013): 309–21.

Bubel, Katharine. 'Nature and Wise Vision in the Poetry of Gerard Manley Hopkins'. *Renascence: Essays on Values in Literature* 62, no. 2 (2010): 117–40.

Bucciantini, Massimo, Michele Camerota, and Franco Giudice. *Galileo's Telescope: A European Story*. Cambridge, MA: Harvard University Press, 2015.

Bunge, Mario. *Scientific Materialism*. Dordrecht: Reidel, 1981.

Burke, Peter. *The Polymath: A Cultural History from Leonardo Da Vinci to Susan Sontag*. New Haven, CT: Yale University Press, 2020.

Burnett, Charles. 'The Introduction of Aristotle's Natural Philosophy into Great Britain: A Preliminary Survey of the Manuscript Evidence'. In *Aristotle in Britain during the Middle Ages*, edited by John Marenbon, 21–50. Turnhout: Brepols, 1996.

Burrow, Colin. 'C. S. Lewis and the Allegory of Love'. *Essays in Criticism* 53, no. 3 (2003): 284–94.

Byl, Simon. *De la médecine magique et religieuse à la médecine rationnelle: Hippocrates*. Paris: Harmattan, 2011.

Calloway, Katherine. '"His Footstep Trace": The Natural Theology of Paradise Lost'. *Milton Studies* 55 (2014): 53–85.

Calloway, Katherine. *Natural Theology in the Scientific Revolution: God's Scientists*. London: Routledge, 2016.

Cambier, Hubert. 'The Evolutionary Meaning of World 3'. *Philosophy of the Social Sciences* 46, no. 3 (2016): 242–64.

Campana, Joseph. 'On Not Defending Poetry: Spenser, Suffering, and the Energy of Affect'. *PMLA* 120, no. 1 (2005): 33–48.

Canceran, Delfo C. 'Social Imaginary in Social Change'. *Philippine Sociological Review* 57 (2009): 21–36.

Cannon, Walter F. 'John Herschel and the Idea of Science'. *Journal of the History of Ideas* 22, no. 2 (1961): 215–39.

Carbonell, Curtis D. 'Wilson and Gould: The Engagement of the Sciences and the Humanities'. *Mediterranean Journal of Social Sciences* 2, no. 2 (2011): 345–59.

Cardona, Carlos Alberto. 'Kepler: Analogies in the Search for the Law of Refraction'. *Studies in History and Philosophy of Science Part A* 59 (2016): 22–35.

Carey, Daniel. *Locke, Shaftesbury, and Hutcheson: Contesting Diversity in the Enlightenment and Beyond*. Cambridge: Cambridge University Press, 2006.

Carlin, Laurence. 'Boyle's Teleological Mechanism and the Myth of Immanent Teleology'. *Studies in History and Philosophy of Science Part A* 43, no. 1 (2012): 54–63.

Carlin, Laurence. 'Boyle on Explanation and Causality'. In *The Bloomsbury Companion to Robert Boyle*, edited by Jan-Erik Jones, 141–68. London: Bloomsbury, 2020.

Carman, Christián, and Gonzalo Recio. 'Ptolemaic Planetary Models and Kepler's Laws'. *Archive for History of Exact Sciences* 73, no. 1 (2019): 39–124.

Carrier, Martin. 'Values and Objectivity in Science: Value-Ladenness, Pluralism and the Epistemic Attitude'. *Science & Education* 22 (2013): 2547–68.

Carroll, Anthony J. 'Disenchantment, Rationality, and the Modernity of Max Weber'. *Forum Philosophicum* 16, no. 1 (2011): 117–37.

Carroll, William. 'Galileo and the Interpretation of the Bible'. *Science & Education* 8, no. 3 (1999): 151–87.

Cartwright, Nancy. *The Dappled World: A Study of the Boundaries of Science*. Cambridge: Cambridge University Press, 1999.

Carusi, Paola. 'Alchimia Islamica e religione: La legittimazione difficile di una scienza della natura'. *Oriente Moderno* 19 (2000): 461–502.

Casey, Edward S. 'How to Get from Space to Place in a Fairly Short Stretch of Time: Phenomenological Prolegomena'. In *Senses of Place*, edited by Steven Feld and Keith H. Basso, 13–52. Santa Fe, NM: School of American Research Press, 1996.

Cat, Jordi. 'Into the "Regions of Physical and Metaphysical Chaos": Maxwell's Scientific Metaphysics and Natural Philosophy of Action (Agency, Determinacy and Necessity from Theology, Moral Philosophy and History to Mathematics, Theory and Experiment)'. *History and Philosophy of Science* 43, no. 1 (2012): 91–104.

Cat, Jordi, Nancy Cartwright, and Hasok Chang. 'Otto Neurath: Politics and the Unity of Science'. In *The Disunity of Science: Boundaries, Contexts, and Power*, edited by Peter Galison and David J. Stump, 347–69. Stanford, CA: Stanford University Press, 1996.

Catana, Leo. *The Historiographical Concept 'System of Philosophy': Its Origin, Nature, Influence and Legitimacy*. Leiden: Brill, 2008.

Catana, Leo. 'Doxographical or Philosophical History of Philosophy: On Michael Frede's Precepts for Writing the History of Philosophy'. History of European Ideas 42, no. 2 (2016): 170–7.

Celenza, Christopher S. *The Lost Italian Renaissance: Humanists, Historians, and Latin's Legacy*. Baltimore, MD: Johns Hopkins University Press, 2004.

Celenza, Christopher S. 'Lorenzo Valla and the Traditions and Transmissions of Philosophy'. *Journal of the History of Ideas* 66 (2005): 483–506.

Celenza, Christopher S. 'What Counted as Philosophy in the Italian Renaissance? The History of Philosophy, the History of Science, and Styles of Life'. *Critical Inquiry* 39, no. 2 (2013): 367–401.

Celenza, Christopher S. *The Intellectual World of the Italian Renaissance: Language, Philosophy, and the Search for Meaning*. Cambridge: Cambridge University Press, 2017.
Cha, Seong Hwan. 'Modern Chinese Confucianism: The Contemporary Neo-Confucian Movement and Its Cultural Significance'. *Social Compass* 50, no. 4 (2003): 481–91.
Chakravartty, Anjan. 'Stance Relativism: Empiricism versus Metaphysics'. *Studies in History and Philosophy of Science Part A* 35, no. 1 (2004): 173–84.
Chalmers, Alan. 'Viewing Past Science from the Point of View of Present Science, Thereby Illuminating Both: Philosophy versus Experiment in the Work of Robert Boyle'. *Studies in History and Philosophy of Science Part A* 55 (2016): 27–35.
Chalmers, David J. 'Why Isn't There More Progress in Philosophy?' *Philosophy* 90, no. 351 (2015): 3–31.
Chang, Hao. 'New Confucianism and the Intellectual Crisis of Contemporary China'. In *The Limits of Change: Essays on Conservative Alternatives in Republican China*, edited by Charlotte Furth, 276–302. Cambridge, MA: Harvard University Press, 1976.
Chapman, Allan. '"Micrographia" on the Moon'. *Astronomy & Geophysics* 56, no. 5 (2015): 23–9.
Charles, J. Daryl. 'Natural Law and Protestant Reform: Lessons from the Forgotten Reformer'. *Pro Ecclesia* 28, no. 3 (2019): 301–19.
Chater, Nick, and George Loewenstein. 'The Under-Appreciated Drive for Sense-Making'. *Journal of Economic Behavior & Organization* 126, Part B (2016): 137–54.
Chauviré, Christiane. 'Peirce, Popper, Abduction, and the Idea of Logic of Discovery'. *Semiotica* 153 (2005): 209–21.
Chen, Hongyan, and Yuhua Bu. 'Anthropocosmic Vision, Time, and Nature: Reconnecting Humanity and Nature'. *Educational Philosophy and Theory* 51, no. 11 (2019): 1130–40.
Chen-Morris, Raz. *Measuring Shadows: Kepler's Optics of Invisibility*. University Park, PA: Pennsylvania State University Press, 2016.
Chenu, M. D. 'La découverte de la nature'. In *La théologie au douzième siècle*, 21–30. Paris: Vrin, 1957.
Chenyang, Li. 'The Confucian Ideal of Harmony'. *Philosophy East and West* 56, no. 4 (2006): 583–603.
Chesterton, G. K. *Autobiography*. San Francisco: Ignatius, 2006.
Chiu, King Pong. *Thomé H. Fang, Tang Junyi and Huayan Thought: A Confucian Appropriation of Buddhist Ideas in Response to Scientism in Twentieth-Century China*. Leiden: Brill, 2016.
Christoffersen, Svein Aage, Geir Tryggve Hellemo, Leonora Onarheim, Nils Holger Petersen, and Margunn Sandall, eds. *Transcendence and Sensoriness: Perceptions, Revelation, and the Arts*. Leiden: Brill, 2015.
Chua, Daniel K. L. *Absolute Music and the Construction of Meaning*. Cambridge: Cambridge University Press, 1999.
Chua, Daniel K. L. 'Vincenzo Galilei, Modernity and the Division of Nature'. In *Music Theory and Natural Order from the Renaissance to the Early Twentieth Century*, edited by Suzannah Clark and Alexander Rehding, 17–29. Cambridge: Cambridge University Press, 2001.
Chudnoff, Elijah. 'Intuition in Mathematics'. In *Rational Intuition: Philosophical Roots, Scientific Investigations*, edited by Lisa M. Osbeck and Barbara S. Held, 174–91. Cambridge: Cambridge University Press, 2014.
Clark, J. C. D. 'Providence, Predestination and Progress: Or, Did the Enlightenment Fail?' *Albion* 35, no. 4 (2003): 559–89.

Clark, J. C. D. 'Secularization and Modernization: The Failure of a "Grand Narrative"'. *Historical Journal* 55, no. 1 (2012): 161–94.

Clark, Jonathan Owen. 'The Voice and Early Modern Historiography: Reading Johannes Kepler's Harmony of the World'. *The Opera Quarterly* 29, no. 3 (2013): 307–27.

Clarke, Steve, and Adrian Walsh. 'Scientific Imperialism and the Proper Relations between the Sciences'. *International Studies in the Philosophy of Science* 23, no. 2 (2009): 195–207.

Cleary, John J. *Aristotle and Mathematics: Aporetic Method in Cosmology and Metaphysics*. Leiden: Brill, 1995.

Cleland, Carol E. 'Methodological and Epistemic Differences between Historical Science and Experimental Science'. *Philosophy of Science* 69 (2002): 447–51.

Clemens, Justin. 'Galileo's Telescope in John Milton's *Paradise Lost*: The Modern Origin of the Critique of Science as Instrumental Rationality?' *Filozofski Vestnik* 33, no. 2 (2012): 163–94.

Clericuzio, Antonio. 'Boyle's Chemistry'. In *The Bloomsbury Companion to Robert Boyle*, edited by Jan-Erik Jones, 65–96. London: Bloomsbury, 2020.

Clody, Michael. 'Deciphering the Language of Nature: Cryptography, Secrecy, and Alterity in Francis Bacon'. *Configurations* 19, no. 1 (2011): 117–42.

Cohen, H. Floris. *The Scientific Revolution: A Historiographical Inquiry*. Chicago: University of Chicago Press, 1994.

Cohen, H. Floris. *How Modern Science Came into the World: Four Civilizations, One 17th-Century Breakthrough*. Amsterdam: Amsterdam University Press, 2010.

Collier, Andrew. *Critical Realism: An Introduction to Roy Bhaskar's Philosophy*. London: Verso, 1994.

Collins, David J. 'Albertus, Magnus or Magus? Magic, Natural Philosophy, and Religious Reform in the Late Middle Ages'. *Renaissance Quarterly* 63, no. 1 (2010): 1–44.

Collins, Harry. 'Actors' and Analysts' Categories in the Social Analysis of Science'. In *Clashes of Knowledge: Orthodoxies and Heterodoxies in Science and Religion*, edited by Peter Meusburger, Michael Welker, and Edgar Wunder, 101–10. Heidelberg: Springer, 2008.

Colombo, Matteo. 'Experimental Philosophy of Explanation Rising: The Case for a Plurality of Concepts of Explanation'. *Cognitive Science* 41, no. 2 (2017): 503–17.

Connell, Philip. *Secular Chains: Poetry and the Politics of Religion from Milton to Pope*. Oxford: Oxford University Press, 2020.

Connolly, Patrick J. 'Metaphysics in Richard Bentley's Boyle Lectures'. *History of Philosophy Quarterly* 34, no. 2 (2017): 155–74.

Conrad, Sebastian. 'Enlightenment in Global History: A Historiographical Critique'. *American Historical Review* 117, no. 4 (2012): 999–1027.

Cook, A. H. 'The Inverse Square Law of Gravitation'. *Contemporary Physics Bulletin* 28, no. 2 (1987): 159–75.

Cook, Margaret G. 'Divine Artifice and Natural Mechanism: Robert Boyle's Mechanical Philosophy of Nature'. *Osiris* 16 (2001): 133–50.

Cooper, David E. 'Animals, Attitudes and Moral Theories'. In *Science and the Self: Animals, Evolution, and Ethics – Essays in Honour of Mary Midgley*, edited by Ian James Kidd and Liz McKinnell, 19–30. New York: Routledge, 2016.

Cooper, L. J. 'William Blake's Aesthetic Reclamation: Newton, Newtonianism, and Absolute Space in the Book of Urizen and Milton'. *European Romantic Review* 29, no. 2 (2018): 247–69.

Copenhaver, Brian. *Magic in Western Culture: From Antiquity to the Enlightenment*. Cambridge: Cambridge University Press, 2015.

Cormack, Lesley B. 'The Role of Mathematical Practitioners and Mathematical Practice in Developing Mathematics as the Language of Nature'. In *The Language of Nature: Reassessing the Mathematization of Natural Philosophy in the 17th Century*, edited by Geoffrey Gorham, Benjamin Hill, Edward Slowik, and C. Kenneth Waters, 205–28. Minneapolis: University of Minnesota Press, 2016.

Corneanu, Sorana. *Regimens of the Mind: Boyle, Locke, and the Early Modern Cultura Animi Tradition*. Chicago: University of Chicago Press, 2011.

Corneanu, Sorana. 'Passions, Providence, and the Cure of the Mind: Robinson Crusoe Meets the Christian Virtuoso'. In *Le corps et ses images dans l'Europe du dix-huitième siècle*, edited by Sabine Arnaud and Helge Jordheim, 260–76. Paris: Honoré Champion, 2012.

Corrales Rodrigáñez, Capi. 'The Use of Mathematics to Read the Book of Nature: About Kepler and Snowflakes'. *Contributions to Science* 6, no. 1 (2010): 27–34.

Corsi, Pietro. *The Age of Lamarck: Evolutionary Theory in France, 1790–1830*. Berkeley, CA: University of California Press, 1988.

Corsi, Pietro. 'Before Darwin: Transformist Concepts in European Natural History'. *Journal of the History of Biology* 38 (2005): 67–83.

Costin, Vlad, and Vivian L. Vignoles. 'Meaning Is About Mattering: Evaluating Coherence, Purpose, and Existential Mattering as Precursors of Meaning in Life Judgments'. *Journal of Personality and Social Psychology* 118, no. 4 (2020): 864–84.

Cottingham, John. *On the Meaning of Life*. London: Routledge, 2003.

Cottingham, John. 'Why Should Analytic Philosophers Do History of Philosophy?' In *History of Philosophy and Analytic Philosophy*, edited by Tom Sorell and G. A. J. Rogers, 25–41. Oxford: Clarendon Press, 2005.

Cottingham, John. 'Meaningfulness, Eternity and Theism'. In *On Meaning in Life*, edited by Beatrix Himmelmann, 99–112. Berlin: de Gruyter, 2013.

Cottingham, John. *Philosophy of Religion: Towards a More Humane Approach*. Cambridge: Cambridge University Press, 2014.

Cowles, Henry M. 'The Age of Methods: William Whewell, Charles Peirce, and Scientific Kinds'. *Isis* 107, no. 4 (2016): 722–37.

Cowles, Henry M. 'History Naturalized'. *Historical Studies in the Natural Sciences* 47, no. 1 (2017): 107–16.

Cowles, Henry M. *The Scientific Method: An Evolution of Thinking from Darwin to Dewey*. Cambridge, MA: Harvard University Press, 2020.

Coyne, George V. 'The Jesuits and Galileo: Fidelity to Tradition and the Adventure of Discovery'. *Forum Italicum* 49, no. 1 (2015): 154–65.

Craig, Martin. *Subverting Aristotle: Religion, History, and Philosophy in Early Modern Science*. Baltimore: Johns Hopkins University Press, 2014.

Craver, Carl, and William Bechtel. 'Top-Down Causation without Top-Down Causes'. *Biology & Philosophy* 22, no. 4 (2007): 547–63.

Creath, Richard. 'The Unity of Science: Carnap, Neurath and Beyond'. In *The Disunity of Science: Boundaries, Contexts, and Power*, edited by Peter Galison and David J. Stump, 158–69. Stanford, CA: Stanford University Press, 1996.

Creed, Walter G. 'René Wellek and Karl Popper on the Mode of Existence of Ideas in Literature and Science'. *Journal of the History of Ideas* 44, no. 4 (1983): 639–56.

Crockatt, Richard. *Einstein and Twentieth-Century Politics: A Salutary Moral Influence*. Oxford: Oxford University Press, 2016.

Crombie, A. C. *Augustine to Galileo: The History of Science A.D. 400–1650*. London: Falcon Press, 1952.

Crowe, Jonathan. 'Metaphysical Foundations of Natural Law Theories'. In *The Cambridge Companion to Natural Law Jurisprudence*, edited by George Duke and Robert P. George, 103–30. Cambridge: Cambridge University Press, 2017.

Crowther-Heyck, Kathleen. 'Wonderful Secrets of Nature: Natural Knowledge and Religious Piety in Reformation Germany'. *Isis* 94 (2003): 253–73.

Cunningham, Andrew. 'How the *Principia* Got Its Name; or, Taking Natural Philosophy Seriously'. *History of Science* 39 (1991): 377–92.

Cunningham, Andrew. 'The Identity of Natural Philosophy: A Response to Edward Grant'. *Early Science and Medicine* 5, no. 3 (2000): 259–78.

Cunningham, Richard. 'Virtual Witnessing and the Role of the Reader in a New Natural Philosophy'. *Philosophy & Rhetoric* 34, no. 3 (2001): 207–24.

Dahm, John J. 'Science and Apologetics in the Early Boyle Lectures'. *Church History* 39 (1970): 172–86.

Dales, Richard C. *Medieval Discussions of the Eternity of the World*. Leiden: Brill, 1990.

Danielson, Dennis R. 'Myth #6: That Copernicanism Demoted Humans from the Center of the Cosmos'. In *Galileo Goes to Jail: And Other Myths About Science and Religion*, edited by Ronald L. Numbers, 50–8. Cambridge, MA: Harvard University Press, 2009.

Darwin, Charles. *The Descent of Man*. 2 vols. London: John Murray, 1871.

Darwin, Charles. *On the Origin of Species by Means of Natural Selection*. 6th edition. London: John Murray, 1872.

Darwin, Charles. *The Autobiography of Charles Darwin, 1809–1882: With Original Omissions Restored*. Edited by Norah Barlow. London: Collins, 1958.

Darwin, Francis, ed. *The Life and Letters of Charles Darwin*. 3 vols. London: John Murray, 1887.

Daston, Lorraine J. 'Objectivity and the Escape from Perspective'. *Social Studies of Science* 44, no. 2 (1992): 597–618.

Daston, Lorraine. 'Attention and the Values of Nature in the Enlightenment'. In *The Moral Authority of Nature*, edited by Lorraine Daston and Fernando Vidal, 100–26. Chicago: University of Chicago Press, 2004.

Daston, Lorraine. 'Scientific Error and the Ethos of Belief'. *Social Research* 72, no. 1 (2005): 1–28.

Daston, Lorraine and Peter Gallison. 'The Image of Objectivity'. *Representations* 40 (1992): 81–128.

Daston, Lorraine, and Peter Galison. *Objectivity*. New York: Zone Books, 2010.

Davies, Martin L. *The Enlightenment and the Fate of Knowledge: Essays on the Transvaluation of Values*. London: Routledge, 2019.

Davies, David. 'Explanatory Disunities and the Unity of Science'. *International Studies in the Philosophy of Science* 10, no. 1 (1996): 5–21.

Davis, Edward B. 'Newton's Rejection of the "Newtonian World View": The Role of Divine Will in Newton's Natural Philosophy'. *Science and Christian Belief* 3, no. 2 (1991): 103–17.

Davis, Edward B. 'Boyle's Philosophy of Religion'. In *The Bloomsbury Companion to Robert Boyle*, edited by Jan-Erik Jones, 257–82. London: Bloomsbury, 2020.

Davis, Nick. 'Rethinking Narrativity: A Return to Aristotle and Some Consequences'. *Storyworlds: A Journal of Narrative Studies* 4 (2012): 1–24.

Davis, William Hatcher. *Peirce's Epistemology*. The Hague: Nijhoff, 1972.

Davis, William S. *Romanticism, Hellenism, and the Philosophy of Nature*. Cham: Palgrave Macmillan, 2018.

Dawson, Gowan, and Bernard V. Lightman, eds. *Victorian Scientific Naturalism: Community, Identity, Continuity*. Chicago: University of Chicago Press, 2014.

De Cruz, Helen. 'Where Philosophical Intuitions Come From'. *Australasian Journal of Philosophy* 93 (2015): 233–49.

De Cruz, Helen, and Johan De Smedt. 'Delighting in Natural Beauty: Joint Attention and the Phenomenology of Nature Aesthetics'. *European Journal for Philosophy of Religion* 5, no. 4 (2015): 167–86.

De Cruz, Helen, and Johan De Smedt. 'Naturalizing Natural Theology'. *Religion, Brain & Behavior* 6, no. 4 (2016): 355–61.

De Cruz, Helen, and Johan De Smedt. 'Intuitions and Arguments: Cognitive Foundations of Argumentation in Natural Theology'. *European Journal for Philosophy of Religion* 9, no. 2 (2017): 57–82.

De Haan, Daniel. 'McGilchrist's Hemispheric Homunculi'. *Religion, Brain & Behavior* 9, no. 4 (2019): 368–79.

De Jong, Willem, and Arianna Betti. 'The Classical Model of Science: A Millennia-Old Model of Scientific Rationality'. *Synthese* 174, no. 2 (2010): 185–203.

de Regt, Henk W., and Dennis Dieks. 'A Contextual Approach to Scientific Understanding'. *Synthese* 144, no. 1 (2005): 137–70.

de Rycker, Kate. 'A World of One's Own: Margaret Cavendish and the Science of Self-Fashioning'. In *English Literature and the Disciplines of Knowledge, Early Modern to Eighteenth Century*, edited by Jorge Bastos da Silva and Miguel Ramalhete Gomes, 76–93. Leiden: Brill Rodopi, 2017.

Dear, Peter. 'From Truth to Disinterestedness in the Seventeenth Century'. *Social Studies of Science* 22 (1992): 619–31.

Dear, Peter. *Discipline & Experience: The Mathematical Way in the Scientific Revolution*. Chicago: University of Chicago Press, 1995.

Dear, Peter. 'Method and the Study of Nature'. In *The Cambridge History of Seventeenth-Century Philosophy*, edited by D. Garber and M. Ayers, vol. 1, 147–77. Cambridge: Cambridge University Press, 1998.

Dear, Peter. 'What Is the History of Science the History Of? Early Modern Roots of the Ideology of Modern Science'. *Isis* 96, no. 3 (2005): 390–406.

Dear, Peter. *The Intelligibility of Nature: How Science Makes Sense of the World*. Chicago: University of Chicago Press, 2006.

Dear, Peter. *Revolutionizing the Sciences: European Knowledge and Its Ambitions, 1500–1700*. 2nd edition. Princeton, NJ: Princeton University Press, 2009.

Dear, Peter. 'Darwin's Sleepwalkers: Naturalists, Nature, and the Practices of Classification'. *Historical Studies in the Natural Sciences* 44, no. 4 (2014): 297–318.

Del Soldato, Eva. *Early Modern Aristotle: On the Making and Unmaking of Authority*. Philadelphia: University of Pennsylvania Press, 2020.

Dellsén, Finnur. 'Certainty and Explanation in Descartes' Philosophy of Science'. *Journal of the International Society for the History of Philosophy of Science* 7, no. 2 (2017): 302–27.

Demeter, Tamás. 'Natural Theology as Superstition: David Hume and the Changing Ideology of Natural Inquiry'. In *Conflicting Values of Inquiry: Ideologies of Epistemology in Early Modern Europe*, edited by Tamás Demeter, Kathryn Murphy, and Claus Zittel, 176–99. Leiden: Brill, 2014.

Demeter, Tamás. *David Hume and the Culture of Scottish Newtonianism: Methodology and Ideology in Enlightenment Inquiry*. Leiden: Brill, 2016.

Denham, Helen. 'The Cunning of Unreason and Nature's Revolt: Max Horkheimer and William Leiss on the Domination of Nature'. *Environment and History* 3, no. 2 (1997): 149–75.

Dennett, Daniel C. *Darwin's Dangerous Idea: Evolution and the Meaning of Life*. London: Penguin, 1995.

Des Chene, Dennis. *Physiologia: Natural Philosophy in Late Aristotelian and Cartesian Thought*. Ithaca, NY: Cornell University Press, 2000.

DeSalvo, Louise A. 'Popper in the Realm of Literary Criticism'. In *In Pursuit of Truth*, edited by Paul Levinson, 175–91. Atlantic Highlands, NJ: Humanities Press, 1982.

Desmond, Adrian J. *Huxley: From Devil's Disciple to Evolution's High Priest*. London: Penguin, 1998.

Dethier, Corey. 'William Whewell's Semantic Account of Induction'. *Journal of the International Society for the History of Philosophy of Science* 8, no. 1 (2018): 141–56.

Deusen, Nancy van. 'On the Usefulness of Music: Motion, Music, and the Thirteenth-Century Reception of Aristotle's Physics'. *Viator* 29 (1998): 167–88.

DeVun, Leah. *Prophecy, Alchemy, and the End of Time: John of Rupescissa in the Late Middle Ages*. New York: Columbia University Press, 2013.

Dewey, John. *The Quest for Certainty*. New York: Capricorn Books, 1960.

DeYoung, Gregg. 'Astronomy in Ancient Egypt'. In *Astronomy across Cultures: The History of Non-Western Astronomy*, edited by Helaine Selin, 475–508. Dordrecht: Springer, 2000.

Dicken, Paul, and Peter Lipton. 'What Can Bas Believe? Musgrave and Van Fraassen on Observability'. *Analysis* 66 (2006): 226–33.

Dieter, Theodor. *Der junge Luther und Aristoteles: Eine historisch-systematische Untersuchung zum Verhältnis von Theologie und Philosophie*. Berlin: de Gruyter, 2001.

Dion, Sonia Maria. 'Pierre Duhem and the Inconsistency between Instrumentalism and Natural Classification'. *Studies in History and Philosophy of Science Part A* 44, no. 1 (2013): 12–19.

Dirac, Paul A. M. 'Logic or Beauty?' *Scientific Monthly* 79, no. 4 (1954): 268–9.

Dittrich, Walter. 'The Eastward Displacement of a Freely Falling Body on the Rotating Earth: Newton and Hooke's Debate of 1679'. *Annalen der Physik* 522, no. 8 (2010): 601–7.

Dixon, Philip. *Nice and Hot Disputes: The Doctrine of the Trinity in the Seventeenth Century*. London: T&T Clark, 2003.

Doby, John T. 'Logic and Levels of Scientific Explanation'. *Sociological Methodology* 1 (1969): 137–54.

Dolnick, Edward. *The Clockwork Universe: Isaac Newton, the Royal Society, and the Birth of the Modern World*. New York: HarperCollins, 2011.

Donahue, William H. 'Kepler's Approach to the Oval of 1602 from the Mars Notebook'. *Journal for the History of Astronomy* 27 (1996): 281–95.

Dongen, Jeroen van. *Einstein's Unification*. Cambridge: Cambridge University Press, 2010.

Doody, Aude. *Pliny's Encyclopedia: The Reception of Natural Philosophy*. Cambridge: Cambridge University Press, 2010.

Downing, Lisa. 'Are Corpuscles Unobservable in Principle for Locke?' *Journal of the History of Philosophy* 30, no. 1 (1992): 33–52.

Downing, Lisa. 'Berkeley's Natural Philosophy and Philosophy of Science'. In *The Cambridge Companion to Berkeley*, edited by Kenneth Winkler, 230–65. Cambridge: Cambridge University Press, 2006.

Drake, Stillman. 'Galileo's Language: Mathematics and Poetry in a New Science'. *Yale French Studies* 49 (1973): 13–27.

Ducheyne, Steffen. 'Reid's Adaptation and Radicalization of Newton's Natural Philosophy'. *History of European Ideas* 32, no. 2 (2006): 173–89.

Ducheyne, Steffen. *'The Main Business of Natural Philosophy': Isaac Newton's Natural-Philosophical Methodology*. Dordrecht: Springer, 2012.

Ducheyne, Steffen. 'The Status of Theory and Hypotheses'. In *The Oxford Handbook of British Philosophy in the Seventeenth Century*, edited by Peter R. Anstey, 169–91. Oxford: Oxford University Press, 2013.

Dudley, Jaj. 'Das betrachtende Leben (Bios Theoretikos) bei Platon und Aristoteles: Ein kritischer Ansatz'. *Neue Zeitschrift für Systematische Theologie und Religionsphilosophie* 37, no. 1 (1995): 20–40.

Duffy, Simon. 'The Difference between Science and Philosophy: The Spinoza–Boyle Controversy Revisited'. *Paragraph* 29, no. 2 (2006): 115–38.

Duhem, Pierre. *Sozein ta phainomena: Essai sur la notion de théorie physique de Platon à Galilée*. Paris: Hermann et Fils, 1908.

Dupré, John. *The Disorder of Things: Metaphysical Foundations of the Disunity of Science*. Cambridge, MA: Harvard University Press, 1993.

Dupré, John. 'Metaphysical Disorder and Scientific Disunity'. In *The Disunity of Science: Boundaries, Contexts, and Power*, edited by Peter Galison and David J. Stump, 101–17. Stanford, CA: Stanford University Press, 1996.

Dupré, John. 'The Miracle of Monism'. In *Processes of Life: Essays in the Philosophy of Biology*, 21–39. Oxford: Oxford University Press, 2012.

Dupré, Sven. 'Kepler's Optics without Hypotheses'. *Synthese* 185, no. 3 (2012): 501–25.

Dustin, Christopher A., and Joanna A. Ziegler. *Practicing Morality: Art, Philosophy, and Contemplative Seeing*. New York: Palgrave Macmillan, 2007.

Dyck, John. 'Sovereign Selves'. In *Whose Will Be Done? Essays on Sovereignty and Religion*, edited by John Dyck, Paul S. Rowe, and Jens Zimmermann, 137–51. Lanham, MD: Lexington Books, 2015.

Eastwood, Bruce S. 'Astronomical Images and Planetary Theory in Carolingian Studies of Martianus Capella'. *Journal for the History of Astronomy* 31, no. 1 (2000): 1–28.

Eddington, Arthur S. *The Nature of the Physical World*. New York: Macmillan, 1928.

Edelstein, Dan. *The Enlightenment: A Genealogy*. Chicago: University of Chicago Press, 2010.

Edelstein, Dan, and Biliana Kassabova. 'How England Fell Off the Map of Voltaire's Enlightenment'. *Modern Intellectual History* 17, no. 1 (2020): 29–53.

Edgerton, Samuel Y. *The Mirror, the Window and the Telescope: How Renaissance Linear Perspective Changed Our Vision of the Universe*. Ithaca, NY: Cornell University Press, 2009.

Edwards, L. Clifton. 'Artful Creation and Aesthetic Rationality: Toward a Creational Theology of Revelatory Beauty'. *Theology Today* 69 (2012): 56–72.

Edwards, Michael. 'Philosophy, Early Modern Intellectual History, and the History of Philosophy'. *Metaphilosophy* 43, no. 1–2 (2012): 82–95.

Einstein, Albert. 'Entwicklung unserer Anschauungen über das Wesen und die Konstitution der Strahlung'. *Physikalische Zeitschrift* 10 (1909): 817–25.

Einstein, Albert. 'Erklärung der Perihelbewegung des Merkur aus der allgemeinen Relativitätstheorie'. *Sitzungsberichte der Preußischen Akademie der Wissenschaften* 47 (1915): 831–9.

Einstein, Albert. *Ideas and Opinions*. New York: Crown Publishers, 1954.

Elder-Vass, Dave. 'Realist Critique without Ethical Naturalism and Moral Realism'. *Journal of Critical Realism* 9, no. 1 (2010): 33–58.

Elders, Leo J. *The Philosophy of Nature of St Thomas Aquinas*. Frankfurt am Main: Peter Lang, 1997.

Elders, Leo J. 'St Thomas Aquinas's Commentary on Aristotle's "Physics"'. *Review of Metaphysics* 66, no. 4 (2013): 713–48.

Ellis, George F. R. 'Physics, Complexity, and Causality'. *Nature* 435 (2005): 743.
Elman, Benjamin A. *On Their Own Terms: Science in China, 1550–1900*. Cambridge, MA: Harvard University Press, 2005.
Elman, Benjamin A. 'Toward a History of Modern Science in Republican China'. In *Science and Technology in Modern China, 1880s–1940s*, edited by Jing Tsu and Benjamin A. Elman, 15–38. Leiden: Brill, 2014.
Elson, Rebecca. *A Responsibility to Awe*. Manchester: Carcanet, 2018.
Emery, Gilles, and Matthew Levering, eds. *Aristotle in Aquinas's Theology*. Oxford: Oxford University Press, 2015.
Encyclopaedia Britannica. 4th edition. 20 vols. Edinburgh: Macfarquhar & Bell, 1810.
Enders, Markus. *Natürliche Theologie im Denken der Griechen*. Frankfurt am Main: Josef Knecht, 2000.
Engelmann, Edward M. 'Scientific Demonstration in Aristotle, "Theoria", and Reductionism'. *Review of Metaphysics* 60, no. 3 (2007): 479–506.
Engvold, Oddbjørn, and Jack B. Zirker. 'The Parallel Worlds of Christoph Scheiner and Galileo Galilei'. *Journal for the History of Astronomy* 47, no. 3 (2016): 322–45.
Enskat, Raine. 'Ist Wissen der Paradoxe epistemische Fall von Wahrheit ohne Wissen? Platon, Gettier, Sartwell und die Folgen'. *Zeitschrift für Philosophische Forschung* 57, no. 3 (2003): 431–45.
Eriksson, Monica, and Bengt Lindström. 'Antonovsky's Sense of Coherence Scale and Its Relation with Quality of Life: A Systematic Review'. *Journal of Epidemiology and Community Health* 61, no. 11 (2007): 938–44.
Ernst, Germana. *Tommaso Campanella: The Book and the Body of Nature*. Dordrecht: Springer, 2010.
Falcon, Andrea. *Aristotle and the Science of Nature: Unity without Uniformity*. Cambridge: Cambridge University Press, 2005.
Fallon, Stephen M. 'Milton, Newton, and the Implications of Arianism'. In *Milton in the Long Restoration*, edited by Blair Hoxby and Ann Baynes Coiro, 319–34. Oxford: Oxford University Press, 2016.
Fara, Patricia. *Newton: The Making of Genius*. New York: Columbia University Press, 2002.
Farkas, J. I., and J. J Sarbo. 'Interpreting Nature'. *Cognitive Systems Research* 49 (2018): 1–8.
Faucher, Luc. 'Unity of Science and Pluralism: Cognitive Neurosciences of Racial Prejudice as a Case Study'. In *Special Sciences and the Unity of Science*, edited by Olga Pombo, Juan Manuel Torres, John Symons, and Shahid Rahman, 177–204. Dordrecht: Springer, 2012.
Feingold, Mordechai. 'Science as a Calling? The Early Modern Dilemma'. *Science in Context* 15, no. 1 (2002): 79–119.
Feingold, Mordechai. *The Newtonian Moment: Isaac Newton and the Making of Modern Culture*. New York: Oxford University Press, 2004.
Feingold, Mordechai. '"Experimental Philosophy": Invention and Rebirth of a Seventeenth-Century Concept'. *Early Science and Medicine* 21, no. 1 (2016): 1–28.
Feldman, Seymour. 'On Plural Universes: A Debate in Medieval Jewish Philosophy and the Duhem–Pines Thesis'. *Aleph* 12, no. 2 (2012): 329–66.
Fend, Michael. 'Historical Alternatives to a Poststructuralist Reading of Johannes Kepler's Harmony of the World: Response to Jonathan Owen Clark'. *The Opera Quarterly* 29, no. 3 (2013): 328–34.
Fend, Michael. 'Probleme mit der Idee der "Harmonia Universalis" in der frühen Neuzeit'. *Archiv für Musikwissenschaft* 71, no. 4 (2014): 307–34.

Feser, Edward. 'Introduction: An Aristotelian Revival?' *In Aristotle on Method and Metaphysics*, edited by Edward Feser, 1–6. Basingstoke: Palgrave Macmillan, 2013.

Fine, Gail. 'Knowledge and Belief in Plato's Republic 5–7'. In *Cambridge Companions to Ancient Thought: I. Epistemology*, edited by S. Everson, 85–115. Cambridge: Cambridge University Press, 1990.

Fine, Gail. 'Knowledge and True Belief in the *Meno*'. *Oxford Studies in Ancient Philosophy* 27 (2004): 41–81.

Fink-Jensen, Morten. 'Medicine, Natural Philosophy, and the Influence of Melanchthon in Reformation Denmark and Norway'. *Bulletin of the History of Medicine* 80, no. 3 (2006): 439–64.

Finlay, Robert. 'China, the West, and World History in Joseph Needham's *Science and Civilisation in China*'. *Journal of World History* 11 no. 2 (2000): 265–303.

Finlay, Robert. 'Weaving the Rainbow: Visions of Color in World History'. *Journal of World History* 18, no. 4 (2007): 383–431.

Finocchiaro, Maurice A. *Defending Copernicus and Galileo: Critical Reasoning in the Two Affairs*. New York: Springer, 2010.

Fiorato, Sidia. *The Relationship between Literature and Science in John Banville's Scientific Tetralogy*. Frankfurt am Main: Peter Lang, 2007.

Fisch, Harold. 'The Scientist as Priest: A Note on Robert Boyle's Natural Theology'. *Isis* 44 (1953): 252–65.

Fisher, Philip. *Wonder, the Rainbow, and the Aesthetics of Rare Experiences*. Cambridge, MA: Harvard University Press, 1998.

Fiske, Ingrid. 'The Poetic Mystery of Dark Matter'. *Nature* 414, no. 6866 (2001): 845–6.

Flórez, Jorge Alejandro. 'Peirce's Theory of the Origin of Abduction in Aristotle'. *Transactions of the Charles S. Peirce Society* 50, no. 2 (2014): 265–80.

Flyvbjerg, Bent. *Rationality and Power: Democracy in Practice*. Chicago: University of Chicago Press, 1998.

Fodor, Jerry A. 'Special Sciences (Or: The Disunity of Science as a Working Hypothesis)'. *Synthese* 28 (1974): 97–115.

Foltz, Bruce V. *The Noetics of Nature: Environmental Philosophy and the Holy Beauty of the Visible*. New York: Fordham University Press, 2013.

Forcada, Miquel. 'Astrology in Al-Andalus during the 11th and 12th Centuries: Between Religion and Philosophy'. In *From Masha'allah to Kepler: The Theory and Practice of Astrology in the Middle Ages and the Renaissance*, edited by Charles Burnett and Dorian Gieseler Greenbaum, 149–76. Ceredigion: Sophia Centre Press, 2015.

Forcada, Miquel. 'Saphaeae and Hay'āt: The Debate between Instrumentalism and Realism in Al-Andalus'. *Medieval Encounters* 23 (2017): 263–86.

Force, James E. 'Newton's God of Dominion: The Unity of Newton's Theological, Scientific, and Political Thought'. In *Essays on the Context, Nature, and Influence of Isaac Newton's Theology*, edited by James E. Force and Richard H. Popkin, 75–102. Dordrecht: Kluwer, 1990.

Force, James E. 'Biblical Interpretation, Newton and English Deism'. In *Scepticism and Irreligion in the Seventeenth and Eighteenth Centuries*, edited by Richard H. Popkin and Arjo J. Vanderjagt, 282–305. Leiden: Brill, 1993.

Force, Pierre. 'The Teeth of Time: Pierre Hadot on Meaning and Misunderstanding in the History of Ideas'. *History and Theory* 50, no. 1 (2011): 20–40.

Ford, Thomas. 'Ruskin's Storm-Cloud: Heavenly Messages and Pathetic Fallacies in a Denatured World'. *International Social Science Journal* 62 (2011): 287–99.

Foss, Jeffrey. 'Subjectivity, Objectivity, and Nagel on Consciousness'. *Dialogue: Canadian Philosophical Review* 32, no. 4 (1993): 725–36.

Fossheim, Hallvard J. 'Individual, Society, and Teleology: An Aristotelian Conception of Meaning in Life'. In *On Meaning in Life*, edited by Beatrix Himmelmann, 45–64. Berlin: de Gruyter, 2013.

Fowler, Robert L. 'Mythos and Logos'. *Journal of Hellenic Studies* 131 (2011): 45–66.

Frances, Bryan. 'Extensive Philosophical Agreement and Progress'. *Metaphilosophy* 48, no. 1–2 (2017): 47–57.

Frank, Günther, and Stefan Rhein, eds. *Melanchthon und die Naturwissenschaften seiner Zeit*. Sigmaringen: Thorbecke, 1998.

Frank, Roberta. 'Interdisciplinarity: The First Half Century'. *Issues in Interdisciplinary Studies* 6 (1988): 139–51.

French, Robert, Hugo Gaggiotti, and Peter Simpson. 'Journeying and the Experiential Gaze in Research: Theorizing as a Form of Knowing'. *Culture and Organization* 20, no. 3 (2014): 185–95.

French, Roger. *Medicine before Science: The Business of Medicine from the Middle Ages to the Enlightenment*. Cambridge: Cambridge University Press, 2003.

Freudenthal, Gad. '"Instrumentalism" and "Realism" as Categories in the History of Astronomy: Duhem vs. Popper, Maimonides vs. Gersonides'. *Centaurus* 45 (2003): 227–48.

Fudge, Robert S. 'Imagination and the Science-Based Aesthetic Appreciation of Unscenic Nature'. *Journal of Aesthetics and Art Criticism* 59, no. 3 (2001): 275–85.

Fuentes, Agustin, and Aku Visala, eds. *Verbs, Bones, and Brains: Interdisciplinary Perspectives on Human Nature*. Notre Dame, IN: University of Notre Dame Press, 2017.

Führer, Markus L. 'Albertus Magnus' Theory of Divine Illumination'. In *Albertus Magnus: Zum Gedenken nach 800 Jahren: Neue Zugänge, Aspekte und Perspektiven*, edited by Walter Senner, 141–56. Berlin: de Gruyter, 2001.

Fumagalli, Roberto. 'Eliminating "Life Worth Living"'. *Philosophical Studies* 175 (2018): 769–92.

Fung, Yu-Lan. 'Why China Has No Science: An Interpretation of the History and Consequences of Chinese Philosophy'. *International Journal of Ethics* 32, no. 3 (1922): 237–63.

Funkenstein, Amos. *Theology and the Scientific Imagination from the Middle Ages to the Seventeenth Century*. Princeton, NJ: Princeton University Press, 1986.

Fyfe, Aileen. 'The Reception of William Paley's Natural Theology in the University of Cambridge'. *British Journal for the History of Science* 30 (1997): 321–35.

Fyfe, Aileen. 'Publishing and the Classics: Paley's Natural Theology and the Nineteenth-Century Scientific Canon'. *Studies in the History and Philosophy of Science* 33 (2002): 433–55.

Gadamer, Hans-Georg. 'Praise of Theory'. In *Praise of Theory: Speeches and Essays*, 16–36. New Haven, CT: Yale University Press, 1998.

Gal, Ofer, and Raz Chen-Morris. *Science in the Age of Baroque*. Dordrecht: Springer, 2013.

Gal, Ofer, and Cindy Hodoba Eric. 'Between Kepler and Newton: Hooke's "Principles of Congruity and Incongruity" and the Naturalization of Mathematics'. *Annals of Science* 76, no. 3/4 (2019): 241–66.

Gale, Monica. *Virgil on the Nature of Things: The Georgics, Lucretius, and the Didactic Tradition*. Cambridge: Cambridge University Press, 2000.

Galera, Andrés. 'The Impact of Lamarck's Theory of Evolution before Darwin's Theory'. *Journal of the History of Biology* 50, no. 1 (2017): 53–70.

Galilei, Galileo, *Le opere di Galileo Galilei*. 20 vols. Florence: Barbèra, 1890–1909.

Gambarotto, Andrea. *Vital Forces, Teleology and Organization: Philosophy of Nature and the Rise of Biology in Germany*. Cham: Springer, 2018.

Gane, Nicholas. *Max Weber and Postmodern Theory: Rationalization versus Re-Enchantment*. New York: Palgrave Macmillan, 2002.

Gaonkar, Dilip. 'Toward New Imaginaries: An Introduction'. *Public Culture* 14, no. 1 (2002): 1–19.

Garber, Daniel. 'What's Philosophical About the History of Philosophy?' In *History of Philosophy and Analytic Philosophy*, edited by Tom Sorell and G. A. J. Rogers, 129–46. Oxford: Clarendon Press, 2005.

Garber, Daniel. 'Remarks on the Pre-History of the Mechanical Philosophy'. In *The Mechanization of Natural Philosophy*, edited by Daniel Garber and Sophie Roux, 3–26. Heidelberg: Springer, 2013.

Garber, Daniel. 'Laws of Nature and the Mathematics of Motion'. In *The Language of Nature: Reassessing the Mathematization of Natural Philosophy in the 17th Century*, edited by Geoffrey Gorham, Benjamin Hill, Edward Slowik, and C. Kenneth Waters, 134–59. Minneapolis: University of Minnesota Press, 2016.

García Castillo, Pablo. 'La filosofía como diálogo permanente con la tradición'. *Disputatio* 5, no. 6 (2016): 377–93.

Gascoigne, John. 'From Bentley to the Victorians: The Rise and Fall of British Newtonian Natural Theology'. *Science in Context* 2 (1988): 219–56.

Gasser-Wingate, Marc. 'Conviction, Priority, and Rationalism in Aristotle's Epistemology'. *Journal of the History of Philosophy* 58, no. 1 (2020): 1–27.

Gaukroger, Stephen. *Francis Bacon and the Transformation of Early-Modern Philosophy*. Cambridge: Cambridge University Press, 2001.

Gaukroger, Stephen. *Descartes' System of Natural Philosophy*. Cambridge: Cambridge University Press, 2002.

Gaukroger, Stephen. 'The Autonomy of Natural Philosophy: From Truth to Impartiality'. In *The Science of Nature in the Seventeenth Century: Patterns of Change in Early Modern Natural Philosophy*, edited by Peter R. Anstey and John A. Schuster, 131–63. Dordrecht: Springer, 2005.

Gaukroger, Stephen. 'Science, Religion and Modernity'. *Critical Quarterly* 47, no. 4 (2005): 1–31.

Gaukroger, Stephen. *The Emergence of a Scientific Culture: Science and the Shaping of Modernity 1210–1685*. Oxford: Oxford University Press, 2006.

Gaukroger, Stephen. 'The Persona of the Natural Philosopher'. In *The Philosopher in Early Modern Europe: The Nature of a Contested Identity*, edited by Conal Condren, Stephen Gaukroger, and Ian Hunter, 17–34. New York: Cambridge University Press, 2006.

Gaukroger, Stephen. *The Collapse of Mechanism and the Rise of Sensibility: Science and the Shaping of Modernity, 1680–1760*. Oxford: Oxford University Press, 2012.

Gaukroger, Stephen. 'The Early Modern Idea of Scientific Doctrine and Its Early Christian Origins'. *Journal of Medieval and Early Modern Studies* 44, no. 1 (2014): 95–112.

Gaukroger, Stephen. 'The Challenges of Empirical Understanding in Early Modern Theology'. In *The Oxford Handbook of Early Modern Theology, 1600–1800*, edited by Ulrich L. Lehner, Richard A. Muller, and A. G. Roebe, 564–75. Oxford: Oxford University Press, 2016.

Gaukroger, Stephen. *The Natural and the Human: Science and the Shaping of Modernity, 1739–1841*. Oxford: Oxford University Press, 2016.

Gaukroger, Stephen. 'The Unity of Science and the Search for a Unity of Understanding in the Modern Era'. *Rivista di Storia della Filosofia* 72, no. 4 (2017): 553–73.

Gaukroger, Stephen. *Civilization and the Culture of Science: Science and the Shaping of Modernity, 1795–1935*. Oxford: Oxford University Press, 2020.

Gaukroger, Stephen. *The Failures of Philosophy: An Historical Essay*. Princeton, NJ: Princeton University Press, 2020.

Gaukroger, Stephen, John Andrew Schuster, and John Sutton, eds. *Descartes' Natural Philosophy*. London: Routledge, 2000.

Gayon, Jean. 'De la biologie à la philosophie de la biologie'. In *Questions vitales: Vie biologique, vie psychique*, edited by Françoise Monnoyeur, 83–95. Paris: Kimé, 2008.

Gayon, Jean. 'Defining Life: Synthesis and Conclusions'. *Origins of Life and Evolution of the Biosphere* 40, no. 2 (2010): 231–44.

Geertz, Clifford. 'Common Sense as a Cultural System'. *Antioch Review* 33, no. 1 (1983): 5–26.

Geldhof, Joris. 'Romantische Metaphysik als natürliche Theologie? Franz von Baader über Gott, die Welt und den Menschen'. In *Idealismus und natürliche Theologie*, edited by Margit Wasmaier-Sailer and Benedikt Paul Göcke, 213–37. Freiburg im Breisgau: Verlag Karl Alber, 2011.

George, L. S., and Crystal L. Park. 'Existential Mattering: Bringing Attention to a Neglected but Central Aspect of Meaning'. In *Meaning in Positive and Existential Psychology*, edited by Alexander Batthyany and Pninit Russo-Netzer, 39–51. New York: Springer, 2014.

Germann, Nadja. 'Natural Philosophy in Earlier Latin Thought'. In *The Cambridge History of Medieval Philosophy*, edited by Robert Pasnau, 219–31. Cambridge: Cambridge University Press, 2014.

Gerson, Lloyd P. 'Platonic Knowledge and the Standard Analysis'. *International Journal of Philosophical Studies* 14, no. 4 (2006): 455–74.

Gettier, Edmund. 'Is Justified True Belief Knowledge?' *Analysis* 23, no. 121–3 (1963).

Giere, Ronald N. *Scientific Perspectivism*. Chicago: University of Chicago Press, 2006.

Gies, David Thatcher, and Cynthia Wall. *The Eighteenth Centuries: Global Networks of Enlightenment*. Charlottesville and London: University of Virginia Press, 2018.

Giglioni, Guido. 'The Hidden Life of Matter: Techniques for Prolonging Life in the Writings of Francis Bacon'. In *Francis Bacon and the Refiguring of Early Modern Thought*, edited by J. R. Solomon and C. G. Martin, 129–44. Aldershot: Ashgate, 2005.

Giglioni, Guido. 'Learning to Read Nature: Francis Bacon's Notion of Experiential Literacy (*Experientia Literata*)'. *Early Science and Medicine* 18, no. 4–5 (2013): 405–34.

Gijsbers, Victor. 'Understanding, Explanation, and Unification'. *Studies in History and Philosophy of Science* 44 (2013): 516–22.

Giletti, Ann. 'Aristotle in Medieval Spain: Writers of the Christian Kingdoms Confronting the Eternity of the World'. *Journal of the Warburg and Courtauld Institutes* 67 (2004): 23–47.

Giletti, Ann. 'The Journey of an Idea: Maimonides, Albertus Magnus, Thomas Aquinas and Ramon Martí (c. 1220–c. 1284/5) on the Undemonstrability of the Eternity of the World'. *Estudos e Textos de Filosofia Medieval* 2 (2011): 239–67.

Gillespie, Neal C. 'Natural History, Natural Theology, and Social Order: John Ray and the "Newtonian Ideology"'. *Journal of the History of Biology* 20 (1987): 1–49.

Gillespie, Neal C. 'Divine Design and the Industrial Revolution: William Paley's Abortive Reform of Natural Theology'. *Isis* 81 (1990): 214–29.

Gilpin, George H. 'William Blake and the World's Body of Science'. *Studies in Romanticism* 43, no. 1 (2004): 35–56.

Gingerich, Owen. 'Is There a Role for Natural Theology Today?' In *Science and Theology: Questions at the Interface*, edited by M. Rae, H. Regan, and J. Stenhouse, 29–48. Edinburgh: T&T Clark, 1994.

Gingerich, Owen. *The Book Nobody Read: Chasing the Revolutions of Nicolaus Copernicus*. New York: Walker & Co., 2004.

Gingerich, Owen. 'The Great Martian Catastrophe and How Kepler Fixed It'. *Physics Today* 64 (2011): 50–4.

Gingerich, Owen, and Albert Van Helden. 'How Galileo Constructed the Moons of Jupiter'. *Journal for the History of Astronomy* 42 (2011): 259–64.

Gingras, Bruno. 'Johannes Kepler's *Harmonices Mundi*: A "Scientific" Version of the Harmony of the Spheres'. *Journal of the Royal Astronomical Society of Canada* 97, no. 6 (2003): 259–65.

Gingras, Yves. 'The Collective Construction of Scientific Memory: The Einstein–Poincaré Connection and Its Discontents, 1905–2005'. *History of Science* 46 (2008): 75–114.

Gintis, Herbert. *The Bounds of Reason: Game Theory and the Unification of the Behavioral Sciences*. Princeton, NJ: Princeton University Press, 2014.

Glasner, Ruth. *Averroes' Physics: A Turning Point in Medieval Natural Philosophy*. Oxford: Oxford University Press, 2009.

Glasner, Ruth. 'The Peculiar History of Aristotelianism among Spanish Jews'. In *Studies in the History of Culture and Science: A Tribute to Gad Freudenthal*, edited by Resianne Fontaine, Ruth Glasner, Reimund Leicht, and Giuseppe Veltri, 361–81. Leiden: Brill, 2010.

Göcke, Benedikt Paul. *After Physicalism*. Notre Dame, IN: University of Notre Dame Press, 2012.

Goddu, André. *Copernicus and the Aristotelian Tradition: Education, Reading, and Philosophy in Copernicus's Path to Heliocentrism*. Leiden: Brill, 2010.

Godfrey-Smith, Peter. 'Metaphysics and the Philosophical Imagination'. *Philosophical Studies* 160 (2012): 97–113.

Godin, Benoît. 'Representation of "Innovation" in Seventeenth-Century England: A View from Natural Philosophy'. *Contributions to the History of Concepts* 11, no. 2 (2016): 24–42.

Goethe, Johann Wolfgang von. *Scientific Studies*. Princeton, NJ: Princeton University Press, 1995.

Goldstein, Bernard R. 'Saving the Phenomena: The Background to Ptolemy's Planetary Theory'. *Journal for the History of Astronomy* 28 (1997): 2–12.

Goldstein, Bernard R. 'Copernicus and the Origin of His Heliocentric System'. *Journal for the History of Astronomy* 33, no. 112 (2002): 219–35.

Goldstein, Bernard R., and Giora Hon. 'Kepler's Move from Orbs to Orbits: Documenting a Revolutionary Scientific Concept'. *Perspectives on Science* 13 (2005): 74–111.

Golinski, Jan. 'Is It Time to Forget Science? Reflections on Singular Science and Its History'. *Osiris* 27, no. 1 (2012): 19–36.

Goodstein, Elizabeth. *Georg Simmel and the Disciplinary Imaginary*. Stanford, CA: Stanford University Press, 2017.

Gorham, Geoffrey, and Benjamin Hill, eds. *The Language of Nature: Reassessing the Mathematization of Natural Philosophy in the 17th Century*. Minneapolis: University of Minnesota Press, 2016.

Gorski, Philip S. 'What Is Critical Realism? And Why Should You Care?' *Contemporary Sociology: A Journal of Reviews* 42, no. 5 (2013): 658–70.

Gotthelf, Allan. *Teleology, First Principles, and Scientific Method in Aristotle's Biology*. Oxford: Oxford University Press, 2012.

Gouk, Penelope. 'Transforming Matter, Refining the Spirit: Alchemy, Music and Experimental Philosophy around 1600'. *European Review* 21, no. 2 (2013): 146–57.

Gould, Stephen Jay. *The Structure of Evolutionary Theory.* Cambridge, MA: Harvard University Press, 2002.

Gould, Stephen Jay. *The Hedgehog, the Fox, and the Magister's Pox: Mending the Gap between Science and the Humanities*. New York: Three Rivers Press, 2003.

Goulding, Robert. 'Binocular Vision and Image Location before Kepler'. *Archive for History of Exact Sciences* 72 (2018): 497–546.

Grady, Hugh. *John Donne and Baroque Allegory: The Aesthetics of Fragmentation.* Cambridge: Cambridge University Press, 2017.

Graff, Harvey J. *Undisciplining Knowledge: Interdisciplinarity in the Twentieth Century.* Baltimore: Johns Hopkins University Press, 2015.

Graff, Harvey J. 'The "Problem" of Interdisciplinarity in Theory, Practice, and History'. *Social Science History* 40, no. 4 (2016): 775–803.

Grafton, Anthony. 'Chronology, Controversy, and Community in the Republic of Letters: The Case of Kepler'. In *Worlds Made by Words: Scholarship and Community in the Modern West*, 114–36. Cambridge, MA: Harvard University Press, 2009.

Graham, Gordon. 'Hume and Smith on Natural Religion'. *Philosophy* 91, no. 357 (2016): 345–60.

Granada, Miguel A. '"A quo moventur planetae?" Kepler et la question de l'agent du mouvement planétaire après la disparition des orbes solides'. *Galilaeana: Journal of Galilean Studies* 7 (2010): 111–41.

Grant, Edward. 'Ways to Interpret the Terms "Aristotelian" and "Aristotelianism" in Medieval and Renaissance Natural Philosophy'. *History of Science* 25 (1987): 335–58.

Grant, Edward. 'God and Natural Philosophy: The Late Middle Ages and Sir Isaac Newton'. *Early Science and Medicine* 5, no. 3 (2000): 279–98.

Grant, Edward. *A History of Natural Philosophy: From the Ancient World to the Nineteenth Century*. Cambridge: Cambridge University Press, 2007.

Grant, Edward. *The Nature of Natural Philosophy in the Late Middle Ages*. Washington, DC: Catholic University of America Press, 2010.

Grant, Edward. 'How Theology, Imagination, and the Spirit of Inquiry Shaped Natural Philosophy in the Late Middle Ages'. *History of Science* 49, no. 1 (2011): 89–108.

Grant, Iain H. *Philosophies of Nature after Schelling*. London: Continuum, 2006.

Grantham, Todd. 'Conceptualizing the (Dis)Unity of Science'. *Philosophy of Science* 71 (2004): 133–55.

Greenham, Paul. 'Clarifying Divine Discourse in Early Modern Science: Divinity, Physico-Theology, and Divine Metaphysics in Isaac Newton's Chymistry'. *The Seventeenth Century* 32, no. 2 (2017): 191–215.

Guicciardini, Niccolò. 'Reconsidering the Hooke–Newton Debate on Gravitation: Recent Results'. *Early Science and Medicine* 10, no. 4 (2005): 510–17.

Guilherme, Alexandre. 'Schelling's *Naturphilosophie* Project: Towards a Spinozian Conception of Nature'. *South African Journal of Philosophy* 29, no. 4 (2010): 373–90.

Gunnoe, Charles D., and Dane T. Daniel. 'Anti-Paracelsianism from Conrad Gessner to Robert Boyle'. *Daphnis* 48, no. 1–2 (2020): 104–39.

Gunster, Shane. 'Fear and the Unknown: Nature, Culture, and the Limits of Reason'. In *Critical Ecologies: The Frankfurt School and Contemporary Environmental Crises*, edited by Andrew Biro, 206–28. Toronto: University of Toronto Press, 2011.

Gussow, Alan. 'Beauty in the Landscape: An Ecological Viewpoint'. In *Landscape in America*, edited by George F. Thompson, 223–40. Austin, TX: University of Texas Press, 1995.

Haack, Susan. *Defending Science—within Reason: Between Scientism and Cynicism*. Amherst, NY: Prometheus Books, 2007.

Haakonssen, Knud. *Natural Law and Moral Philosophy: From Grotius to the Scottish Enlightenment*. Cambridge: Cambridge University Press, 1996.

Haakonssen, Knud. 'Early Modern Natural Law Theories'. In *The Cambridge Companion to Natural Law Jurisprudence*, edited by George Duke and Robert P. George, 76–102. Cambridge: Cambridge University Press, 2017.

Hacker, P. M. S. 'Philosophy: A Contribution, Not to Human Knowledge, but to Human Understanding'. *Royal Institute of Philosophy Supplement* 84, no. 65 (2009): 129–53.

Hacking, Ian. 'The Disunities of Science'. In *The Disunity of Science: Boundaries, Contexts, and Power*, edited by Peter Galison and David J. Stump, 37–74. Stanford, CA: Stanford University Press, 1996.

Hadot, Pierre. *What Is Ancient Philosophy?* Cambridge, MA: Harvard University Press, 2002.

Hadot, Pierre. 'La philosophie antique: Une éthique ou une pratique?' In *Études de Philosophie Antique*, 207–32. Paris: Les Belles Lettres, 2010.

Haldane, John. 'Scientism and Its Challenge to Humanism'. *New Blackfriars* 93, no. 1048 (2012): 671–86.

Hale, Piers J. *Political Descent: Malthus, Mutualism, and the Politics of Evolution in Victorian England*. Chicago: University of Chicago Press, 2014.

Hall, Bryan. 'Kant on Newton, Genius, and Scientific Discovery'. *Intellectual History Review* 24, no. 4 (2014): 539–56.

Hall, A. Rupert. *All Was Light: An Introduction to Newton's Opticks*. Oxford: Clarendon Press, 1993.

Hall, A. Rupert. 'Newton versus Leibniz: From Geometry to Metaphysics'. In *The Cambridge Companion to Newton*, edited by I. Bernard Cohen and George E. Smith, 431–54. Cambridge: Cambridge University Press, 2002.

Hall, Marie Boas. *Promoting Experimental Learning: Experiment and the Royal Society 1660–1727*. Cambridge: Cambridge University Press, 1991.

Halmi, Nicholas. *The Genealogy of the Romantic Symbol*. Oxford: Oxford University Press, 2007.

Hamilton, Christopher. '"Frail Worms of the Earth": Philosophical Reflections on the Meaning of Life'. *Religious Studies* 54, no. 1 (2018): 55–71.

Hamilton, Ross. 'Deep History: Association and Natural Philosophy in Wordsworth's Poetry'. *European Romantic Review* 18, no. 4 (2007): 459–81.

Hamou, Philippe. *La mutation du visible: Essai sur la portée épistémologique des instruments d'optique au XVIIe siècle*. Lille: Presses Universitaires du Septentrion, 1999.

Hankins, Thomas L. *Science and the Enlightenment*. Cambridge: Cambridge University Press, 1985.

Hankinson, R. J. *Cause and Explanation in Ancient Greek Thought*. Oxford: Clarendon Press, 2001.

Haralambous, Bronwen, and Thomas W. Nielsen. 'Wonder as a Gateway Experience'. In *Wonderful Education: The Centrality of Wonder in Teaching and Learning*, edited by Kieran Egan, Annabella Cant, and Gillian Judson, 219–38. London: Routledge, 2013.

Hardin, Jeff, Ronald L. Numbers, and Ronald A. Binzley, eds. *The Warfare between Science and Religion: The Idea That Wouldn't Die*. Baltimore: Johns Hopkins University Press, 2018.

Harman, P. M. *The Natural Philosophy of James Clerk Maxwell*. Cambridge: Cambridge University Press, 1998.

Harper, William. 'Newton's Methodology and Mercury's Perihelion before and after Einstein'. *Philosophy of Science* 74, no. 5 (2007): 932–42.

Harper, William L. *Isaac Newton's Scientific Method: Turning Data into Evidence About Gravity and Cosmology*. Oxford: Oxford University Press, 2014.

Harrington, Michael. 'Creation and Natural Contemplation in Maximus the Confessor's *Ambiguum* 10:19'. In *Divine Creation in Ancient, Medieval, and Early Modern Thought*, edited by Willemien Otten, Walter Hannam, and Michael Treschow, 191–212. Leiden: Brill, 2007.

Harrison, Carol. *On Music, Sense, Affect, and Voice*. London: T&T Clark, 2019.

Harrison, Peter. *The Bible, Protestantism, and the Rise of Natural Science*. Cambridge: Cambridge University Press, 1998.

Harrison, Peter. 'Subduing the Earth: Genesis 1, Early Modern Science, and the Exploitation of Nature'. *Journal of Religion* 79, no. 1 (1999): 86–109.

Harrison, Peter. 'Curiosity, Forbidden Knowledge, and the Reformation of Natural Philosophy in Early Modern England'. *Isis* 92, no. 2 (2001): 265–90.

Harrison, Peter. '"Priests of the Most High God, with Respect to the Book of Nature": The Vocational Identity of the Early Modern Naturalist'. In *Reading God's World*, edited by Angus Menuge, 55–80. St Louis, MO: Concordia, 2004.

Harrison, Peter. 'Physico-Theology and the Mixed Sciences: The Role of Theology in Early Modern Natural Philosophy'. In *The Science of Nature in the Seventeenth Century* edited by Peter Anstey and John Schuster, 165–83. Dordrecht: Springer, 2005.

Harrison, Peter. 'The "Book of Nature" and Early Modern Science'. In *The Book of Nature in Early Modern and Modern History*, edited by Klaas van Berkel and Arie Johan Vanderjagt, 1–26. Louvain: Peeters, 2006.

Harrison, Peter. 'The Natural Philosopher and the Virtues'. In *The Philosopher in Early Modern Europe: The Nature of a Contested Identity*, edited by Conal Condren, Stephen Gaukroger, and Ian Hunter, 202–28. New York: Cambridge University Press, 2006.

Harrison, Peter. 'Natural Theology, Deism, and Early Modern Science'. In *Science, Religion, and Society: An Encyclopedia of History, Culture and Controversy*, edited by Arri Eisen and Gary Laderman, 426–33. New York: Sharp, 2006.

Harrison, Peter. *The Fall of Man and the Foundations of Science*. Cambridge: Cambridge University Press, 2007.

Harrison, Peter. 'Was There a Scientific Revolution?' *European Review* 15, no. 4 (2007): 445–57.

Harrison, Peter. 'The Development of the Concept of Laws of Nature'. In *Creation: Law and Probability*, edited by Fraser Watts, 13–36. Aldershot: Ashgate, 2008.

Harrison, Peter. 'Religion, the Royal Society, and the Rise of Science'. *Theology and Science* 6 (2008): 255–71.

Harrison, Peter. *The Territories of Science and Religion*. Chicago: University of Chicago Press, 2015.

Harrison, Peter. 'Laws of God or Laws of Nature? Natural Order in the Early Modern Period'. In *Science without God? Rethinking the History of Scientific Naturalism*, edited by Peter Harrison and Jon Roberts, 59–77. Oxford: Oxford University Press, 2019.

Harrison, Victoria. 'The Pragmatics of Defining Religion in a Multi-Cultural World'. *International Journal for Philosophy of Religion* 59 (2006): 133–52.

Harvey, Steven. 'The Hebrew Translation of Averroes' *Prooemium* to his Long Commentary on Aristotle's *Physics*'. *Proceedings of the American Academy for Jewish Research* 52 (1985): 55–84.

Haskins, Charles H. *The Renaissance of the Twelfth Century*. Cambridge, MA: Harvard University Press, 1927.

Hatch, Laurie Camp. 'Gerard Manley Hopkins and Victorian Approaches to the Problems of Perception: Affirming the Metaphysical in the Physical'. *Christianity & Literature* 65, no. 2 (2016): 170–94.

Hatfield, Gary. 'The History of Philosophy as Philosophy'. In *Analytic Philosophy and History of Philosophy*, edited by Tom Sorell and G. A. J. Rogers, 83–128. Oxford: Clarendon Press, 2005.

Hawking, Stephen, and Leonard Mlodinow. *The Grand Design*. New York: Bantam Books, 2010.

Hedley, Douglas. 'Gods and Giants: Cudworth's Platonic Metaphysics and His Ancient Theology'. *British Journal for the History of Philosophy* 25, no. 5 (2017): 932–54.

Heilmann, Anja. *Boethius' Musiktheorie und das Quadrivium: Eine Einführung in den neuplatonischen Hintergrund von 'de Institutione Musica'*. Göttingen: Vandenhoeck & Ruprecht, 2007.

Heinecke, Berthold. 'Naturphilosophie bei Georg Philipp Harsdörffer'. In *Georg Philipp Harsdörffers Universalität: Beiträge zu einem Uomo Universale des Barock*, edited by Stefan Keppler-Tasaki and Ursula Kocher, 247–78. Berlin: de Gruyter, 2011.

Heisenberg, Werner. 'Die Goethe'sche und Newtonische Farbenlehre im Lichte der modernen Physik'. In *Wandlungen in den Grundlagen der Naturwissenschaft*, 105–25. Stuttgart: Hirtzel, 2014.

Henderson, Andrea K. *Romantic Identities: Varieties of Subjectivity, 1774–1830*. Cambridge: Cambridge University Press, 1996.

Henry, John. 'Metaphysics and the Origins of Modern Science: Descartes and the Importance of Laws of Nature'. *Early Science and Medicine* 9 (2004): 73–114.

Henry, John. *The Scientific Revolution and the Origins of Modern Science: Studies in European History*. 3rd edition. Basingstoke: Palgrave Macmillan, 2008.

Henry, John. 'Gravity and *De Gravitatione*: The Development of Newton's Ideas on Action at a Distance'. *Studies in History and Philosophy of Science Part A*, 42, no. 1 (2011): 11–27.

Herkommer, Hubert. 'Buch der Schrift und Buch der Natur: Zur Spiritualität der Welterfahrung im Mittelalter, mit einem Ausblick auf ihren Wandel in der Neuzeit'. *Zeitschrift für schweizerische Archäologie und Kunstgeschichte* 43 (1986): 167–78.

Hermann, Marc. 'A Critical Evaluation of Fang Dongmei's Philosophy of Comprehensive Harmony'. *Journal of Chinese Philosophy* 34, no. 1 (2007): 59–97.

Herschel, John F. W. *A Preliminary Discourse on the Study of Natural Philosophy*. London: Longman, Rees, Orme, Brown & Green, 1830.

Hersey, George L. 'Ruskin as an Optical Thinker'. In *The Ruskin Polygon: Essays on the Imagination of John Ruskin*, edited by John Dixon Hunt and Faith M. Holland, 44–64. Manchester: Manchester University Press, 1982.

Hesse, Hermann. *Mit dem Erstaunen fängt es an. Herkunft und Heimat. Natur und Kunst*. Frankfurt am Main: Suhrkamp, 2000.

Heuschling, Sophie. '"Don't Ask the Questions You've Been Taught by Science": Rebecca Elson's Astronomical Poetry'. *Journal of Literature and Science* 12, no. 2 (2019): 43–61.

Hicks, Joshua A., and Laura A. King. 'Meaning in Life and Seeing the Big Picture: Positive Affect and Global Focus'. *Cognition and Emotion* 21, no. 7 (2007): 1577–84.

Hill, Greg. 'Solidarity, Objectivity, and the Human Form of Life: Wittgenstein vs. Rorty'. *Critical Review* 11, no. 4 (1997): 555–80.

Hintikka, Jaakko. 'Aristotelian Induction'. *Revue Internationale de Philosophie* 34 (1980): 422–39.

Hirai, Hiro. *Medical Humanism and Natural Philosophy: Renaissance Debates on Matter, Life, and the Soul*. Leiden: Brill, 2011.

Hittinger, Russell. *A Critique of the New Natural Law Theory*. Notre Dame, IN: University of Notre Dame Press, 1987.

Hobson, Marian. *Diderot and Rousseau: Networks of Enlightenment*. Oxford: Voltaire Foundation, 2011.

Hoekstra, Kinch. 'Disarming the Prophets: Thomas Hobbes and Predictive Power'. *Rivista di storia della filosofia* 1 (2004): 97–153.

Holcomb, Justin S., ed. *Christian Theologies of Scripture: A Comparative Introduction*. New York: New York University Press, 2006.

Holden, Thomas. 'Robert Boyle on Things above Reason'. *British Journal for the History of Philosophy* 15, no. 2 (2007): 283–312.

Hollander, John. *The Untuning of the Sky: Ideas of Music in English Poetry, 1500–1700*. Princeton, NJ: Princeton University Press, 1961.

Hollinger, David A. 'The Unity of Knowledge and the Diversity of Knowers: Science as an Agent of Cultural Integration in the United States between the Two World Wars'. *Pacific Historical Review* 80, no. 2 (2011): 211–23.

Holmes, John, and Paul Smith. 'Visions of Nature: Reviving Ruskin's Legacy at the Oxford University Museum'. *Journal of Art Historiography* 22 (2020): 1–15.

Hon, Giora. 'Kepler's Revolutionary Astronomy: Theological Unity as a Comprehensive View of the World'. In *Conflicting Values of Inquiry: Ideologies of Epistemology in Early Modern Europe*, edited by Tamás Demeter, Kathryn Murphy, and Claus Zittel, 155–75. Leiden: Brill, 2014.

Hopkins, Gerard Manley. *The Major Works*. Edited by Catherine Phillips. Oxford: Oxford University Press, 2002.

Horkheimer, Max. *Eclipse of Reason*. New York: Oxford University Press, 1947.

Horn, Christoph. 'Happiness and the Meaning of Life'. In *On Meaning in Life*, edited by Beatrix Himmelmann, 9–22. Berlin: de Gruyter, 2013.

Hoskin, Michael. *The Construction of the Heavens: William Herschel's Cosmology*. Cambridge: Cambridge University Press, 2012.

Hösle, Vittorio. 'How Did the Western Culture Subdivide Its Various Forms of Knowledge and Justify Them? Historical Reflections on the Metamorphoses of the Tree of Knowledge'. In *Forms of Truth and the Unity of Knowledge*, edited by Vittorio Hösle, 29–69. Notre Dame, IN: University of Notre Dame Press, 2014.

Hostettler, Nick, and Alan Norrie. 'Are Critical Realist Ethics Foundationalist?' In *Critical Realism: The Difference It Makes*, edited by Justin Cruickshank, 30–53. Abingdon: Routledge, 2003.

Howell, Kenneth J. 'The Hermeneutics of Nature and Scripture in Early Modern Science and Theology'. In *Nature and Scripture in the Abrahamic Religions: Up to 1700*, edited by Jitse M. van der Meer and Scott Mandelbrote, vol. 1, 275–98. 2 vols. Leiden: Brill, 2008.

Huber, Irmtraud. 'Competing for Eternity: Tracing the Relation between Poetry and Science in Victorian Discourse'. *Journal of Literature and Science* 12, no. 1 (2019): 1–20.

Hübner, Wolfgang. 'The Professional Astrologos'. In *Hellenistic Astronomy*, edited by Paul T. Keyser, John Scarborough, and Alan C. Bowen, 297–320. Oxford: Oxford University Press, 2019.

Hughes, Austin L. 'The Folly of Scientism'. *The New Atlantis* 37 (2012): 32–50.

Hughes, J. T. 'The Medical Education of Sir Thomas Browne, a Seventeenth-Century Student at Montpellier, Padua, and Leiden'. *Journal of Medical Biography* 9 (2001): 70–6.

Hughes, Jonathan. *The Rise of Alchemy in Fourteenth-Century England: Plantagenet Kings and the Search for the Philosopher's Stone*. London: Continuum, 2012.
Hui, Wang. 'The Fate of "Mr. Science" in China: The Concept of Science and Its Application in Modern Chinese Thought'. *Positions: East Asia Cultures Critique* 3, no. 1 (1995): 1–68.
Huijgen, Arnold. *Divine Accommodation in John Calvin's Theology: Analysis and Assessment*. Göttingen: Vandenhoeck & Ruprecht, 2011.
Hume, David. *Enquiries Concerning Human Understanding and Concerning the Principles of Morals*. Edited by P. H. Nidditch, and L. A. Selby-Bigge. 3rd edition. Oxford: Clarendon Press, 1975.
Huneman, Philippe. 'From the Critique of Judgment to the Hermeneutics of Nature: Sketching the Fate of Philosophy of Nature after Kant'. *Continental Philosophy Review* 39, no. 1 (2006): 1–34.
Hünemörder, Christian. 'Traditionelle Naturkunde, realistische Naturbeobachtung und theologische Naturdeutung in Enzyklopädien des hohen Mittelalters'. In *Mittelalter. Konzeptionen—Erfahrungen—Wirkungen*, edited by Peter Dilg, 124–35. Berlin: Akademie Verlag, 2003.
Hunt, G. M. K. 'Is Philosophy a "Theory of Everything"?' *Royal Institute of Philosophy* Supplement 33 (1992): 219–31.
Hunt, Robert. *The Poetry of Science: Or, Studies of the Physical Phenomena of Nature*. 3rd edition. London: Henry G. Bohn, 1854.
Hunter, Ian. Rival Enlightenments: Civil and Metaphysical Philosophy in Early Modern Germany. Cambridge: Cambridge University Press, 2001.
Hunter, Ian. 'The History of Philosophy and the Persona of the Philosopher'. *Modern Intellectual History* 4 (2007): 571–600.
Hunter, Michael. 'Latitudinarianism and the "Ideology" of the Early Royal Society: Thomas Sprat's History of the Royal Society (1667) Reconsidered'. In *Philosophy, Science, and Religion in England 1640–1700*, edited by Richard W. F. Kroll, Richard Ashcraft, and Perez Zagorin, 199–229. Cambridge: Cambridge University Press, 1992.
Hunter, Michael. *Robert Boyle (1627–91): Scrupulosity and Science*. Woodbridge: Boydell Press, 2002.
Hunter, Michael. 'Robert Boyle and the Early Royal Society: A Reciprocal Exchange in the Making of Baconian Science'. *British Journal for the History of Science* 40, no. 1 (2007): 1–23.
Hunter, Michael. *Boyle: Between God and Science*. New Haven, CT: Yale University Press, 2009.
Huxley, Thomas H. 'On Certain Zoological Arguments Commonly Adduced in Favour of the Hypothesis of the Progressive Development of Animal Life in Time'. *Proceedings of the Royal Institution* 2 (1858): 82–5.
Huxley, Thomas H. *On Our Knowledge of the Causes of the Phenomena of Organic Nature*. London: Robert Hardwicke, 1863.
Huxley, Thomas H. *Hume*. London: Macmillan, 1879.
Huxley, Thomas H. *Introductory Science Primer*. London: Appleton, 1887.
Huxley, Thomas H. *Science & Education*. London: Macmillan, 1893.
Huxley, Thomas H. *Science and Hebrew Tradition: Essays*. London: Macmillan, 1893.
Huxley, Thomas H. *Evolution and Ethics and Other Essays*. London: Macmillan, 1894.
Huxley, Thomas H. 'The Progress of Science'. In *Method and Results*, 42–129. New York: Appleton, 1911.
Ichikawa, Jonathan Jenkins. 'Intuition in Contemporary Philosophy'. In *Rational Intuition: Philosophical Roots, Scientific Investigations*, edited by Lisa M. Osbeck and Barbara S. Held, 192–210. Cambridge: Cambridge University Press, 2014.

Iliffe, Robert. '"Is He Like Other Men?" The Meaning of the *Principia Mathematica* and the Author as Idol'. In *Culture and Society in the Stuart Restoration*, edited by Gerald MacLean, 159–76. Cambridge: Cambridge University Press, 1995.

Iliffe, Robert. 'Abstract Considerations: Disciplines and the Incoherence of Newton's Natural Philosophy'. *Studies in History and Philosophy of Science* 35A, no. 3 (2004): 427–54.

Iliffe, Robert. *Priest of Nature: The Religious Worlds of Isaac Newton*. Oxford: Oxford University Press, 2017.

Ilnitchi, Gabriela. '"Musica Mundana", Aristotelian Natural Philosophy and Ptolemaic Astronomy'. *Early Music History* 21 (2002): 37–74.

Irving, Sarah. 'Public Knowledge, Natural Philosophy, and the Eighteenth-Century Republic of Letters'. *Early American Literature* 49, no. 1 (2014): 67–88.

Irwin, Terence. *Aristotle's First Principles*. Oxford: Clarendon Press, 1988.

Jackson, Michael. 'Phénomènes limites: Un essai sur l'ambiguïté de la nature'. *Anthropological Quarterly* 51, no. 1 (2009): 133–43.

Jackson, Myles W. 'Music and Science during the Scientific Revolution'. *Perspectives on Science* 9, no. 1 (2001): 106–15.

Jacob, J. R. 'The Ideological Origins of Robert Boyle's Natural Philosophy'. *Journal of European Studies* 2, no. 1 (1972): 1–21.

Jacob, Margaret C. *The Newtonians and the English Revolution 1689–1720*. Ithaca, NY: Cornell University Press, 1976.

Jacob, Margaret C. *Scientific Culture and the Making of the Industrial West*. Oxford: Oxford University Press, 1997.

Jacob, Margaret C., and Larry Stewart. *Practical Matter: Newton's Science in the Service of Industry and Empire, 1687–1851*. Cambridge, MA: Harvard University Press, 2004.

Jacobs, Jerry. *In Defense of Disciplines: Interdisciplinarity and Specialization in the Research University*. Chicago: University of Chicago Press, 2014.

James, Frank A. J. L., and Robert Bud. 'Epilogue: Science after Modernity'. In *Being Modern: The Cultural Impact of Science in the Early Twentieth Century*, edited by Robert Bud, Paul Greenhalgh, Frank James, and Morag Shiach, 386–93. London: UCL Press, 2018.

James, William. *The Varieties of Religious Experience: A Study in Human Nature*. London: Longmans Green & Co., 1911.

James, William. *Essays in Radical Empiricism*. Cambridge, MA: Harvard University Press, 1976.

Jammer, Max. *Einstein and Religion: Physics and Theology*. Princeton, NJ: Princeton University Press, 1999.

Janiak, Andrew. *Newton as Philosopher*. Cambridge: Cambridge University Press, 2008.

Janiak, Andrew. 'Metaphysics and Natural Philosophy in Descartes and Newton'. *Foundations of Science* 18 (2012): 1–15.

Janiak, Andrew. 'Newton and Descartes: Theology and Natural Philosophy'. *Southern Journal of Philosophy* 50, no. 3 (2012): 414–35.

Jaworski, William. *Structure and the Metaphysics of Mind: How Hylomorphism Solves the Mind–Body Problem*. Oxford: Oxford University Press, 2016.

Jay, Martin. *Reason after Its Eclipse: On Late Critical Theory*. Madison, WI: University of Wisconsin Press, 2016.

Jenkins, Richard. 'Disenchantment, Enchantment and Re-Enchantment: Max Weber at the Millennium'. *Max Weber Studies* 1, no. 1 (2000): 11–32.

Joas, Hans. *Die Macht des Heiligen: Eine alternative zur Geschichte von der Entzauberung*. Berlin: Suhrkamp 2017.

Johansen, Thomas. 'Timaeus in the Cave'. In *The Platonic Art of Philosophy*, edited by George Boys-Stones, Dimitri El Murr, and Christopher Gill, 90–109. Cambridge: Cambridge University Press, 2013.

Johansen, Thomas K. *Plato's Natural Philosophy. A Study of the Timaeus-Critias.* Cambridge: Cambridge University Press, 2004.

Johnson, Christopher. 'Bricoleur and Bricolage: From Metaphor to Universal Concept'. *Paragraph* 35, no. 3 (2012): 355–72.

Johnson, Monte Ransome. *Aristotle on Teleology*. Oxford: Clarendon Press, 2005.

Johnson, Nuala C. 'Grand Design(er)s: David Moore, Natural Theology and the Royal Botanic Gardens in Glasnevin, Dublin, 1838–1879'. *Cultural Geographies* 14, no. 1 (2007): 29–55.

Joly, Bernard. 'Francis Bacon, the Reformer of Alchemy: Alchemic Tradition and Scientific Invention in the Seventeenth Century'. *Revue philosophique de la France et de l'étranger* 128, no. 1 (2003): 23–40.

Jonas, Hans. 'Die Freiheit des Bildens: Homo Pictor und die Differentia des Menschen'. In *Zwischen Nichts und Ewigkeit: Zur Lehre vom Menschen*, 26–43. Göttingen: Vandenhoeck & Ruprecht, 1961.

Jordan, Karen, and Kristján Kristjánsson. 'Sustainability, Virtue Ethics, and the Virtue of Harmony with Nature'. *Environmental Education Research* 23, no. 9 (2017): 1205–29.

Jorink, Eric. *Reading the Book of Nature in the Dutch Golden Age, 1575–1715*. Leiden: Brill, 2010.

Josephson-Storm, Jason A. 'Why Do We Think We Are Disenchanted?' *The New Atlantis* 56 (2018): 3–13.

Judson, Lindsay. 'Aristotle and Crossing the Boundaries between the Sciences'. *Archiv für Geschichte der Philosophie* 101, no. 2 (2019): 177–204.

Kail, P. J. E. *Projection and Realism in Hume's Philosophy*. Oxford: Oxford University Press, 2007.

Kaiser, Christopher B. 'Humanity as the Exegete of Creation with Reference to the Work of Natural Scientists'. *Horizons in Biblical Theology* 14, no. 1 (1992): 112–28.

Kaiser, Marie I. *Reductive Explanation in the Biological Sciences*. Cham: Springer, 2015.

Kakkuri-Knuuttila, Marja-Liisa, and Simo Knuuttila. 'Induction and Conceptual Analysis in Aristotle'. In *Language, Knowledge, and Intentionality: Perspectives on the Philosophy of Jaakko Hintikka*, edited by L. Haaparanta, M. Kusch, and I. Niiniluoto, 294–303. Helsinki: Philosophical Society of Finland, 1990.

Kang, Xiaoguang. 'A Study of the Renaissance of Traditional Confucian Culture in Contemporary China'. In *Confucianism and Spiritual Traditions in Modern China and Beyond*, edited by Fenggang Yang and Joseph B. Tamney, 33–74. Leiden: Brill, 2012.

Kanterian, Edward. 'Naturalism, Involved Philosophy, and the Human Predicament'. In *New Models of Religious Understanding*, edited by Fiona Ellis, 59–78. Oxford: Oxford University Press, 2018.

Kaplan, Lawrence J. 'Time, History, Space, and Place: Abraham Joshua Heschel on the Religious Significance of the Land of Israel'. *Journal of Modern Jewish Studies* 17, no. 4 (2018): 496–504.

Karasmanis, Vassilis. 'Plato's Republic: The Line and the Cave'. *Apeiron* 21 (1988): 147–71.

Karbowski, Joseph. 'Justification "by Argument" in Aristotle's Natural Science'. *Oxford Studies in Ancient Philosophy* 51 (2016): 119–60.

Kargon, Robert H. 'The Testimony of Nature: Boyle, Hooke and Experimental Philosophy'. *Albion* 3, no. 2 (1971): 72–81.

Katsikadelis, John. 'Derivation of Newton's Law of Motion from Kepler's Laws of Planetary Motion'. *Archive of Applied Mechanics* 88, no. 1 (2018): 27–38.

Kawall, Jason. 'Meaningful Lives, Ideal Observers, and Views from Nowhere'. *Journal of Philosophical Research* 37 (2012): 73–97.

Keats, John. *Poetry and Prose*. Edited by Jeffrey N. Cox. London: W. W. Norton, 2009.

Kekes, John. 'Meaning and Narratives'. In *On Meaning in Life*, edited by Beatrix Himmelmann, 65–82. Berlin: de Gruyter, 2013.

Keller, Eve. 'Producing Petty Gods: Margaret Cavendish's Critique of Experimental Science'. *ELH* 64, no. 2 (1997): 447–71.

Kelter, Irving A. 'Reading the Book of God as the Book of Nature: The Case of the Louvain Humanist Cornelius Valerius (1512–1578)'. In *The Word and the World: Biblical Exegesis and Early Modern Science*, edited by Kevin Killeen and Peter J. Forshaw, 174–87. Basingstoke: Palgrave Macmillan, 2007.

Kepler, Johann. *Gesammelte Werke*. Edited by Max Caspar. 22 vols. Munich: C. H. Beck, 1937–83.

Keppler-Tasaki, Stefan, and Ursula Kocher, eds. *Georg Philipp Harsdörffers Universalität: Beiträge zu einem Uomo Universale des Barock*. Berlin: de Gruyter, 2011.

Kessler, Eckhard. 'Metaphysics or Empirical Science? The Two Faces of Aristotelian Natural Philosophy in the Sixteenth Century'. In *Renaissance Readings of Corpus Aristotelicum*, edited by Marianne Pade, 79–101. Copenhagen: Museum Tusculanum Press, 2001.

Kessler, Eckhard. '*O vitae experientia dux*. Die Rolle der Erfahrung im theoretischen und praktischen Weltbezug des frühen Humanismus und ihre Konsequenzen'. *Das Mittelalter* 17, no. 2 (2012): 60–74.

Ketcham, Michael G. 'Scientific and Poetic Imagination in James Thomson's "Poem Sacred to the Memory of Sir Isaac Newton"'. *Philological Quarterly* 61, no. 1 (1982): 33–50.

Keymer, Thomas. 'The Subjective Turn'. In *The Oxford Handbook of British Romanticism*, edited by David Duff, 312–27. Oxford: Oxford University Press, 2018.

Keynes, R. D. *Fossils, Finches and Fuegians: Charles Darwin's Adventures and Discoveries on the Beagle, 1832–1836*. London: HarperCollins, 2002.

Kidd, Ian James. 'Doing Science an Injustice: Midgley on Scientism'. In *Science and the Self: Animals, Evolution, and Ethics*, edited by Ian James Kidd and Liz McKinnell, 151–67. New York: Routledge, 2016.

Kidd, Ian James. 'Reawakening to Wonder: Wittgenstein, Feyerabend, and Scientism'. In *Wittgenstein and Scientism*, edited by Ian James Kidd and Jonathan Beale, 101–15. London: Routledge, 2017.

Kieckhefer, Richard. 'The Specific Rationality of Medieval Magic'. *American Historical Review* 99 (1994): 813–36.

Kiefer, Thomas. 'Hermeneutical Understanding as the Disclosure of Truth: Hans-Georg Gadamer's Distinctive Understanding of Truth'. *Philosophy Today* 57, no. 1 (2013): 42–60.

Kim, Yung Sik. *The Natural Philosophy of Chu Tsi, 1130–1200*. Philadelphia: Memoirs of the American Philosophic Society, 2000.

Kinch, Michael Paul. 'Geographical Distribution and the Origin of Life: The Development of Early Nineteenth Century British Explanations'. *Journal of the History of Biology* 13, no. 1 (1980): 91–119.

Kind, Amy. 'Imagination and the Imaginary, by Kathleen Lennon'. *Mind* 125, no. 500 (2016): 1244–51.

King, Amy M. *The Divine in the Commonplace: Reverent Natural History and the Novel in Britain*. Cambridge: Cambridge University Press, 2019.

Kipperman, Mark. *Beyond Enchantment: German Idealism and English Romantic Poetry*. Philadelphia: University of Pennsylvania Press, 1986.
Kitcher, Philip. 'The Trouble with Scientism'. *New Republic* 243, no. 8 (2012): 20–5.
Klaassen, Frank F. *The Transformations of Magic: Illicit Learned Magic in the Later Middle Ages and Renaissance*. University Park, PA: Pennsylvania State University Press, 2013.
Klein, Julie Thompson. *Crossing Boundaries: Knowledge, Disciplinarities, and Interdisciplinarities*. Charlottesville, VA: University Press of Virginia, 1996.
Kleineberg, Michael. 'From Linearity to Co-Evolution: On the Architecture of Nicolai Hartmann's Levels of Reality'. In *New Research on the Philosophy of Nicolai Hartmann*, edited by Keith R. Peterson and Roberto Poli, 81–108. Berlin: de Gruyter, 2016.
Klepper, Deeana Copeland. 'Theories of Interpretation: The Quadriga and Its Successors'. In *The New Cambridge History of the Bible*, edited by Euan Cameron, 418–38. Cambridge: Cambridge University Press, 2016.
Knight, Harriet. 'Boyle's Baconianism'. In *The Bloomsbury Companion to Robert Boyle*, edited by Jan-Erik Jones, 9–38. London: Bloomsbury, 2020.
Knox, Dilwyn. 'Ficino and Copernicus'. In *Marsilio Ficino: His Theology, His Philosophy, His Legacy*, edited by Michael Allen and Valery Rees, 399–418. Leiden: Brill, 2002.
Kock, Christoph. *Natürliche Theologie: Ein evangelischer Streitbegriff*. Neukirchen-Vluyn: Neukirchener, 2001.
Koestler, Arthur. *The Ghost in the Machine*. London: Hutchinson, 1967.
Kogan, Barry S. 'Judah Halevi and His Use of Philosophy in the *Kuzari*'. In *The Cambridge Companion to Medieval Jewish Philosophy*, edited by Daniel H. Frank and Oliver Leaman, 111–35. Cambridge: Cambridge University Press, 2003.
Konchak, William. 'Gadamer's "Practice" of Theoria'. *Epoché* 24, no. 2 (2020): 453–65.
Koopman, Colin. 'Pragmatist Resources for Experimental Philosophy: Inquiry in Place of Intuition'. *Journal of Speculative Philosophy* 26, no. 1 (2012): 1–24.
Kosso, Peter. 'The Omniscienter: Beauty and Scientific Understanding'. *International Studies in the Philosophy of Science* 16, no. 1 (2002): 39–48.
Kozhamthadam, Job. 'The Religious Foundations of Kepler's Science'. *Revista Portuguesa de Filosofia* 58, no. 4 (2002): 887–901.
Kragh, Helge. *Conceptions of Cosmos: From Myths to the Accelerating Universe – A History of Cosmology*. Oxford: Oxford University Press, 2007.
Kragh, Helge. *Higher Speculations: Grand Theories and Failed Revolutions in Physics and Cosmology*. Oxford: Oxford University Press, 2015.
Kretzmann, Norman. *The Metaphysics of Theism: Aquinas's Natural Theology in Summa Contra Gentiles I*. Oxford: Clarendon Press, 1997.
Kretzmann, Norman. *The Metaphysics of Creation: Aquinas's Natural Theology in Summa Contra Gentiles II*. Oxford: Clarendon Press, 1999.
Krichauff, Skye. *Memory, Place and Aboriginal-Settler History*. London: Anthem Press, 2018.
Kriesel, Howard. *Judaism as Philosophy: Studies in Maimonides and the Medieval Jewish Philosophers of Provence*. Boston: Academic Studies Press, 2015.
Krohn, Wolfgang. 'Interdisciplinary Cases and Disciplinary Knowledge: Epistemic Challenges of Interdisciplinary Research'. In *The Oxford Handbook of Interdisciplinarity*, edited by Robert Frodeman, Julie Thompson Klein, and Roberto C. S. Pacheco, 40–52. Oxford: Oxford University Press, 2017.
Krüger, Malte Dominik. 'Mehr als notwendig – Natürliche Theologie nach Schelling'. In *Idealismus und natürliche Theologie*, edited by Margit Wasmaier-Sailer and Benedikt Paul Göcke, 135–46. Freiburg im Breisgau: Verlag Karl Alber, 2011.

Kuhn, Thomas S. *The Road since Structure: Philosophical Essays, 1970–1993*. Chicago: University of Chicago Press, 2000.

Kulp, Christopher B. *The End of Epistemology: Dewey and His Current Allies on the Spectator Theory of Knowledge*. Westport, CT: Greenwood Press, 1992.

Kusukawa, Sachiko. *The Transformation of Natural Philosophy: The Case of Philip Melanchthon*. Cambridge: Cambridge University Press, 1995.

Kusukawa, Sachiko. 'The Natural Philosophy of Melanchthon and His Followers'. In *Sciences et religions: De Copernic à Galilée (1540–1610)*, 443–53. Rome: École Française de Rome, 1999.

Kuukkanen, Jouni-Matti. 'Making Sense of Conceptual Change'. *History and Theory* 47, no. 3 (2008): 351–72.

Kuukkanen, Jouni-Matti. 'Autonomy and Objectivity of Science'. *International Studies in the Philosophy of Science* 26, no. 3 (2012): 309–34.

Kwa, Chunglin. *Styles of Knowing: A New History of Science from Ancient Times to the Present*. Pittsburgh: University of Pittsburgh Press, 2011.

Ladyman, James. 'Does Physics Answer Metaphysical Questions?' *Royal Institute of Philosophy* Supplement 61 (2007): 179–201.

Ladyman, James. 'Science, Metaphysics and Method'. *Philosophical Studies* 160, no. 1 (2012): 31–51.

Ladyman, James. 'An Apology for Naturalized Metaphysics'. In *Metaphysics and the Philosophy of Science: New Essays*, edited by Matthew Slater and Zanja Yudell, 141–62. Oxford: Oxford University Press, 2017.

Ladyman, James. 'Scientism with a Humane Face'. In *Scientism: Prospects and Problems*, edited by Jeroen de Ridder, Rik Peels, and René van Woudenberg, 106–26. New York: Oxford University Press, 2018.

Ladyman, James, and Don Ross. *Every Thing Must Go: Metaphysics Naturalized*. Oxford: Oxford University Press, 2007.

Lærke, Mogens. 'The Anthropological Analogy and the Constitution of Historical Perspectivism'. In *Philosophy and Its History*, edited by Mogens Lærke, Justin E. H. Smith, and Eric Schliesser, 7–29. Oxford: Oxford University Press, 2013.

Lagerlund, Henrik. 'The Changing Face of Aristotelian Empiricism in the Fourteenth Century'. *Quaestio* 10 (2010): 315–27.

Lancaster, James A. T. 'Natural Knowledge as a Propaedeutic to Self-Betterment: Francis Bacon and the Transformation of Natural History'. *Early Science and Medicine* 17, no. 1–2 (2012): 181–96.

Landau, Iddo. 'Why Has the Question of the Meaning of Life Arisen in the Last Two and a Half Centuries?' *Philosophy Today* 41 (1997): 263–70.

Landau, Iddo. 'Conceptualizing Great Meaning in Life: Metz on the Good, the True, and the Beautiful'. *Religious Studies* 49, no. 4 (2013): 505–14.

Lane, Belden C. 'Thomas Traherne and the Awakening of Want'. *Anglican Theological Review* 81, no. 4 (1999): 651–64.

Langermann, Y. Tzvi. 'Moses Maimonides and Judah Halevi on Order and Law in the World of Nature, and Beyond'. *Studies in History and Philosophy of Science Part A* 81 (2020): 39–45.

Laurand, Valéry. 'La contemplation chez Philon d'Alexandrie'. In *Theoria, Praxis and the Contemplative Life after Plato and Aristotle*, edited by Thomas Bénatouïl and Mauro Bonazzi, 121–38. Leiden: Brill, 2012.

Laurent, Catherine. 'Plurality of Science and Rational Integration of Knowledge'. In *Special Sciences and the Unity of Science*, edited by Olga Pombo, Juan Manuel Torres, John Symons, and Shahid Rahman, 219–31. Dordrecht: Springer, 2012.

Lawson, Ian. 'Bears in Eden, or, This Is Not the Garden You're Looking For: Margaret Cavendish, Robert Hooke and the Limits of Natural Philosophy'. *British Journal for the History of Science* 48, no. 4 (2015): 583–605.

Lawson, Ian. 'Crafting the Microworld: How Robert Hooke Constructed Knowledge About Small Things'. *Notes and Records of the Royal Society of London* 70, no. 1 (2016): 23–44.

Lazari-Radek, Katarzyna de, and Peter Singer. *The Point of View of the Universe: Sidgwick and Contemporary Ethics*. Oxford: Oxford University Press, 2014.

Leahy, Angela. 'Natural Law as Early Social Thought: The Recovery of Natural Law for Sociology'. *History of the Human Sciences* 33, no. 2 (2020): 72–90.

Léchot, Pierre-Olivier. 'Calvin et la connaissance naturelle de Dieu: Une relecture'. *Études théologiques et religieuses* 93, no. 2 (2018): 271–99.

Leeds, Stephen. 'Constructive Empiricism'. *Synthese* 101, no. 2 (1994): 187–221.

Lefèvre, Eckhard. 'Der Tithonus Aristons von Chios und Ciceros Cato: Von der philosophischen Theoria zur politischen Betätigung'. *Hermes* 135 (2007): 43–65.

Leftow, Brian. 'The Ontological Argument'. In *The Oxford Handbook of Philosophy of Religion*, edited by William J. Wainwright, 80–115. Oxford: Oxford University Press, 2005.

Leigh-Choate, Tova, William T. Flynn, and Margot E. Fassler, 'Hearing the Heavenly Symphony: An Overview of Hildegard's Musical Oeuvre with Case Studies'. In *A Companion to Hildegard of Bingen*, edited by Debra Stoudt, George Ferzoco, and Beverly Kienzle, 163–92. Leiden: Brill, 2014.

Leiss, William. 'Modern Science, Enlightenment, and the Domination of Nature: No Exit?' In *Critical Ecologies: The Frankfurt School and Contemporary Environmental Crises*, edited by Andrew Biro, 23–42. Toronto: University of Toronto Press, 2011.

Leitgeb, Hannes. 'Scientific Philosophy, Mathematical Philosophy, and All That'. *Metaphilosophy* 44, no. 3 (2013): 267–75.

Lennon, Kathleen. 'Imaginary Bodies and Worlds'. *Inquiry* 47, no. 2 (2004): 107–22.

Lennon, Kathleen. *Imagination and the Imaginary*. London: Routledge, 2015.

Lennox, James G. 'Nature Does Nothing in Vain'. In *Beiträge zur Antiken Philosophie: Festschrift für Wolfgang Kullmann*, edited by Hans-Christian Günther, Antonios Rengakos, and Ernst Vogt, 199–214. Stuttgart: Franz Steiner Verlag, 1997.

Lennox, James G. 'Bios and Explanatory Unity in Aristotle's Biology'. In *Definition in Greek Philosophy*, edited by David Charles, 329–58. Oxford: Oxford University Press, 2010.

Leoni, Simona Boscani. 'Conrad Gessner and a Newly Discovered Enthusiasm for Mountains in the Renaissance'. In *Conrad Gessner: Die Renaissance der Wissenschaften*, edited by Urs Leu and Peter Opitz, 119–26. Berlin: de Gruyter, 2019.

Leoni, Simona Boscani. 'A Hybrid Physico-Theology: The Case of the Swiss Confederation'. In *Physico-Theology: Religion and Science in Europe, 1650–1750*, edited by Ann Blair and Kaspar von Greyerz, 222–34. Baltimore: Johns Hopkins University Press, 2020.

Leroi, Armand Marie. *The Lagoon: How Aristotle Invented Science*. London: Bloomsbury, 2014.

Leunissen, Mariska. *Explanation and Teleology in Aristotle's Science of Nature*. Cambridge: Cambridge University Press, 2010.

Leuschner, Anna. 'Pluralism and Objectivity: Exposing and Breaking a Circle'. *Studies in History and Philosophy of Science Part A* 43, no. 1 (2012): 191–8.

Levitin, Dmitri. 'The Experimentalist as Humanist: Robert Boyle on the History of Philosophy'. *Annals of Science* 71, no. 2 (2014): 149–82.

Lewis, C. S. *Studies in Words*. Cambridge: Cambridge University Press, 1960.

Lewis, C. S. *An Experiment in Criticism*. Cambridge: Cambridge University Press, 1992.

Lewis, C. S. *The Discarded Image: An Introduction to Medieval and Renaissance Literature*. Cambridge: Cambridge University Press, 1994.

Lewis, C. S. *Essay Collection and Other Short Pieces*. London: HarperCollins, 2000.

Lewis, C. S. *The Allegory of Love: A Study in Medieval Tradition*. Cambridge: Cambridge University Press, 2013.

Licitra Rosa, Carmelo, Carla Antonucci, Alberto Siracusano, and Diego Centonze. 'From the Imaginary to Theory of the Gaze in Lacan'. *Frontiers in Psychology* 12 (2021): 578277. doi: 10.3389/fpsyg.2021.578277

Lightman, Bernard. 'Huxley and Scientific Agnosticism: The Strange History of a Failed Rhetorical Strategy'. *British Journal for the History of Science* 35, no. 3 (2002): 271–89.

Lightman, Bernard V. *Victorian Popularizers of Science: Designing Nature for New Audiences*. Chicago: University of Chicago Press, 2007.

Lightman, Bernard V., and Michael S. Reidy, eds. *The Age of Scientific Naturalism: Tyndall and His Contemporaries*. London: Pickering & Chatto, 2014.

Lim, Paul Chang-Ha. *Mystery Unveiled: The Crisis of the Trinity in Early Modern England*. New York: Oxford University Press, 2012.

Lindberg, David C. 'The Transmission of Greek and Arabic Learning to the West'. In *Science in the Middle Ages*, 52–90. Chicago: University of Chicago Press, 1978.

Lindholm, Philip. '"At the Mere Touch of Cold Philosophy": Science, Sensation and Synaesthesia in John Keats's "Lamia"'. *European Journal of English Studies* 22, no. 3 (2018): 258–72.

Liu, Ting, Liuna Geng, Lijuan Ye, and Kexin Zhou. '"Mother Nature" Enhances Connectedness to Nature and Pro-Environmental Behavior'. *Journal of Environmental Psychology* 61 (2019): 37–45.

Llewelyn, John. *Gerard Manley Hopkins and the Spell of John Duns Scotus*. Edinburgh: Edinburgh University Press, 2015.

Lloyd, G. E. R. 'Saving the Appearances'. *Classical Quarterly* 28 (1978): 202–22.

Lloyd, G. E. R. *Ancient Worlds, Modern Reflections: Philosophical Perspectives on Greek and Chinese Science and Culture*. Oxford: Clarendon Press, 2004.

Lloyd, G. E. R. *Disciplines in the Making: Cross-Cultural Perspectives on Elites, Learning, and Innovation*. Oxford: Oxford University Press, 2009.

Locke, John. *An Essay Concerning Human Understanding*. Edited by P. H. Nidditch. Oxford: Oxford University Press, 1975.

Loeb, Jacques. *The Mechanistic Conception of Life*. Cambridge, MA: Belknap Press, 1964.

Loesberg, Jonathan. 'Darwin, Natural Theology, and Slavery: A Justification of Browning's Caliban'. *ELH* 75 (2008): 871–97.

Longino, Helen E. *The Fate of Knowledge*. Princeton, NJ: Princeton University Press, 2002.

Louth, Andrew. 'Theology, Contemplation and the University'. *Studies in Christian Ethics* 17, no. 1 (2004): 69–79.

Lovejoy, Arthur O. *The Great Chain of Being: A Study of the History of an Idea*. Cambridge, MA: Harvard University Press, 1936.

Lowe, E. J. 'The Rationality of Metaphysics'. *Synthese* 178, no. 1 (2011): 99–109.

Lugg, Andrew. 'History, Discovery and Induction: Whewell on Kepler on the Orbit of Mars'. In *An Intimate Relation: Studies in the History and Philosophy of Science*, edited by James R. Brown and Jürgen Mittelstrass, 283–98. Dordrecht: Reidel, 1989.

Lund, Roger D. 'Laughing at Cripples: Ridicule, Deformity and the Argument from Design'. *Eighteenth Century Studies* 39 (2005): 91–114.

Luscombe, David. 'Crossing Philosophical Boundaries *c.* 1150–*c.* 1250'. In *Crossing Boundaries at Medieval Universities*, edited by Spencer E. Young, 7–27. Leiden: Brill, 2010.

Lüthy, Christoph. 'What to Do with Seventeenth-Century Natural Philosophy? A Taxonomic Problem'. *Perspectives on Science* 8, no. 2 (2000): 164–95.

Lynch, William T. 'A Society of Baconians? The Collective Development of Bacon's Method in the Royal Society of London'. In *Francis Bacon and the Refiguring of Early Modern Thought*, edited by J. R. Solomon and C. G. Martin, 173–202. Aldershot: Ashgate, 2005.

McAllister, James W. 'Is Beauty a Sign of Truth in Scientific Theories?' *American Scientist* 86 (1998): 174–83.

McCaskey, John P. 'Freeing Aristotelian *Epagōgē* from *Prior Analytics* II 23'. *Apeiron* 40, no. 4 (2007): 345–74.

McCosh, James. *Agnosticism of Hume and Huxley*. New York: Charles Scribner's Sons, 1884.

McDonald, Grantley. *Biblical Criticism in Early Modern Europe: Erasmus, the Johannine Comma, and Trinitarian Debate*. New York: Cambridge University Press, 2016.

McDonald, Matthew, Stephen L. Wearing, and Jess Ponting. 'The Nature of Peak Experience in Wilderness'. *The Humanistic Psychologist* 37, no. 4 (2009): 370–85.

MacDougal, Douglas W. *Newton's Gravity: An Introductory Guide to the Mechanics of the Universe*. New York: Springer, 2012.

McGilchrist, Iain. *The Master and His Emissary: The Divided Brain and the Making of the Western World*. New Haven, CT: Yale University Press, 2012.

McGinnis, Jon. *Avicenna*. Oxford: Oxford University Press, 2010.

McGinnis, Jon. 'The Eternity of the World: Proofs and Problems in Aristotle, Avicenna, and Aquinas'. *American Catholic Philosophical Quarterly* 88, no. 3 (2014): 271–88.

McGrath, Alister E. *The Intellectual Origins of the European Reformation*. Oxford: Blackwell, 2003.

McGrath, Alister E. '"Schläft ein Lied in allen Dingen"? Gedanken über die Zukunft der natürlichen Theologie'. *Theologische Zeitschrift* 65 (2009): 246–60.

McGrath, Alister E. 'Chance and Providence in the Thought of William Paley'. In *Abraham's Dice: Chance and Providence in the Monotheistic Traditions*, edited by Karl Giberson, 240–59. Oxford: Oxford University Press, 2016.

McGrath, Alister E. *Re-Imagining Nature: The Promise of a Christian Natural Theology*. Oxford: Wiley-Blackwell, 2016.

McGrath, Alister E. 'Natürliche Theologie: Ein Plädoyer für eine neue Definition und Bedeutungserweiterung'. *Neue Zeitschrift für Systematische Theologie und Religionsphilosophie* 59, no. 3 (2017): 297–310.

McGrath, Alister E. 'The Famous Stone: The Alchemical Tropes of George Herbert's "The Elixir" in their Late Renaissance Context'. *George Herbert Journal* 42, no. 1–2 (Fall 2018/Spring 2019): 114–27.

McGrath, Alister E. *The Territories of Human Reason: Science and Theology in an Age of Multiple Rationalities*. Oxford: Oxford University Press, 2019.

McGrath, Alister E. 'The Owl of Minerva: Reflections on the Theological Significance of Mary Midgley'. *Heythrop Journal* 61, no. 5 (2020): 852–64.

McGrath, Alister E. 'A Consilience of Equal Regard: Stephen Jay Gould on the Relation of Science and Religion'. *Zygon* 56, no. 3 (2021): 547–65.

McGuire, James E. 'The Rhetoric of Sprat's Defence of the Royal Society'. *Archives internationals d'histoire des sciences* 55 (2005): 203–10.

McHenry, Leemon B., ed. *Science and the Pursuit of Wisdom: Studies in the Philosophy of Nicholas Maxwell*. Frankfurt: Ontos, 2009.

MacIntosh, J. J. 'Boyle's Epistemology: The Interaction between Scientific and Religious Knowledge'. In *The Bloomsbury Companion to Robert Boyle*, edited by Jan-Erik Jones, 97–140. London: Bloomsbury, 2020.

MacIntyre, Alasdair. 'The Relationship of Philosophy to Its Past'. In *Philosophy in History*, edited by Richard Rorty, J. B. Schneewind, and Quentin Skinner, 31–48. Cambridge: Cambridge University Press, 1984.

MacIntyre, Alasdair. *After Virtue*. London: Duckworth, 1985.

MacIntyre, Alasdair. *Three Rival Versions of Moral Enquiry: Encyclopedia, Genealogy, and Tradition*. Notre Dame, IN: University of Notre Dame Press, 1990.

McKaughan, Daniel J. 'From Ugly Duckling to Swan: C. S. Peirce, Abduction, and the Pursuit of Scientific Theories'. *Transactions of the Charles S. Peirce Society* 44, no. 3 (2008): 446–68.

McKnight, Patrick E., and Todd B. Kashdan. 'Purpose in Life as a System That Creates and Sustains Health and Well-Being: An Integrative, Testable Theory'. *Review of General Psychology* 13, no. 3 (2009): 242–51.

McKnight, Stephen A. 'Religion and Francis Bacon's Scientific Utopianism'. *Zygon* 42, no. 2 (2007): 463–86.

McLeish, Tom. 2019. *The Poetry and Music of Science: Comparing Creativity in Science and Art*. Oxford: Oxford University Press, 2019.

McLeish, Tom. 'The Re-discovery of Contemplation through Science'. *Zygon* 56, no. 3 (2021): 758–76.

McMullin, Ernan. 'The Impact of Newton's *Principia* on the Philosophy of Science'. *Philosophy of Science* 68, no. 3 (2001): 279–310.

McMullin, Ernan. 'Galileo's Theological Venture'. In *The Church and Galileo*, edited by Ernan McMullin, 88–116. Notre Dame, IN: University of Notre Dame Press, 2005.

Maffie, James. 'Naturalism, Scientism and the Independence of Epistemology'. *Erkenntnis* 43, no. 1 (1995): 1–27.

Maher, Patrick. 'Prediction, Accommodation, and the Logic of Discovery'. *Philosophy of Science Association* 1 (1988): 273–85.

Mahoney, Michael. 'Changing Canons of Mathematical and Physical Intelligibility in the Later Seventeenth Century'. *Historia Mathematica* 11 (1984): 417–23.

Mahoney, Michael. 'The Mathematical Realm of Nature'. In *The Cambridge History of Seventeenth-Century Philosophy*, edited by Daniel Garber and Michael Ayers, 702–55. Cambridge: Cambridge University Press, 1998.

Malpas, Jeff E. *Place and Experience: A Philosophical Topology*. Cambridge: Cambridge University Press, 1999.

Mancosu, Paolo. *Philosophy of Mathematics and Mathematical Practice in the Seventeenth Century*. New York: Oxford University Press, 1996.

Mandelbrote, Scott. 'Eighteenth-Century Reactions to Newton's Anti-Trinitarianism'. In *Newton and Newtonianism: New Studies*, edited by J. E. Force and S. Hutton, 93–112. Dordrecht: Kluwer, 2004.

Mandelbrote, Scott. 'The Uses of Natural Theology in Seventeenth-Century England'. *Science in Context* 20 (2007): 451–80.

Mandelbrote, Scott. 'Biblical Hermeneutics and the Sciences, 1700–1900: An Overview'. In *Nature and Scripture in the Abrahamic Religions: 1700–Present*, edited by Scott H. Mandelbrote and J. M. van der Meer, 1–37. Leiden: Brill, 2008.

Mandelbrote, Scott. 'What Was Physico-Theology For?' In *Physico-Theology: Religion and Science in Europe, 1650–1750*, edited by Ann Blair and Kaspar von Greyerz, 67–77. Baltimore: Johns Hopkins University Press, 2020.

Mann, Mark H. 'Wesley and the Two Books: John Wesley, Natural Philosophy, and Christian Faith'. In *Connecting Faith and Science: Philosophical and Theological Inquiries*, edited by Matthew Nelson Hill and W. Curtis Holtzen, 11–30. Claremont, CA: Claremont Press, 2017.

Margolis, Joseph. *The Unraveling of Scientism: American Philosophy at the End of the Twentieth Century*. Ithaca, NY: Cornell University Press, 2003.

Markley, Robert. 'Objectivity as Ideology: Boyle, Newton, and the Languages of Science'. *Genre* 16 (1983): 355–72.

Marmodoro, Anna. 'Aristotle's Hylomorphism without Reconditioning'. *Philosophical Inquiry* 37, no. 1–2 (2013): 5–22.

Marmodoro, Anna. *Everything in Everything: Anaxagoras's Metaphysics*. Oxford: Oxford University Press, 2017.

Marmodoro, Anna. 'Whole, but Not One'. In *Ontology, Modality, Mind: Themes from the Metaphysics of E. J. Lowe*, edited by John Heil, Sophie Gibb, and Alex Carruth, 60–72. Oxford: Oxford University Press, 2018.

Marrow, Stanley B. '*Κόσμος* in John'. *Catholic Biblical Quarterly* 64, no. 1 (2002): 90–102.

Marshall, John. *John Locke, Toleration and Early Enlightenment Culture: Religious Intolerance and Arguments for Religious Toleration in Early Modern and 'Early Enlightenment' Europe*. Cambridge: Cambridge University Press, 2009.

Marshall, Peter H. *The Magic Circle of Rudolf II: Alchemy and Astrology in Renaissance Prague*. New York: Walker & Company, 2006.

Martens, Rhonda. *Kepler's Philosophy and the New Astronomy*. Princeton, NJ: Princeton University Press, 2000.

Martens, Rhonda. 'Harmony and Simplicity: Aesthetic Virtues and the Rise of Testability'. *Studies in History and Philosophy of Science Part A* 40, no. 3 (2009): 258–66.

Martin, Catherine Gimelli. '"What If the Sun Be Centre to the World?": Milton's Epistemology, Cosmology, and Paradise of Fools Reconsidered'. *Modern Philology* 99, no. 2 (2001): 231–65.

Martin, Craig. *Subverting Aristotle: Religion, History, and Philosophy in Early Modern Science*. Baltimore: Johns Hopkins University Press, 2014.

Marx, Karl. *Der achtzehnte Brumaire des Louis Bonaparte*. 2nd edition. Hamburg: Otto Meissner, 1869.

Mascolo, Michael F. 'Beyond Objectivity and Subjectivity: The Intersubjective Foundations of Psychological Science'. *Integrative Psychological & Behavioral Science* 50, no. 4 (2016): 543–54.

Maslow, Abraham H. *Religions, Values and Peak Experiences*. Columbus, OH: Ohio State University Press, 1964.

Massimi, Michela. 'Galileo's Mathematization of Nature at the Crossroad between the Empiricist and the Kantian Tradition'. *Perspectives on Science* 18 (2010): 152–88.

Mavrodes, George I. 'Self-Referential Incoherence'. *American Philosophical Quarterly* 22, no. 1 (1985): 65–72.

Maxwell, James Clerk. *The Scientific Papers of James Clerk Maxwell*. 2 vols. Cambridge: Cambridge University Press, 1890.

Maxwell, Nicholas. *From Knowledge to Wisdom*. London: Pentire Press, 2007.

Maxwell, Nicholas. 'In Praise of Natural Philosophy: A Revolution for Thought and Life'. *Philosophia* 40, no. 4 (2012): 705–15.

Maxwell, Nicholas. 'Popper's Paradoxical Pursuit of Natural Philosophy'. In *The Cambridge Companion to Popper*, edited by Jeremy Shearmur and Geoffrey Stoke, 170–207. Cambridge: Cambridge University Press, 2012.

Maxwell, Nicholas. *In Praise of Natural Philosophy: A Revolution for Thought and Life*. Montréal: McGill-Queen's University Press, 2017.

Mayer, Claude-Hélène, Rian Viviers, Aden-Paul Flotman, and Detlef Schneider-Stengel. 'Enhancing Sense of Coherence and Mindfulness in an Ecclesiastical, Intercultural Group'. *Journal of Religion and Health* 55, no. 6 (2016): 2023–38.

Mayne, Michael. *This Sunrise of Wonder: Letters for the Journey*. London: Darton, Longman, and Todd, 2008.
Mayr, Ernst. *The Growth of Biological Thought*. Cambridge, MA: Belknap Press, 1982.
Mensch, Jennifer. 'Intuition and Nature in Kant and Goethe'. *European Journal of Philosophy* 19, no. 3 (2011): 431–53.
Merchant, Carolyn. '"The Violence of Impediments": Francis Bacon and the Origins of Experimentation'. *Isis* 99 (2008): 731–60.
Merkur, Daniel. 'The Study of Spiritual Alchemy: Mysticism, Gold-Making, and Esoteric Hermeneutics'. *Ambix* 37 (1990): 35–45.
Merleau-Ponty, Jacques. *Philosophie et théorie physique chez Eddington*. Paris: Les Belles Lettres, 1965.
Methuen, Charlotte. 'This Comet or New Star: Theology and the Interpretation of the Nova of 1572'. *Perspectives on Science* 5 (1997): 499–509.
Methuen, Charlotte. *Kepler's Tübingen: Stimulus to a Theological Mathematics*. Aldershot: Ashgate, 1998.
Metz, Thaddeus. 'The Good, the True, and the Beautiful: Toward a Unified Account of Great Meaning in Life'. *Religious Studies* 47 (2011): 389–409.
Metz, Thaddeus. *Meaning in Life*. Oxford: Oxford University Press, 2013.
Mews, Constant J. 'The World as Text: The Bible and the Book of Nature in Twelfth-Century Theology'. In *Scripture and Pluralism: Reading the Bible in the Religiously Plural Worlds of the Middle Ages and Renaissance*, edited by Thomas J. Heffernan and Thomas E. Burman, 95–122. Leiden: Brill, 2005.
Middleton, J. Richard. *The Liberating Image: The Imago Dei in Genesis 1*. Grand Rapids, MI: Brazos Press, 2005.
Midgley, Mary. *Animals and Why They Matter*. Harmondsworth: Penguin, 1983.
Midgley, Mary. 'De-Dramatizing Darwin'. *The Monist* 67, no. 2 (1984): 200–15.
Midgley, Mary. *Science as Salvation: A Modern Myth and Its Meaning*. London: Routledge, 1992.
Midgley, Mary. 'One World but a Big One'. *Journal of Consciousness Studies* 3, no. 5–6 (1996): 500–14.
Midgley, Mary. 'A Well-Meaning Cannibal [review of *Consilience* by Edward O. Wilson]'. *Commonweal* 125, no. 13 (1998): 23–4.
Midgley, Mary. *Science and Poetry*. London: Routledge, 2001.
Midgley, Mary. *Beast and Man: The Roots of Human Nature*. Revised edition. London: Routledge, 2002.
Midgley, Mary. 'Pluralism: The Many Maps Model'. *Philosophy Now* 35 (2002): 10–11.
Midgley, Mary. *The Myths We Live By*. London: Routledge, 2004.
Midgley, Mary. 'Mapping Science: In Memory of John Ziman'. *Interdisciplinary Science Reviews* 30, no. 3 (2005): 195–7.
Midgley, Mary. 'Dover Beach: Understanding the Pains of Bereavement'. *Philosophy* 81, no. 316 (2006): 209–30.
Midgley, Mary. *Are You an Illusion?* Durham: Acumen, 2014.
Midgley, Mary. *What Is Philosophy For?* London: Bloomsbury Academic, 2018.
Mieg, Harald A., and Julia Evetts. 'Professionalism, Science, and Expert Roles: A Social Perspective'. In *The Cambridge Handbook of Expertise and Expert Performance*, edited by K. A. Ericsson, R. R. Hoffman, A. Kozbelt, and A. M. William, 127–48. Cambridge: Cambridge University Press, 2008.
Miller, Arthur I. 'Cultures of Creativity: Mathematics and Physics'. *Diogenes* 45, no. 177 (1997): 53–72.

Miller, David M. '*O male factum*: Rectilinearity and Kepler's Discovery of the Ellipse'. *Journal for the History of Astronomy* 39, no. 1 (2008): 43–63.

Miller, Henry. *On Writing*. New York: New Directions, 1964.

Miller, Jerome. *In the Throe of Wonder: Intimations of the Sacred in a Post-Modern World*. Albany, NY: State University of New York Press, 1992.

Miller, Jon, ed. *The Reception of Aristotle's Ethics*. Cambridge: Cambridge University Press, 2012.

Minkov, Svetozar. 'Baconian Science and the Intelligibility of Human Experience: The Case of Love'. *Review of Politics* 71, no. 3 (2009): 389–410.

Minkowski, Hermann. 'Raum und Zeit'. *Jahresbericht der deutschen Mathematiker-Vereinigung* 18 (1909): 75–88.

Mitcham, Carl, and Nan Wang. 'Interdisciplinarity in Ethics'. In *The Oxford Handbook of Interdisciplinarity*, edited by Robert Frodeman, Julie Thompson Klein, and Roberto C. S. Pacheco, 241–54. Oxford: Oxford University Press, 2017.

Mitchell, S. D., and M. R. Dietrich. 'Integration without Unification: An Argument for Pluralism in the Biological Sciences'. *American Naturalist* 168 (2006): S73–S79.

Miteva, Evelina. 'Intellect, Natural Philosophy, Finality: Albertus Magnus' Attempt at a Universal System of Sciences'. *Philobiblon* 22, no. 2 (2017): 37–50.

Mittelstrass, Jürgen. *Theoria: Chapters in the Philosophy of Science*. Berlin: de Gruyter, 2018.

Mitter, Rana. *A Bitter Revolution: China's Struggle with the Modern World*. Oxford: Oxford University Press, 2004.

Modrak, D. K. W. 'Aristotle on the Difference between Mathematics and Physics and First Philosophy'. *Apeiron* 22, no. 4 (1989): 121–40.

Montiglio, Silvia. 'The (Cultural) Harmony of Nature: Music, Love, and Order in "Daphnis and Chloe"'. *Transactions of the American Philological Association* 142, no. 1 (2012): 133–56.

Moody, Todd C. 'Progress in Philosophy'. *American Philosophical Quarterly* 23, no. 1 (1986): 35–46.

Moravcsik, Julius M. 'What Makes Reality Intelligible? Reflections on Aristotle's Theory of Aitia'. In *Aristotle's Physics: A Collection of Essays*, edited by Lindsay Judson, 31–48. Oxford: Clarendon Press, 1991.

Morejón, Gil. 'Differentiation and Distinction: On the Problem of Individuation from Scotus to Deleuze'. *Deleuze and Guattari Studies* 12, no. 3 (2018): 353–73.

Morello, Gustavo. 'Charles Taylor's "Imaginary" and "Best Account" in Latin America'. *Philosophy & Social Criticism* 33, no. 5 (2007): 617–39.

Morgan, Kathryn A. *Myth and Philosophy from the Pre-Socratics to Plato*. Cambridge: Cambridge University Press, 2000.

Morison, Benjamin. *On Location: Aristotle's Concept of Place*. Oxford: Clarendon Press, 2002.

Morrison, Margaret. *Unifying Scientific Theories: Physical Concepts and Mathematical Structures*. Cambridge: Cambridge University Press, 2000.

Morrisson, Mark S. *Modern Alchemy: Occultism and the Emergence of the Atomic Theory*. Oxford: Oxford University Press, 2007.

Morse, Marston. 'Mathematics and the Arts'. In *Musings of the Masters: An Anthology of Mathematical Reflections*, edited by Raymond G. Ayoub, 81–96. Washington DC: Mathematical Association of America, 2004.

Moss, Jean Dietz. 'Rhetoric and Science'. In *The Oxford Handbook of Rhetorical Studies*, edited by Michael J. MacDonald, 423–36. Oxford: Oxford University Press, 2017.

Motion, Andrew. *Keats*. London: Faber and Faber, 1998.Moulin, Isabelle. 'Beauty as Natural Order: The Legacy of Antiquity to Bonaventure's Symbolical Theology and

Nicholas of Cusa's Spiritual Theophany'. *Studies in History and Philosophy of Science Part A* 81 (2020): 32–8.
Moyer, Ann E. 'Music, Mathematics, and Aesthetics: The Case of the Visual Arts in the Renaissance'. In *Music and Mathematics in Late Medieval and Early Modern Europe*, edited by Philippe Vendrix, 111–46. Turnhout: Brepols, 2008.
Mueller, Olaf L. 'Prismatic Equivalence: A New Case of Underdetermination – Goethe vs. Newton on the Prism Experiments'. *British Journal for the History of Philosophy* 24, no. 2 (2016): 323–47.
Müller, Götz. *Jean Pauls Ästhetik und Naturphilosophie*. Berlin: Springer, 1983.
Münch, Richard. 'Max Webers These der Entzauberung der Welt: Versuch einer Verteidigung'. *Soziologische Revue* 42, no. 2 (2019): 177–87.
Murdoch, Iris. *The Sovereignty of Good*. London: Macmillan, 1970.
Murdoch, Iris. *Metaphysics as a Guide to Morals*. London: Penguin, 1992.
Murphy, Kathryn, and Anita Traninger, eds. *The Emergence of Impartiality*. Leiden: Brill, 2014.
Naddaf, Gerard. 'Plato: The Creator of Natural Theology'. *International Studies in Philosophy* 36, no. 1 (2004): 103–27.
Naddaf, Gerard. *The Greek Concept of Nature*. Albany, NY: State University of New York Press, 2005.
Nagel, Fritz. 'Scientia Experimentalis: Zur Cusanus-Rezeption in England'. *Mitteilungen und Forschungsbeiträge der Cusanus-Gesellschaft* 29 (2005): 95–109.
Nagel, Thomas. *The View from Nowhere*. New York: Oxford University Press, 1986.
Nagel, Thomas. *Mortal Questions*. Cambridge: Cambridge University Press, 2012.
Naour, Paul. *E.O. Wilson and B.F. Skinner: A Dialogue between Sociobiology and Radical Behaviorism*. New York: Springer, 2009.
Nassar, Dalia. 'From a Philosophy of Self to a Philosophy of Nature: Goethe and the Development of Schelling's *Naturphilosophie*'. *Archiv für Geschichte der Philosophie* 92, no. 3 (2010): 304–21.
Nauenberg, Michael. 'Robert Hooke's Seminal Contribution to Orbital Dynamics'. *Physics in Perspective* 7, no. 1 (2005): 4–34.
Naylor, Ron. 'Galileo, Copernicanism and the Origins of the New Science of Motion'. *British Journal for the History of Science* 36, no. 2 (2003): 151–81.
Nehamas, Alexander. *The Art of Living: Socratic Reflections from Plato to Foucault*. Berkeley, CA: University of California Press, 1998.
Nelson, Eric S. 'Revisiting the Dialectic of Environment: Nature as Ideology and Ethics in Adorno and the Frankfurt School'. *Telos*, no. 155 (2011): 105–26.
Nestle, Wilhelm. *Vom Mythos zum Logos: Die Selbstentfaltung des griechischen Denkens von Homer bis auf die Sophistik und Sokrates*. 2nd edition. Stuttgart: Kröner, 1942.
Neumann, Hanns-Peter. *Natura Sagax—Die geistige Natur: Zum Zusammenhang von Naturphilosophie und Mystik in der Frühen Neuzeit am Beispiel Johann Arndts*. Berlin: de Gruyter, 2004.
Neves, J. C. S. 'Einstein contra Aristotle: The Sound from the Heavens'. *Physics Essays* 30, no. 3 (2017): 279–80.
Newman, William R. *Promethean Ambitions: Alchemy and the Quest to Perfect Nature*. Chicago: University of Chicago Press, 2004.
Newman, William R. 'Mercury and Sulphur among the High Medieval Alchemists: From Rāzī and Avicenna to Albertus Magnus and Pseudo-Roger Bacon'. *Ambix* 61, no. 4 (2014): 327–44.

Newman, William R. *Newton the Alchemist: Science, Enigma, and the Quest for Nature's 'Secret Fire'*. Princeton, NJ: Princeton University Press, 2019.
Newton, Isaac. 'An Account of the Book Entituled *Commercium Epistolicum*'. *Philosophical Transactions of the Royal Society* 29 (1714–15): 173–224.
Newton, Isaac. *The Correspondence of Isaac Newton*. Edited by H. W. Turnbull. Cambridge: Cambridge University Press, 1959–78.
Newton-Smith, W. H. 'Explanation'. In *A Companion to the Philosophy of Science*, edited by W. H. Newton-Smith, 127–33. Oxford: Blackwell, 2000.
Nichols, Ryan. 'Natural Philosophy and Its Limits in the Scottish Enlightenment'. *The Monist* 90, no. 2 (2007): 233–50.
Nightingale, Andrea Wilson. 'On Wondering and Wandering: Theoria in Greek Philosophy and Culture'. *Arion: A Journal of Humanities and the Classics* 9 (2001): 111–46.
Nightingale, Andrea Wilson. *Spectacles of Truth in Classical Greek Philosophy: Theoria in Its Cultural Context*. Cambridge: Cambridge University Press, 2004.
Niiniluoto, Ilkka. 'Hintikka and Whewell on Aristotelian Induction'. *Grazer Philosophische Studien* 49 (1994): 40–61.
Nola, Robert. 'Darwin's Arguments in Favour of Natural Selection and against Special Creationism'. *Science & Education* 22, no. 3 (2013): 149–71.
North, Gerald. *Mastering Astronomy*. Basingstoke: Macmillan Education, 1988.
Nuccetelli, Susana, and Gary Seay. 'Reasoning, Normativity, and Experimental Philosophy'. *American Philosophical Quarterly* 49, no. 2 (2012): 151–63.
Nummedal, Tara. 'Alchemy and Religion in Christian Europe'. *Ambix* 60, no. 4 (2014): 311–22.
Nussbaum, Martha C. *The Fragility of Goodness: Luck and Ethics in Greek Tragedy and Philosophy*. Cambridge: Cambridge University Press, 1986.
Nussbaum, Martha C. *Love's Knowledge: Essays on Philosophy and Literature*. Oxford: Oxford University Press, 1990.
O'Beirne, Emer. 'Mapping the Non-Lieu in Marc Augé's Writings'. *Forum for Modern Language Studies* 42, no. 1 (2006): 38–50.
O'Gorman, Francis. 'Religion'. In *Cambridge Companion to John Ruskin*, edited by Francis O'Gorman, 144–56. Cambridge: Cambridge University Press, 2015.
Ochs, Elinor. 'Transcriptions as Theory'. In *Developmental Pragmatics*, edited by Elinor Ochs and Bambi B. Schieffelin, 43–72. New York: Academic Press, 1979.
Oddie, Graham. 'Non-Naturalist Moral Realism, Autonomy and Entanglement'. *Topoi* 37, no. 4 (2018): 607–20.
Odom, Herbert H. 'The Estrangement of Celestial Mechanics and Religion'. *Journal of the History of Ideas* 27 (1966): 533–58.
Ogilvie, Brian W. 'The Many Books of Nature: Renaissance Naturalists and Information Overload'. *Journal of the History of Ideas* 64 (2003): 29–40.
Ogilvie, Brian W. 'Natural History, Ethics, and Physico-Theology'. In *Historia: Empiricism and Erudition in Early Modern Europe*, edited by Gianna Pomata, 75–104. Cambridge, MA: MIT Press, 2005.
Ogilvie, Brian W. *The Science of Describing: Natural History in Renaissance Europe*. Chicago: University of Chicago Press, 2006.
Oldstone-Moore, Jennifer. 'Scientism and Modern Confucianism'. In *The Sage Returns: Confucian Revival in Contemporary China*, edited by Kenneth J. Hammond and Jeffrey L. Richey, 39–63. Albany, NY: State University of New York Press, 2015.
Olson, Richard. *Science and Scientism in Nineteenth-Century Europe*. Urbana, IL: University of Illinois Press, 2007.

Omodeo, Pietro Daniel. *Copernicus in the Cultural Debates of the Renaissance: Reception, Legacy, Transformation*. Leiden: Brill, 2014.

Oppenheim, Peter, and Hilary Putnam. 'The Unity of Science as a Working Hypothesis'. In *Minnesota Studies in the Philosophy of Science*, vol. 2, edited by Herbert Feigl, 3–36. Minneapolis: University of Minnesota Press, 1958.

Orange, A. D. 'The Origins of the British Association for the Advancement of Science'. *British Journal for the History of Science* 6, no. 3 (1972): 152–76.

Orringer, Nelson R. 'Ortega y Gasset's Sportive Vision of Plato'. *MLN* 88, no. 2 (1973): 264–80.Osbeck, Lisa M., and Barbara S. Held. 'Introduction'. In *Rational Intuition: Philosophical Roots, Scientific Investigations*, edited by Lisa M. Osbeck and Barbara S. Held, 1–36. Cambridge: Cambridge University Press, 2014.

Osler, Margaret J. 'Mixing Metaphors: Science and Religion or Natural Philosophy and Theology in Early Modern Europe'. *History of Science* 36 (1998): 91–113.

Osler, Margaret J. 'The Canonical Imperative: Rethinking the Scientific Revolution'. In *Rethinking the Scientific Revolution*, edited by Margaret J. Osler, 3–22. Cambridge: Cambridge University Press, 2000.

Ouyang, Guangwei. 'Scientism, Technocracy, and Morality in China'. *Journal of Chinese Philosophy* 30, no. 2 (2003): 177–93.

Owens, Thomas. '*Nature*'s Motto: Wordsworth and the Macmillans'. *Notes and Queries* 62, no. 3 (2015): 430–5.

Padian, Kevin. 'Charles Darwin's Views of Classification in Theory and Practice'. *Systematic Biology* 48, no. 2 (1999): 352–64.

Page, Sophie. 'Medieval Magic'. In *The Oxford Illustrated History of Witchcraft and Magic*, edited by Owen Davies, 29–64. Oxford: Oxford University Press, 2017.

Pais, Abraham. '*Subtle Is the Lord…*': *The Science and the Life of Albert Einstein*. Oxford: Oxford University Press, 2005.

Palmerino, Carla Rita. 'The Mathematical Characters of Galileo's Book of Nature'. In *The Book of Nature in Early Modern and Modern History*, edited by Klaas van Berkel and Arie Johan Vanderjagt, 27–44. Louvain: Peeters, 2006.

Palmerino, Carla Rita. 'Reading the Book of Nature: The Ontological and Epistemological Underpinnings of Galileo's Mathematical Realism'. In *The Language of Nature: Reassessing the Mathematization of Natural Philosophy in the 17th Century*, edited by Geoffrey Gorham, Benjamin Hill, Edward Slowik, and C. Kenneth Waters, 29–50. Minneapolis: University of Minnesota Press, 2016.

Panayides, Christos. 'A Note on Aristotelian First Principles'. *Hermathena* 184 (2008): 19–51.

Paradis, James G. *T. H. Huxley: Man's Place in Nature*. Lincoln: University of Nebraska Press, 1978.

Paradis, James. '*Evolution and Ethics* in Its Victorian Context'. In *Evolution and Ethics: T. H. Huxley's Evolution and Ethics in Its Victorian and Sociobiological Context*, edited by James Paradis and George C. Williams, 3–55. Princeton, NJ: Princeton University Press, 1989.

Park, Crystal L. 'Religion as a Meaning-Making Framework in Coping with Life Stress'. *Journal of Social Issues* 61, no. 4 (2005): 707–29.

Park, Katharine. 'Nature in Person: Medieval and Renaissance Allegories and Emblems'. In *The Moral Authority of Nature*, edited by Lorraine Daston and Fernando Vidal. Chicago: University of Chicago Press, 2004.

Parrish, Stephen E. *The Knower and the Known: Physicalism, Dualism, and the Nature of Intelligibility*. South Bend, IN: St Augustine's Press, 2013.

Parshall, Karen H., Michael T. Walton, and Bruce T. Moran, eds. *Bridging Traditions: Alchemy, Chemistry, and Paracelsian Practices in the Early Modern Era*. Kirksville, MO: Truman State University Press.

Pasachoff, Jay M. 'Simon Marius's *Mundus Iovialis* and the Discovery of the Moons of Jupiter'. In *Simon Marius and His Research*, edited by Hans Gaab and Pierre Leich, 191–204. Cham: Springer, 2018.

Pastorino, Cesare. 'Weighing Experience: Experimental Histories and Francis Bacon's Quantitative Program'. *Early Science and Medicine* 16, no. 6 (2011): 542–70.

Paton, Rob. 'The Mutability of Time and Space as a Means of Healing History in an Australian Aboriginal Community'. In *Long History, Deep Time: Deepening Histories of Place*, edited by Ann McGrath and Mary Anne Jebb, 67–82. Acton: ANU Press, 2015.

Peckhaus, Volker. *Logik, Mathesis Universalis und allgemeine Wissenschaft: Leibniz und Die Wiederentdeckung der formalen Logik im 19. Jahrhundert*. Berlin: Akademie-Verlag, 1997.

Pennington, Kenneth. '*Lex Naturalis* and *Ius Naturale*'. In *Crossing Boundaries at Medieval Universities*, edited by Spencer E. Young, 227–53. Leiden: Brill, 2010.

Penrose, Roger. *The Road to Reality: A Complete Guide to the Laws of the Universe*. London: Jonathan Cape, 2004.

Pérez-Ramos, Antonio. *Francis Bacon's Idea of Science and the Maker's Knowledge Tradition*. Oxford: Clarendon Press, 1988.

Perloff, Marjorie. *Wittgenstein's Ladder: Poetic Language and the Strangeness of the Ordinary*. Chicago: University of Chicago Press, 1996.

Perovic, Slobodan. 'Emergence of Complementarity and the Baconian Roots of Niels Bohr's Method'. *Studies in History and Philosophy of Modern Physics* 44, no. 3 (2013): 162–73.

Peruzzi, Alberto. 'Hartmann's Stratified Reality'. *Axiomathes* 21 (2001): 227–60.

Pesic, Peter. 'Desire, Science, and Polity: Francis Bacon's Account of Eros'. *Interpretation* 26, no. 3 (1999): 333–52.

Pesic, Peter. 'Proteus Unbound: Francis Bacon's Successors and the Defense of Experiment'. *Studies in Philology* 98, no. 4 (2001): 428–56.

Pesic, Peter. 'Proteus Rebound: Reconsidering the Torture of Nature'. *Isis* 99, no. 2 (2008): 304–17.

Pesic, Peter. *Music and the Making of Modern Science*. Cambridge, MA: MIT Press, 2014.

Peteet, Julie. 'Words as Interventions: Naming in the Palestine–Israel Conflict'. *Third World Quarterly* 26, no. 1 (2005): 153–72.

Peterfreund, Stuart. 'Imagination at a Distance: Bacon's Epistemological Double-Bind, Natural Theology, and the Way of Scientific Discourse in the Seventeenth and Eighteenth Centuries'. *The Eighteenth Century: Theory and Interpretation* 21 (2000): 110–40.

Peterfreund, Stuart. 'Robert Browning's Decoding of Natural Theology in "Caliban Upon Setebos"'. *Victorian Poetry* 43 (2005): 317–31.

Peterson, Anne S. 'Matter in Biology: An Aristotelian Metaphysics for Contemporary Homology'. *American Catholic Philosophical Quarterly* 92, no. 2 (2018): 353–71.

Peterson, Keith R. 'Nicolai Hartmann's Philosophy of Nature: Realist Ontology and Philosophical Anthropology'. *Scripta Philosophiae Naturalis* 2 (2012): 143–79.

Peterson, Keith R. 'Stratification, Dependence, and Non-Anthropocentrism: Nicolai Hartmann's Critical Ontology'. In *Ontologies of Nature: Continental Perspectives and Environmental Reorientations*, edited by Gerard Kuperus and Marjolein Oele, 159–80. Cham: Springer, 2017.

Peterson, Mark A. *Galileo's Muse: Renaissance Mathematics and the Arts*. Cambridge, MA: Harvard University Press, 2011.

Petitot, Jean, Francisco J. Varela, Bernard Pachoud, and Jean-Michel Roy. 'Beyond the Gap: An Introduction to Naturalizing Phenomenology'. In *Naturalizing Phenomenology: Issues in Contemporary Phenomenology and Cognitive Science*, edited by Jean-Michel Roy, Jean Petitot, Bernard Pachoud, and Francisco J. Varela, 1–79. Stanford, CA: Stanford University Press, 1999.

Petrarca, Francesco. *Invectives*. Translated by David Marsh. Cambridge, MA: Harvard University Press, 2003.

Phillips, Ian, ed. *The Routledge Handbook of Philosophy of Temporal Experience*. London: Routledge, 2017.

Piccolino, Marco, and Marco Bresadola. *Rane, Torpedini e scintilla: Galvani, Volta e l'elettricità animale*. Turin: Bollati-Boringhieri, 2003.

Pickering, David. 'New Directions in Natural Theology'. *Theology* 124, no. 5 (2021): 349–57.

Pietraß, Manuela. 'Towards Systematicity: Comparing from the Perspective of Philosophy of Science'. *Research in Comparative and International Education* 12, no. 3 (2017): 276–88.

Pigden, Charles R. 'Logic and the Autonomy of Ethics'. *Australasian Journal of Philosophy* 67, no. 2 (1989): 127–51.

Pigden, Charles R., ed. *Hume on 'Is' and 'Ought'*. Basingstoke: Palgrave Macmillan, 2010.

Pigliucci, Massimo. 'The Borderlands between Science and Philosophy: An Introduction'. *Quarterly Review of Biology* 83, no. 1 (2008): 7–15.

Pigliucci, Massimo. 'The Demarcation Problem: A (Belated) Response to Laudan'. In *Philosophy of Pseudoscience: Reconsidering the Demarcation Problem*, edited by Massimo Pigliucci and Maarten Boudry, 9–28. Chicago: University of Chicago Press, 2013.

Pigliucci, Massimo. 'New Atheism and the Scientistic Turn in the Atheism Movement'. *Midwest Studies in Philosophy* 37, no. 1 (2013): 142–53.

Pigliucci, Massimo. 'The Limits of Consilience and the Problem of Scientism'. In *Darwin's Bridge: Uniting the Humanities and Sciences*, edited by Joseph Carroll, Dan P. McAdams, and Edward O. Wilson, 247–64. Oxford: Oxford University Press, 2016.

Planck, Max. *Acht Vorlesungen über theoretische Physik*. Leipzig: Hirzel, 1910.

Plantinga, Alvin. *Where the Conflict Really Lies: Science, Religion, and Naturalism*. New York: Oxford University Press, 2011.

Plotnitsky, Arkady. 'On the Reasonable and Unreasonable Effectiveness of Mathematics in Classical and Quantum Physics'. *Foundations of Physics* 41, no. 3 (2011): 466–91.

Pocock, J. G. A. 'Historiography and Enlightenment: A View of Their History'. *Modern Intellectual History* 5 (2008): 83–9.

Poirel, Dominique. *Livre de la nature et débat trinitaire au XIIe siècle: Le 'De Tribus Diebus' de Hugues de Saint-Victor*. Turnhout: Brepols, 2002.

Pollock, Jacob. 'The Voyage Account, the Royal Society and Textual Production, 1687–1707'. *Studies in Travel Writing* 17, no. 3 (2013): 281–99.

Pollock, John L. *Contemporary Theories of Knowledge*. Totowa, NJ: Rowman & Littlefield, 1986.

Ponzio, Paolo. 'Perera, Bellarmino, Galileo e il "Concordismo" tra Sacre Scritture e ricerca scientifica'. *Quaestio* 14 (2014): 257–69.

Pope, Alexander. *An Essay on Man*. London: Knapton, 1745.

Popper, Karl. 'Epistemology without a Knowing Subject'. *Studies in Logic and the Foundations of Mathematics* 52 (1968): 333–73.

Popper, Karl R. 'Natural Selection and the Emergence of Mind'. *Dialectica* 32 (1978): 339–55.

Porter, Dahlia. *Science, Form, and the Problem of Induction in British Romanticism*. Cambridge: Cambridge University Press, 2020.

Portuondo, María M. *The Spanish Disquiet: The Biblical Natural Philosophy of Benito Arias Montano*. Chicago: University of Chicago Press, 2019.

Poulakos, John, and Nathan Crick. 'There Is Beauty Here, Too: Aristotle's Rhetoric for Science'. *Philosophy & Rhetoric* 45, no. 3 (2012): 295–311.

Pozzo, Riccardo. 'Wissenschaft und Reformation: Die Beispiele der Universitäten Königsberg und Helmstedt'. *Berichte zur Wissenschaftsgeschichte* 18, no. 2 (1995): 103–13.

Preston, Claire. *Thomas Browne and the Writing of Early Modern Science*. Cambridge: Cambridge University Press, 2005.

Prigogine, Ilya, and Isabelle Stengers. *Order out of Chaos: Man's New Dialogue with Nature*. London: Heinemann, 1984.

Principe, Lawrence M. *The Aspiring Adept: Robert Boyle and His Alchemical Quest*. Princeton, NJ: Princeton University Press, 1998.

Principe, Lawrence M. 'Transmuting Chymistry into Chemistry: Eighteenth-Century Chrysopoeia and Its Repudiation'. In *Neighbours and Territories: The Evolving Identity of Chemistry*, edited by José Ramón Bertomeu-Sánchez, Duncan Thorburn Burns, and Brigitte van Tiggelen, 21–34. Louvain-la-Neuve: Mémosciences, 2008.

Prior, Arthur N. 'The Autonomy of Ethics'. *Australasian Journal of Philosophy* 38, no. 3 (1960): 199–206.

Proust, Marcel. *La prisonnière*. Paris: Gallimard, 1925.

Puig, Jaume de. *La filosofia de Ramon Sibiuda*. Barcelona: Institut d'Estudis Catalans, 1997.

Pumfrey, Stephen. '"Your Astronomers and Ours Differ Exceedingly": The Controversy over the "New Star" of 1572 in the Light of a Newly Discovered Text by Thomas Digges'. *British Journal for the History of Science* 44, no. 1 (2011): 29–60.

Putnam, Hilary. *The Collapse of the Fact/Value Dichotomy and Other Essays*. Cambridge, MA: Harvard University Press, 2002.

Putnam, Ruth Anna. 'Varieties of Experience and Pluralities of Perspective'. In *Pragmatism as a Way of Life: The Lasting Legacy of William James and John Dewey*, edited by Hilary Putnam and Ruth Anna Putnam, 232–47. Cambridge, MA: Harvard University Press, 2017.

Qiu, Peipei. 'Onitsura's Makoto and the Daoist Concept of the Natural'. *Philosophy East and West* 51, no. 2 (2001): 232–46.

Rabassó, Georgina. '*In Caelesti Gaudio*: Hildegard of Bingen's Auditory Contemplation of the Universe'. *Quaestio* 15 (2015): 393–401.

Rabouin, David. 'Interpretations of Leibniz's Mathesis Universalis at the Beginning of the XXth Century'. In *New Essays on Leibniz Reception*, edited by Ralf Krömer and Yannick Chin-Drian, 187–201. Basel: Springer, 2012.

Radice, Roberto. 'Ordine musica bellezza in Agostino'. *Rivista di Filosofia Neo-Scolastica* 84, no. 4 (1992): 587–607.

Ramachandran, Ayesha. *The Worldmakers: Global Imagining in Early Modern Europe*. Chicago: University of Chicago Press, 2015.

Ramos, Alice M. *Dynamic Transcendentals: Truth, Goodness, and Beauty from a Thomistic Perspective*. Washington, DC: Catholic University of America Press, 2012.

Ratcliffe, Matthew. 'Husserl and Nagel on Subjectivity and the Limits of Physical Objectivity'. *Continental Philosophy Review* 35, no. 4 (2002): 353–77.

Rausch, Hannelore. *Theoria: Von ihrer sakralen zur philosophischen Bedeutung*. Munich: Fink, 1982.

Rawlins, Dennis. 'A Long Lost Observation of Uranus: Flamsteed 1714'. *Publications of the Astronomical Society of the Pacific* 80 (1968): 217–19.

Re Manning, Russell. *God's Scientists. Science, Religion, and the Spectre of Atheism in Restoration England. The Boyle Lectures, 1692–1732*. London: Pickering & Chatto, 2015.

Re Manning, Russell, John Hedley Brooke, and Fraser N. Watts, eds. *The Oxford Handbook of Natural Theology*. Oxford: Oxford University Press, 2013.
Rea, Michael C. 'Hylomorphism Reconditioned'. *Philosophical Perspectives* 25 (2011): 341–58.
Reck, Erich H., ed. *The Historical Turn in Analytic Philosophy*. Basingstoke: Palgrave Macmillan, 2013.
Reeve, C. D. C. *Action, Contemplation, and Happiness: An Essay on Aristotle*. Cambridge, MA: Harvard University Press, 2012.
Reeves, Eileen Adair. *Galileo's Glassworks: The Telescope and the Mirror*. Cambridge, MA: Harvard University Press, 2008.
Reeves, Eileen. 'From Dante's Moonspots to Galileo's Sunspots'. *MLN* 124 (2009): S190–S209.
Reeves, Josh A. 'The Field of Science and Religion as Natural Philosophy'. *Theology and Science* 6, no. 4 (2008): 403–19.
Regier, Jonathan. 'Kepler's Theory of Force and His Medical Sources'. *Early Science and Medicine* 19, no. 1 (2014): 1–27.
Reinhardt, Carsten. 'Disciplines, Research Fields, and Their Boundaries'. In *Chemical Sciences in the 20th Century: Bridging Boundaries*, edited by Carsten Reinhardt, 1–13. Chichester: Wiley, 2001.
Reisch, George. 'How Postmodern Was Neurath's Idea of Unity of Science?' *Studies in History and Philosophy of Science* 28, no. 3 (1997): 439–51.
Repellini, Ferruccio Franco. 'Platone e la salvezza dei fenomeni'. *Rivista di Storia della Filosofia* 44, no. 3 (1989): 419–42.
Rescher, Nicholas. 'Some Issues Regarding the Completeness of Science and the Limits of Scientific Knowledge'. In *The Structure and Development of Science*, edited by G. Radnitzky and G. Andersson, 19–40. Dordrecht: D. Reidel, 1979.
Rescher, Nicholas. *Objectivity: The Obligations of Impersonal Reason*. Notre Dame, IN: University of Notre Dame Press, 1997.
Rescher, Nicholas. *Nature and Understanding: The Metaphysics and Method of Science*. Oxford: Oxford University Press, 2003.
Reutlinger, Alexander. 'Explanation Beyond Causation? New Directions in the Philosophy of Scientific Explanation'. *Philosophy Compass* 12, no. 2 (2017).
Richards, Robert J. 'Darwin's Theory of Natural Selection and Its Moral Purpose'. In *The Cambridge Companion to the 'Origin of Species'*, edited by Michael Ruse and Robert J. Richards, 47–66. Cambridge: Cambridge University Press, 2009.
Ridder, Jeroen de, Rik Peels, and René van Woudenberg, eds. *Scientism: Prospects and Problems*. New York: Oxford University Press, 2018.
Ritchey, Sara. 'Rethinking the Twelfth-Century Discovery of Nature'. *Journal of Medieval and Early Modern Studies* 39, no. 2 (2009): 225–55.
Robichaud, Denis J.-J. 'Ficino on Force, Magic, and Prayers: Neoplatonic and Hermetic Influences in Ficino's Three Books on Life'. *Renaissance Quarterly* 70, no. 1 (2017): 44–87.
Robinson, Timothy A. 'Getting It All Together: The Fragmentation of the Disciplines and the Unity of Knowledge'. *Headwaters* 25 (2008): 102–14.
Rochberg, Francesca. 'Reasoning, Representing, and Modeling in Babylonian Astronomy'. *Journal of Ancient Near Eastern History* 5, no. 1 (2018): 131–47.
Rocke, Alan J. 'In Search of El Dorado: John Dalton and the Origins of the Atomic Theory'. *Social Research* 72, no. 1 (2005): 125–58.
Rockwood, Nathan. 'Locke on Empirical Knowledge'. *History of Philosophy Quarterly* 35, no. 4 (2018): 317–36.

Rodrigues, Cassiano Terra. 'The Method of Scientific Discovery in Peirce's Philosophy: Deduction, Induction, and Abduction'. *Logica Universalis* 5 (2011): 127–64.

Rodríguez-García, José María. 'Scientia potestas est—Knowledge Is Power: Francis Bacon to Michel Foucault'. *Neohelicon* 28 (2001): 109–22.

Roelants, Nienke W. J. 'The Physical Status of Astronomical Models before the 1570s: The Curious Case of Lutheran Astronomer Georg Joachim Rheticus'. *Theology and Science* 10, no. 4 (2012): 367–90.

Rogers, G. A. J. 'Stillingfleet, Locke and the Trinity'. In *Judaeo-Christian Intellectual Culture in the Seventeenth Century: A Celebration of the Library of Narcissus Marsh (1638–1713)*, edited by Allison P. Coudert, Sarah Hutton, Richard H. Popkin, and Gordon M. Weiner, 207–24. Dordrecht: Springer, 1999.

Rolston, Holmes. 'Does Aesthetic Appreciation of Landscapes Need to Be Science-Based?' *British Journal of Aesthetics* 33 (1995): 374–86.

Rolston, Holmes. 'Value in Nature and the Nature of Value'. In *Philosophy and the Natural Environment*, edited by Robin Attfield and Andrew Belsey, 13–30. Cambridge: Cambridge University Press, 2009.

Roochnik, David. 'What Is Theoria? Nicomachean Ethics Book 10.7–8'. *Classical Philology* 104, no. 1 (2009): 69–82.

Roochnik, David. *Retrieving Aristotle in an Age of Crisis*. Albany, NY: SUNY Press, 2013.

Rorty, Amelie Oksenberg. 'The Place of Contemplation in Aristotle's Nicomachean Ethics'. *Mind* 87, no. 347 (1978): 343–58.

Rorty, Richard. 'Solidarity or Objectivity?' In *Objectivity, Relativism and Truth: Philosophical Papers*, 21–34. Cambridge: Cambridge University Press, 1991.

Rose, David, and David Danks. 'In Defense of a Broad Conception of Experimental Philosophy'. *Metaphilosophy* 44, no. 4 (2013): 512–32.

Rose, Hilary, and Steven Rose, eds. *Alas, Poor Darwin: Arguments against Evolutionary Psychology*. London: Jonathan Cape, 2000.

Rose, P. L. 'Universal Harmony in Regiomontanus and Copernicus'. In *Avant, avec, après Copernic: La représentation de l'univers et ses conséquences épistémologiques*, edited by Suzanne Delorme, 153–8. Paris: Blanchard, 1975.

Rose, Steven. 'The Biology of the Future and the Future of Biology'. In *Explanations: Styles of Explanation in Science*, edited by John Cornwell, 125–42. Oxford: Oxford University Press, 2004.

Rose, Steven, and Hilary Rose. 'The Changing Face of Human Nature'. *Daedalus* 138, no. 3 (2009): 7–20.

Rosenberg, Alexander. *The Atheist's Guide to Reality: Enjoying Life without Illusions*. New York: W.W. Norton, 2011.

Ross, Sydney. 'Scientist: The Story of a Word'. *Annals of Science* 18, no. 2 (1962): 65–85.

Ross, W. D. *Aristotle*. London: Methuen, 1923.

Rossi, Pietro. 'Rationalisation, "désenchantement" du monde, modernité'. *Revue européenne des sciences sociales* 33, no. 101 (1995): 81–94.

Rossiter, Elliot. 'Locke, Providence, and the Limits of Natural Philosophy'. *British Journal for the History of Philosophy* 22, no. 2 (2014): 217–35.

Rothman, Aviva. *The Pursuit of Harmony: Kepler on Cosmos, Confession, and Community*. Chicago: University of Chicago Press, 2018.

Roux, Sophie. 'An Empire Divided: French Natural Philosophy (1670–1690)'. In *The Mechanization of Natural Philosophy*, edited by Daniel Garber, 55–95. Dordrecht: Springer, 2013.

Rueger, Alexander. 'Perspectival Models and Theory Unification'. *British Journal for the Philosophy of Science* 56 (2005): 579–94.

Rupke, Nicolaas A. 'Neither Creation nor Evolution: The Third Way in Mid-Nineteenth Century Thinking About the Origin of Species'. *Annals of the History and Philosophy of Biology and Philosophy* 10 (2005): 143–72.

Ruskin, John. *Complete Works*. 39 vols. London: George Allen, 1903.

Rutherford, Ian. *State Pilgrims and Sacred Observers in Ancient Greece: A Study of Theōria and Theōroi*. Cambridge: Cambridge University Press, 2013.

Ryrie, Alec. *Unbelievers: An Emotional History of Doubt*. London: Collins, 2019.

Saccenti, Riccardo. 'The *Ministerium Naturae*: Natural Law in the Exegesis and Theological Discourse at Paris between 1160 and 1215'. *Journal of the History of Ideas* 79, no. 4 (2018): 527–45.

Saliba, George. 'Arabic versus Greek Astronomy: A Debate over the Foundations of Science'. *Perspectives on Science* 8, no. 4 (2000): 328–41.

Sanderson, Marie. 'The Classification of Climates from Pythagoras to Koeppen'. *Bulletin of the American Meteorological Society* 80, no. 4 (1999): 669–73.

Sangiacomo, Andrea. 'Modelling the History of Early Modern Natural Philosophy: The Fate of the Art–Nature Distinction in the Dutch Universities'. *British Journal for the History of Philosophy* 27, no. 1 (2019): 46–74.

Sarasohn, Lisa T. *The Natural Philosophy of Margaret Cavendish: Reason and Fancy during the Scientific Revolution*. Baltimore: Johns Hopkins University Press, 2010.

Sargent, Rose-Mary. *The Diffident Naturalist: Robert Boyle and the Philosophy of Experiment*. Chicago: University of Chicago Press, 1995.

Sargent, Rose-Mary. 'Baconian Experimentalism: Comments on McMullin's History of the Philosophy of Science'. *Philosophy of Science* 68, no. 3 (2001): 311–18.

Sarisky, Darren, ed. *Theologies of Retrieval: An Exploration and Appraisal*. London: Bloomsbury, 2017.

Sauer, Tilman. 'Einstein's Unified Theory Program'. In *The Cambridge Companion to Einstein*, edited by Michel Janssen and Christoph Lehner, 281–305. Cambridge: Cambridge University Press, 2014.

Saumell, Jordi Crespo. 'Aristóteles y la medicina'. *Asclepio: Revista de Historia de la Medicina y de la Ciencia* 69, no. 1 (2017). https://doi.org/10.3989/asclepio.2017.01.

Sawday, Jonathan. *Engines of the Imagination: Renaissance Culture and the Rise of the Machine*. London: Routledge, 2007.

Scalercio, Mauro. 'Dominating Nature and Colonialism. Francis Bacon's View of Europe and the New World'. *History of European Ideas* 44, no. 8 (2018): 1078–91.

Sceski, John H. *Popper, Objectivity and the Growth of Knowledge*. London: Continuum, 2007.

Schaefer, Jame. 'Appreciating the Beauty of Earth'. *Theological Studies* 62, no. 1 (2001): 23–52.

Schaffer, Simon. 'Herschel in Bedlam: Natural History and Stellar Astronomy'. *British Journal for the History of Science* 13, no. 3 (1980): 211–39.

Schaffer, Simon. 'Scientific Discoveries and the End of Natural Philosophy'. *Social Studies of Science* 16, no. 3 (1986): 387–20.

Schaffer, Simon. 'Glass Works: Newton's Prisms and the Uses of Experiment'. In *The Uses of Experiment: Studies in the Natural Sciences*, edited by David Gooding, Trevor Pinch, and Simon Schaffer, 67–104. Cambridge: Cambridge University Press, 1989.

Schilpp, Paul A., ed. *The Philosophy of Rudolf Carnap*. La Salle, IL: Open Court, 1963.

Schipperges, Heinrich. 'Kosmologische Aspekte der Lebensordnung und Lebensführung bei Hildegard von Bingen', in *Kosmos und Mensch aus der Sicht Hildegards von Bingen*,

edited by Adelgundis Führkötter, 1–25. Mainz: Verlag der Gesellschaft für mittelrheinische Kirchengeschichte, 1987.
Schliesser, Eric. 'Newton's Challenge to Philosophy: A Programmatic Essay'. *Journal of the International Society for the History of Philosophy of Science* 1, no. 1 (2011): 101–28.
Schluchter, Wolfgang. *Die Entzauberung der Welt: Sechs Studien zu Max Weber*. Tübingen: Mohr Siebeck, 2009.
Schmaltz, Tad. *Early Modern Cartesianisms: Dutch and French Constructions*. New York: Oxford University Press, 2016.
Schmid, Jelscha. 'Schelling's Method of Darstellung: Presenting Nature through Experiment'. *Studies in History and Philosophy of Science* 69 (2018): 12–22.
Schmidt, James. 'Inventing the Enlightenment: Anti-Jacobins, British Hegelians, and the Oxford English Dictionary'. *Journal of the History of Ideas* 64, no. 3 (2003): 421–43.
Schmidt, James. 'Enlightenment as Concept and Context'. *Journal of the History of Ideas* 75, no. 4 (2014): 677–85.
Schnädelbach, Herbert. *Philosophie in Deutschland, 1831–1933*. Frankfurt am Main: Suhrkamp, 1983.
Schneider, Ulrich J. 'L'historicisation de l'enseignement de la philosophie dans les universités allemandes du XIXème siècle'. *Actes de la Recherche en Sciences Sociales* 109 (1995): 29–40.
Schuler, Robert M. 'Some Spiritual Alchemies of Seventeenth-Century England'. *Journal of the History of Ideas* 41, no. 2 (1980): 293–318.
Schumacher, Lydia. 'The Lost Legacy of Anselm's Argument: Rethinking the Purpose of Proofs for the Existence of God'. *Modern Theology* 27, no. 1 (2011): 87–101.
Schupbach, Jonah N. 'Conjunctive Explanations and Inference to the Best Explanation'. *Teorema* 38, no. 3 (2019): 143–62.
Schupbach, Jonah N., and David H. Glass. 'Hypothesis Competition beyond Mutual Exclusivity'. *Philosophy of Science* 84, no. 5 (2017): 810–24.
Schurz, Gerhard. 'Patterns of Abduction'. *Synthese* 164, no. 2 (2008): 201–34.
Scott, Callum. 'The Frontiers of Empirical Science: A Thomist-Inspired Critique of Scientism'. *HTS Teologiese Studies* 72, no. 3 (2016): 1–10.
Scruton, Roger. 'Scientism in the Arts and Humanities'. *The New Atlantis* 40 (2013): 33–46.
Scruton, Roger. *Conservatism: An Invitation to the Great Tradition*. New York: St Martin's Press, 2017.
Sdegno, Emma. 'The Alps'. In *The Cambridge Companion to John Ruskin*, edited by Francis O'Gorman, 32–48. Cambridge: Cambridge University Press, 2015.
Seaford, Richard. The Origins of Philosophy in Ancient Greece and Ancient India: A Historical Comparison. Cambridge: Cambridge University Press, 2019.
Seager, William. 'The Philosophical and Scientific Metaphysics of David Bohm'. *Entropy* 20, no. 7 (2018): 493.
Sedley, David. *The Midwife of Platonism: Text and Subtext in Plato's Theaetetus*. Oxford: Clarendon Press, 2004.
Segerstrale, Ullica. 'Wilson and the Unification of Science'. *Annals of the New York Academy of Sciences* 1093, no. 1 (2006): 46–73.
Sellars, John. *The Art of Living: The Stoics on the Nature and Function of Philosophy*. Aldershot: Ashgate, 2003.
Serjeantson, Richard. 'Francis Bacon and the "Interpretation of Nature" in the Late Renaissance'. *Isis* 105, no. 4 (2013): 681–705.
Sgarbi, Marco. *The Aristotelian Tradition and the Rise of British Empiricism: Logic and Epistemology in the British Isles (1570–1689)*. Dordrecht: Springer, 2013.

Shank, J. B. *The Newton Wars and the Beginning of the French Enlightenment*. Chicago: University of Chicago Press, 2008.

Shank, J. B. 'Between Isaac Newton and Enlightenment Newtonianism: The "God Question" in the Eighteenth Century'. In *Science without God? Rethinking the History of Scientific Naturalism*, edited by Peter Harrison and Jon Roberts, 78–96. Oxford: Oxford University Press, 2019.

Shapin, Steven. 'The Sciences of Subjectivity'. *Social Studies of Science* 42, no. 2 (2012): 170–84.

Shapiro, Lisa. 'Revisiting the Early Modern Philosophical Canon'. *Journal of the American Philosophical Association* 2, no. 3 (2016): 365–83.

Shattuck, Roger. 'Does It All Fit Together? Evolution, the Arts, and Consilience'. *Academic Questions* 11 (1998): 56–61.

Shaw, W. David. 'The Optical Metaphor: Victorian Poetics and the Theory of Knowledge'. *Victorian Studies* 23, no. 3 (1980): 293–324.

Shea, William. 'Looking at the Moon as Another Earth: Terrestrial Analogies and Seventeenth-Century Telescopes'. In *Metaphor and Analogy in the Sciences*, edited by Fernand Hallyn, 83–104. Dordrecht: Kluwer Academic, 2000.

Shiach, Morag. 'Woolf's Atom, Eliot's Catalyst and Richardson's Waves of Light: Science and Modernism in 1919'. In *Being Modern: The Cultural Impact of Science in the Early Twentieth Century*, edited by Robert Bud, Paul Greenhalgh, Frank James, and Morag Shiach, 58–76. London: UCL Press, 2018.

Shiffrin, Richard M., and Katy Börner. 'Mapping Knowledge Domains'. *Proceedings of the National Academy of Sciences* 101 (2004): 5183–5.

Sitter, John. 'Eighteenth-Century Ecological Poetry and Ecotheology'. *Religion and Literature* 40, no. 1 (2008): 11–37.

Skupin, André, and Sara Irina Fabrikant. 'Spatialization Methods: A Cartographic Research Agenda for Non-Geographic Information Visualization'. *Cartography and Geographic Information Systems* 30, no. 2 (2013): 99–119.

Slater, Michael R. 'Two Rival Interpretations of Xunzi's Views on the Basis of Morality'. *Journal of Religious Ethics* 45, no. 2 (2017): 363–79.

Slatkin, Laura M. 'Measuring Authority, Authoritative Measures: Hesiod's *Works and Days*'. In *The Moral Authority of Nature*, edited by Lorraine Daston and Fernando Vidal, 25–49. Chicago: University of Chicago Press, 2004.

Sloman, Steven A. *Causal Models: How People Think About the World and Its Alternatives*. Oxford: Oxford University Press, 2005.

Slowik, Edward. 'The Fate of Mathematical Place: Objectivity and the Theory of Lived-Space from Husserl to Casey'. In *Space, Time, and Spacetime*, edited by Vesselin Petkov, 291–311. Berlin: Springer, 2010.

Smith, Chris U. M. 'How the Modern World Began: Stephen Gaukroger's Descartes' System of Natural Philosophy'. *Journal of the History of the Neurosciences* 14, no. 1 (2005): 57–63.

Smith, Christian. *Moral, Believing Animals: Human Personhood and Culture*. Oxford: Oxford University Press, 2009.

Smith, Courtney Weiss. *Empiricist Devotions: Science, Religion, and Poetry in Early Eighteenth-Century England*. Charlottesville, VA: University of Virginia Press, 2016.

Smith, Justin. *The Philosopher: A History in Six Types*. Princeton, NJ: Princeton University Press, 2016.

Smith, Pamela H. 'Alchemy as a Language of Mediation at the Habsburg Court'. *Isis* 85, no. 1 (1994): 1–25.

Snobelen, Stephen D. 'Newton, Heretic: The Strategies of a Nicodemite'. *British Journal for the History of Science* 32 (1999): 381–419.

Snobelen, Stephen D. '"God of Gods, and Lord of Lords": The Theology of Isaac Newton's General Scholium to the *Principia*'. *Osiris* 16 (2001): 169–208.

Snobelen, Stephen D. 'To Discourse of God: Isaac Newton's Heterodox Theology and His Natural Philosophy'. In *Science and Dissent in England, 1688–1945*, edited by Paul B. Wood, 39–65. Aldershot: Ashgate, 2004.

Snobelen, Stephen D. 'The Theology of Isaac Newton's *Principia Mathematica*: A Preliminary Survey'. *Neue Zeitschrift für Systematische Theologie und Religionsphilosophie* 52, no. 4 (2010): 377–412.

Snobelen, Stephen D. 'The Myth of the Clockwork Universe: Newton, Newtonianism, and the Enlightenment'. In *The Persistence of the Sacred in Modern Thought*, edited by C. L. Firestone and N. Jacobs, 149–84. Notre Dame, IN: University of Notre Dame Press, 2012.

Snow, C. P. *The Search*. Harmondsworth: Penguin, 1965.

Snyder, Laura J. 'Discoverers' Induction'. *Philosophy of Science* 64 (1997): 580–604.

Snyder, Laura J. 'The Mill–Whewell Debate: Much Ado About Induction'. *Perspectives on Science* 5 (1997): 159–98.

Snyder, Laura J. 'Renovating the "Novum Organum": Bacon, Whewell and Induction'. *Studies in History and Philosophy of Science* 30, no. 4 (1999).

Snyder, Laura J. *Reforming Philosophy: A Victorian Debate on Science and Society*. Chicago: University of Chicago Press, 2006.

Snyder, Laura J. *The Philosophical Breakfast Club: Four Remarkable Friends Who Transformed Science and Changed the World*. New York: Broadway Books, 2011.

Soares, Luiz Carlos. 'John Banks: An Independent and Itinerant Lecturer of Natural and Experimental Philosophy at the Threshold of the English Industrial Revolution'. *Circumscribere* 19 (2017): 18–33.

Solère, Jean Luc. 'Avant-Propos'. In *La servante et la consolatrice: La philosophie dans ses rapports avec la théologie uu Moyen Age*, edited by Jean-Luc Solère and Zénon Kaluza, v–xv. Paris: Vrin, 2002.

Solomon, Julie R. *Objectivity in the Making: Francis Bacon and the Politics of Inquiry*. Baltimore, MD: Johns Hopkins University Press, 1998.

Sorell, Tom. *Scientism: Philosophy and the Infatuation with Science*. London: Routledge, 1991.

Sorell, Tom, and G. A. J. Rogers, eds. *Analytic Philosophy and History of Philosophy*. Oxford: Oxford University Press, 2005.

Sosa, Ernest. 'Experimental Philosophy and Philosophical Intuition'. *Philosophical Studies* 132, no. 1 (2007): 99–107.

Specter, Matthew. 'From Eclipse of Reason to the Age of Reasons? Historicizing Habermas and the Frankfurt School'. *Modern Intellectual History* 16, no. 1 (2017): 321–37.

Speer, Andreas. *Die entdeckte Natur: Untersuchungen zu Begründungsversuchen einer Scientia Naturalis im 12. Jahrhundert*. Leiden: Brill, 1995.

Speer, Andreas. 'Secundum Physicam. Die entdeckte Natur und die Begründung einer Scientia Naturalis im 12. Jahrhundert'. *Documenti e studi sulla tradizione filosofica medievale* 6 (1995): 1–37.

Speer, Andreas. 'Zwischen Naturbeobachtung und Metaphysik. Zur Entwicklung und Gestalt der Naturphilosophie im 12. Jahrhundert'. In *Aufbruch—Wandel—Erneuerung: Beiträge zur sogenannten Renaissance des 12. Jahrhunderts*, edited by Georg Wieland, 155–80. Stuttgart: Frommann-Holzboog, 1995.

Sperber, Peter. 'There's No Success Like Failure: On the Early Reception of Kant's Most Famous Synthesis'. *Rivista di Storia della Filosofia* 72, no. 4 (2017): 597–606.

Speziali, Pierre, ed. *Albert Einstein–Michele Besso Correspondence, 1903–55*. Paris: Hermann, 1972.

Spiller, Elizabeth A. 'Reading through Galileo's Telescope: Margaret Cavendish and the Experience of Reading'. *Renaissance Quarterly* 53, no. 1 (2000): 192–221.

Spinelli, Emidio. 'Beyond the Theoretikos Bios: Philosophy and Praxis in Sextus Empiricus'. In *Theoria, Praxis and the Contemplative Life after Plato and Aristotle*, edited by Thomas Bénatouïl and Mauro Bonazzi, 101–17. Leiden: Brill, 2012.

Spitzer, Leo. *Classical and Christian Ideas of World Harmony: Prolegomena to an Interpretation of the Word 'Stimmung'*. Baltimore: Johns Hopkins University Press, 1963.

Sprandel, Rolf. 'Vorwissenschaftliches Naturverstehen und Entstehung von Naturwissenschaften'. *Sudhoffs Archiv* 63, no. 4 (1979): 313–25.

Spranzi, Marta. 'Galileo and the Mountains of the Moon: Analogical Reasoning, Models and Metaphors in Scientific Discovery'. *Journal of Cognition and Culture* 4, no. 3–4 (2004): 451–83.

Sprat, Thomas. *History of the Royal Society of London for the Improving of Natural Knowledge*. London: J. Martyn, 1667.

Stabile, Giorgio. 'Linguaggio della natura e linguaggio della Scrittura in Galilei. Dalla istoria sulle Macchie Solari alle lettere Copernicane'. *Nuncius* 9, no. 1 (1994): 37–64.

Staley, Richard. *Einstein's Generation: The Origins of the Relativity Revolution*. Chicago: University of Chicago Press, 2008.

Stanford, P. Kyle. 'Reading Nature: Realist, Instrumentalist, and Quietist Interpretations of Scientific Theories'. In *Physical Theory: Method and Interpretation*, edited by Lawrence Sklar, 94–126. Oxford: Oxford University Press, 2014.

Stanley, Matthew. *Huxley's Church and Maxwell's Demon: From Theistic Science to Naturalistic Science*. Chicago: University of Chicago Press, 2015.

Steane, Andrew M. *Science and Humanity: A Humane Philosophy of Science and Religion*. Oxford: Oxford University Press, 2018.

Steel, Carlos. 'Maximus Confessor on Theory and Praxis: A Commentary on Ambigua Ad Johannem VI (10) 1–19'. In *Theoria, Praxis and the Contemplative Life after Plato and Aristotle*, edited by Thomas Bénatouïl and Mauro Bonazzi, 229–57. Leiden: Brill, 2012.

Steiner, Mark. 'Penrose and Platonism'. In *The Growth of Mathematical Knowledge*, edited by Emily Grosholz and Herbert Breger, 133–41. Dordrecht: Springer, 2000.

Steinicke, Wolfgang. 'William Herschel, Flamsteed Numbers and Harris's Star Maps'. *Journal for the History of Astronomy* 45, no. 3 (2014): 287–303.

Stekeler-Weithofer, Pirmin. *Formen der Anschauung: Eine Philosophie der Mathematik*. Berlin: de Gruyter, 2008.

Stenhouse, John. 'Darwin's Captain: F. W. Hutton and the Nineteenth-Century Darwinian Debates'. *Journal of the History of Biology* 23 (1990): 411–42.

Stenmark, Mikael. *Scientism: Science, Ethics and Religion*. Aldershot: Ashgate, 2001.

Stephenson, Bruce. *The Music of the Heavens: Kepler's Harmonic Astronomy*. Princeton, NJ: Princeton University Press, 2014.

Stewart, Ian G. 'Res, Veluti Per Machinas, Conficiatur: Natural History and the "Mechanical" Reform of Natural Philosophy'. *Early Science and Medicine* 17, no. 1/2 (2012): 87–111.

Stewart, Larry R. *The Rise of Public Science: Rhetoric, Technology, and Natural Philosophy in Newtonian Britain, 1660–1750*. Cambridge: Cambridge University Press, 1992.

Stock, Ady van den. *The Horizon of Modernity: Subjectivity and Social Structure in New Confucian Philosophy*. Leiden: Brill, 2016.

Stoljar, Daniel. *Physicalism*. London: Routledge, 2010.

Stoljar, Daniel. *Philosophical Progress: In Defence of a Reasonable Optimism*. Oxford: Oxford University Press, 2019.

Stout, Jeffrey. *Ethics after Babel: The Languages of Morals and Their Discontents*. 2nd edition. Princeton, NJ: Princeton University Press, 2001.

Strano, Giorgio. 'Galileo's Telescope: History, Scientific Analysis, and Replicated Observations'. *Experimental Astronomy* 25, no. 1 (2009): 17–31.

Strickland, Lloyd. 'Leibniz's Harmony between the Kingdoms of Nature and Grace'. *Archiv für Geschichte der Philosophie* 98, no. 3 (2016): 302–29.

Striner, Richard. 'Political Newtonianism: The Cosmic Model of Politics in Europe and America'. *William and Mary Quarterly* 52, no. 4 (1995): 583–608.

Struck, Peter T. 'The Invention of Mythic Truth in Antiquity'. In *Antike Mythen: Medien, Transformationen und Konstruktionen*, edited by Ueli Dill and Christine Walde, 25–37. Berlin: de Gruyter, 2009.

Stump, Eleonore. *Wandering in Darkness: Narrative and the Problem of Suffering*. Oxford: Clarendon Press, 2010.

Stump, Eleonore, and Norman Kretzmann. 'Being and Goodness'. In *Being and Goodness: The Concept of the Good in Metaphysics and Philosophical Theology*, edited by Scott MacDonald, 281–312. Ithaca, NY: Cornell University Press, 1991.

Sturm, Thomas. 'Intuition in Kahneman and Tversky's Psychology of Rationality'. In *Rational Intuition: Philosophical Roots, Scientific Investigations*, edited by Lisa M. Osbeck and Barbara S. Held, 257–86. Cambridge: Cambridge University Press, 2014.

Subbiondo, Joseph L. 'Preliminary Reflections on a Philosophical Language: A Study of John Wilkins' *Mercury: Or the Secret and Swift Messenger*'. *Language & History* 63, no. 2 (2020): 105–19.

Swerdlow, Noel M. 'Regiomontanus on the Critical Problems of Astronomy'. In *Nature, Experiment, and the Sciences*, edited by Trevor H. Levere and William R. Shea, 165–95. Dordrecht: Kluwer, 1990.

Swerdlow, Noel M. 'The Empirical Foundations of Ptolemy's Planetary Theory'. *Journal for the History of Astronomy* 35, no. 3 (2004): 249–71.

Swerdlow, Noel M., and Otto Neugebauer. *Mathematical Astronomy in Copernicus's De Revolutionibus*. 2 vols. New York: Springer, 1984.

Sytsma, David S. *Richard Baxter and the Mechanical Philosophers*. Oxford: Oxford University Press, 2017.

Sytsma, Justin. 'Two Origin Stories for Experimental Philosophy'. *Teorema: Revista Internacional de Filosofía* 36, no. 3 (2017): 23–43.

Tait, Adrian. 'The "True Philosophy" of Robert Hunt: Forms of Understanding in *Panthea* and *The Poetry of Science*'. *European Journal of English Studies* 22, no. 3 (2018): 273–86.

Takahashi, Adam. 'Nature, Formative Power and Intellect in the Natural Philosophy of Albert the Great'. *Early Science and Medicine* 13, no. 5 (2008): 451–81.

Tallis, Raymond. *In Defence of Wonder and Other Philosophical Reflections*. Durham: Acumen, 2012.

Taminiaux, Jacques. 'Bios Politikos and Bios Theoretikos in the Phenomenology of Hannah Arendt'. *International Journal of Philosophical Studies* 4, no. 2 (1996): 215–32.

Tanzella-Nitti, Giuseppe. 'The Book of Nature and the God of Scientists According to the Encyclical "Fides Et Ratio"'. In *The Human Search for Truth: Philosophy, Science, Faith.*

The Outlook for the Third Millennium, 82–90. Philadelphia: St Joseph's University Press, 2001.

Tanzella-Nitti, Giuseppe. 'The Two Books Prior to the Scientific Revolution'. *Annales Theologici* 18 (2004): 51–83.

Tanzella-Nitti, Giuseppe. 'Theologia Physica: Razionalità scientifica e domanda su Dio'. *Hermeneutica: Annuario di filosofi a e teologia* New Series 2012 (2012): 37–54.

Tasić, Vladimir. *Mathematics and the Roots of Postmodern Thought*. Oxford: Oxford University Press, 2001.

Tate, Gregory. 'Poetry and Science'. In *The Routledge Research Companion to Nineteenth-Century British Literature and Science*, edited by John Holmes and Sharon Ruston, 101–14. London: Routledge, 2017.

Taves, Ann. *Religious Experience Reconsidered*. Princeton, NJ: Princeton University Press, 2009.

Taylor, Charles. *Defining Science: A Rhetoric of Demarcation*. Madison, WI: University of Wisconsin Press, 1996.

Taylor, Charles. *Modern Social Imaginaries*. Durham, NC: Duke University Press, 2004.

Taylor, Charles. *A Secular Age*. Cambridge, MA: Belknap Press, 2007.

Thagard, Paul. 'Creative Intuition: How Eureka Results from Three Neural Mechanisms'. In *Rational Intuition: Philosophical Roots, Scientific Investigations*, edited by Lisa M. Osbeck and Barbara S. Held, 287–306. Cambridge: Cambridge University Press, 2014.

Theunissen, Bert. 'Darwin and His Pigeons: The Analogy between Artificial and Natural Selection Revisited'. *Journal of the History of Biology* 45, no. 2 (2012): 179–212.

Thijssen, J. M. M. H. 'Late Medieval Natural Philosophy: Some Recent Trends in Scholarship'. *Recherches de théologie et philosophie médiévales* 67, no. 1 (2000): 158–90.

Thijssen, J. M. M. H., and H. A. G. Braakhuis. *The Commentary Tradition on Aristotle's De Generatione et Corruptione: Ancient, Medieval and Early Modern*. Turnhout: Brepols, 1999.

Thomasson, Amie L. 'Experimental Philosophy and the Methods of Ontology'. *The Monist* 95, no. 2 (2012): 175–99.

Thoren, Victor E., with John R. Christianson. *The Lord of Uraniborg: A Biography of Tycho Brahe*. Cambridge: Cambridge University Press, 1990.

Tilley, Christopher, and Kate Cameron-Daum. *Anthropology of Landscape: The Extraordinary in the Ordinary*. London: UCL Press, 2017.

Todd, Cain S. 'Unmasking the Truth beneath the Beauty: Why the Supposed Aesthetic Judgements Made in Science May Not Be Aesthetic at All'. *International Studies in the Philosophy of Science* 22 (2008): 61–79.

Topham, Jonathan R. 'Biology in the Service of Natural Theology: Darwin, Paley, and the Bridgewater Treatises'. In *Biology and Ideology: From Descartes to Dawkins*, edited by Denis R. Alexander and Ronald Numbers, 88–113. Chicago: University of Chicago Press, 2010.

Traherne, Thomas, *Poetry and Prose*. Edited by Denise Inge. London: SPCK, 2002.

Trop, Gabriel. 'The Aesthetics of Schelling's *Naturphilosophie*'. *Symposium* 19, no. 1 (2015): 140–52.

Trubowitz, Rachel. 'Reading Milton and Newton in the Radical Reformation: Poetry, Mathematics, and Religion'. *ELH* 84, no. 1 (2017): 33–62.

Tuan, Yi-Fu. *Space and Place: The Perspective of Experience*. London: Edward Arnold, 1977.

Tuominen, Miira. 'Philosophy of the Ancient Commentators on Aristotle'. *Philosophy Compass* 7, no. 12 (2012): 852–95.

Turner, Bryan S. 'Alasdair Macintyre on Morality, Community and Natural Law'. *Journal of Classical Sociology* 13, no. 2 (2013): 239–53.

Upton, Thomas V. 'Infinity and Perfect Induction in Aristotle'. *Proceedings of the Catholic Philosophic Association* 55 (1981): 149–58.

Urroz, Eloy. 'Karl Popper y Mario Vargas Llosa: ¿Igualdad o Libertad?' *Revista de la Universidad Nacional Autónoma de México* 86 (2011): 31–40.

Vallat, Philippe. 'Al-Fārābī's Arguments for the Eternity of the World and the Contingency of Natural Phenomena'. In *Interpreting the Bible and Aristotle in Late Antiquity: The Alexandrian Commentary Tradition between Rome and Baghdad*, edited by Josef Lössl and John W. Watt, 259–86. Aldershot: Ashgate, 2011.

van Buren, John. 'Environmental Hermeneutics: Deep in the Forest'. In *Interpreting Nature: The Emerging Field of Environmental Hermeneutics*, edited by Forrest Clingerman, Brian Treanor, Martin Drenthen, and David Utsler, 17–35. Fordham, NY: Fordham University Press, 2013.

van den Hoek, Annewies. 'Mistress and Servant: An Allegorical Theme in Philo, Clement and Origen'. In *Origeniana Quarta: Die Referate des 4. Internationalen Origeneskongresses*, edited by Lothar Lies, 344–48. Innsbruck: Tyrolia, 1987.

Van Dyck, Maarten, and Koen Vermeir. 'Varieties of Wonder: John Wilkins' Mathematical Magic and the Perpetuity of Invention'. *Historia Mathematica* 41, no. 4 (2014): 463–89.

van Fraassen, Bas C. 'A Re-Examination of Aristotle's Philosophy of Science'. *Dialogue* 19, no. 1 (1980): 20–45.

Van Helden, Albert, Sven Dupré, Rob van Gent, and Huib Zuidervaart, eds. *The Origins of the Telescope*. Amsterdam: KNAW Press, 2010.

Vanderhooft, David. 'Dwelling beneath the Sacred Place: A Proposal for Reading 2 Samuel 7:10'. *Journal of Biblical Literature* 118, no. 4 (1999): 625–33.

Vanzo, Alberto. 'Christian Wolff and Experimental Philosophy'. In *Oxford Studies in Early Modern Philosophy*, Volume VII, edited by Daniel Garber and Donald Rutherford, 225–55. Oxford: Oxford University Press, 2015.

Vargas Llosa, Mario. 'Updating Karl Popper: Suspect Truth, the Closed Society and the Third World, Historicism and Fiction'. *PMLA* 105 (1990): 1018–25.

Vassányi, Miklós *Anima Mundi: The Rise of the World Soul Theory in Modern German Philosophy*. Dordrecht: Springer, 2011.

Verger, Jacques. 'La Faculté des Arts: Le cadre institutionnel'. In *L'enseignement des disciplines à la Faculté des Arts (Paris et Oxford, XIIIe–XIVe siècles)*, edited by Olga Weijers and Louis Holtz, 17–42. Turnhout: Brepols, 1997.

Viano, Cristina. 'Aristote et l'alchimie grecque: La transmutation et le modèle Aristotélicien entre théorie et pratique'. *Revue d'histoire des sciences* 49, no. 2–3 (1996): 189–213.

Viano, Cristina, ed. *L'alchimie et ses racines philosophiques: La tradition grecque et la tradition arabe*. Paris: Vrin, 2005.

Viano, Cristina. 'Mixis and Diagnôsis: Aristotle and the "Chemistry" of the Sublunary World'. *Ambix* 62, no. 3 (2015): 203–14.

Vidal, Fernando, and Bernard Kleeberg. 'Knowledge, Belief, and the Impulse to Natural Theology'. *Science in Context* 20 (2007): 381–400.

Viladesau, Richard. 'Natural Theology and Aesthetics: An Approach to the Existence of God from the Beautiful?' *Philosophy & Theology* 3 (1988): 145–60.

Vincent, Jean-Marie. 'Le désenchantement du monde: Max Weber et Walter Benjamin'. *Revue Européenne des Sciences Sociales* 33, no. 101 (1995): 95–106.

Viola, Enrico. 'The Specificity of Logical Empiricism in the Twentieth-Century History of Scientific Philosophy'. *Journal of the International Society for the History of Philosophy of Science* 3, no. 2 (2013): 191–209.

Voelkel, James R. 'Commentary on Ernan McMullin, "The Impact of Newton's *Principia* on the Philosophy of Science"'. *Philosophy of Science* 68, no. 3 (2001): 319–26.

Walker, Matthew D. *Aristotle on the Uses of Contemplation*. Cambridge: Cambridge University Press, 2018.

Wallace, Dewey D. *Shapers of English Calvinism, 1660–1714: Variety, Persistence, and Transformation*. Oxford University Press: Oxford, 2011.

Wallis, Faith. '*Si Naturam Quaeras*: Reframing Bede's "Science"'. In *Innovation and Tradition in the Writings of the Venerable Bede*, edited by Scott DeGregorio, 65–99. Morgantown, WV: West Virginia University Press, 2006.

Walmsley, Jonathan. 'Locke's Natural Philosophy in Draft A of the Essay'. *Journal of the History of Ideas* 65, no. 1 (2004): 15–37.

Walsham, Alexandra. 'The Reformation and "the Disenchantment of the World" Reassessed'. *Historical Journal* 51, no. 2 (2008): 497–528.

Ward, Eleanor, Giorgio Ganis, and Patric Bach. 'Spontaneous Vicarious Perception of the Content of Another's Visual Perspective'. *Current Biology* 29 (2019): 874–80.

Warner, Brian. 'Charles Darwin and John Herschel'. *South African Journal of Science* 105, no. 11–12 (2009): 432–9.

Waters, C. Kenneth. 'Beyond Theoretical Reduction and Layer-Cake Antireduction: How DNA Retooled Genetics and Transformed Biological Practice'. In *The Oxford Handbook of Philosophy of Biology*, edited by Michael Ruse, 238–62. Oxford: Oxford University Press, 2008.

Watson, Micah. 'Natural Law in the *Abolition of Man*'. In *Contemporary Perspectives on C. S. Lewis: History, Philosophy, Education, and Science*, edited by Tim Mosteller and Gayne John Anacker, 25–46. London: Bloomsbury Academic, 2017.

Watzl, Sebastian. 'The Nature of Attention'. *Philosophy Compass* 6, no. 11 (2011): 842–53.

Wayne, Andrew. 'Theoretical Unity: The Case of the Standard Model'. *Perspectives on Science* 4, no. 4 (1996): 391–407.

Webster, Charles. *The Great Instauration: Science, Medicine and Reform 1626–1660*. London: Duckworth, 1975.

Wegmann, Milene. *Naturwahrnehmung im Mittelalter im Spiegel der lateinischen Historiographie des 12. und 13. Jahrhunderts*. Bern: Peter Lang, 2005.

Weinberg, Steven. *Dreams of a Final Theory: The Search for the Fundamental Laws of Nature*. New York: Pantheon, 1992.

Weinberg, Steven. *Facing Up: Science and Its Cultural Adversaries*. Cambridge, MA: Harvard University Press, 2001.

Weinberg, Steven. *To Explain the World: The Discovery of Modern Science*. London: Allen Lane, 2015.

Weisz, Eduardo. 'Los procesos de secularización y pos-secularización a la luz de la sociología Weberiana de la racionalización'. *Política & Sociedade* 16, no. 36 (2017): 97–127.

Weisz, Eduardo. 'Science, Rationalization, and the Persistence of Enchantment'. *Max Weber Studies* 20, no. 1 (2020): 8–24.

Welch, Edward. 'Marc Augé, Jean Rolin and the Mapping of (Non-)Place in Modern France'. *Irish Journal of French Studies* 9 (2009): 49–68.

Werkmeister, W. H. *Nicolai Hartmann's New Ontology*. Tallahassee, FL: Florida State University Press, 1990.

West, John B. 'Robert Boyle's Landmark Book of 1660 with the First Experiments on Rarified Air'. *Journal of Applied Physiology* 98, no. 1 (2005): 31–9.
Westfall, Richard S. 'The Scientific Revolution of the Seventeenth Century: A New World View'. In *The Concept of Nature*, edited by John Torrance, 63–93. Oxford: Oxford University Press, 1992.
Westman, Robert S. 'The Melanchthon Circle, Rheticus and the Wittenberg Interpretation of the Copernican Theory'. *Isis* 66 (1975): 165–93.
Westman, Robert S. The *Copernican Question: Prognostication, Skepticism, and Celestial Order*. Berkeley, CA: University of California Press, 2011.
Whewell, William. *Philosophy of the Inductive Sciences*. 2 vols. London: John W. Parker, 1847.
Whewell, William. 'Criticism of Aristotle's Account of Induction'. *Transactions of the Cambridge Philosophical Society* 9, no. 1 (1850): 63–72.
Whistler, Daniel. *Schelling's Theory of Symbolic Language: Forming the System of Identity*. Oxford: Oxford University Press, 2013.
White, Lynn. 'The Historical Roots of Our Ecological Crisis'. *Science* 155 (1967): 1203–7.
White, Paul. *Thomas Huxley: Making the 'Man of Science'*. Cambridge: Cambridge University Press, 2003.
White, R. S. *Natural Law in English Renaissance Literature*. Cambridge: Cambridge University Press, 2006.
Whitehouse, Harvey. 'Cognitive Evolution and Religion', in *The Evolution of Religion: Studies, Theories & Critiques*, edited by Joseph Abdul Bulbulia, 19–29. Santa Margarita, CA: Collins Foundation Press, 2008.
Whitman, Walt. *Leaves of Grass*. Boston: James R. Osgood and Co., 1881–2.
Wigelsworth, Jeffrey R. 'Competing to Popularize Newtonian Philosophy: John Theophilus Desaguliers and the Preservation of Reputation'. *Isis* 94, no. 3 (2003): 435–55.
Wigelsworth, Jeffrey R. *Deism in Enlightenment England: Theology, Politics, and Newtonian Public Science*. Manchester: Manchester University Press, 2009.
Wigner, Eugene Paul. *Symmetries and Reflections: Scientific Essays*. Bloomington, IN: Indiana University Press, 1967.
Wiles, Maurice. *Archetypal Heresy: Arianism through the Centuries*. Oxford: Clarendon Press, 1996.
Wilkenfeld, Daniel. 'Functional Explaining: A New Approach to the Philosophy of Explanation'. *Synthese* 191, no. 14 (2014): 3367–91.
Wilkins, John. *Mathematicall Magick, or, The wonders that may be performed by Mechanicall Geometry in Two Books*. London: Gellibrand, 1648.
Willan, Claude. 'The Proper Study of Mankind in Pope and Thomson'. *ELH* 84, no. 1 (2017): 63–90.
Willey, Basil. *The Eighteenth Century Background: Studies on the Idea of Nature in the Thought of the Period*. London: Chatto & Windus, 1940.
Williams, Dee Mack. 'Representations of Nature on the Mongolian Steppe: An Investigation of Scientific Knowledge Construction'. *American Anthropologist* 102, no. 3 (2000): 503–19.
Williams, Michael. 'The Ambiguity of Nature'. *Antipode* 30, no. 1 (1998): 26–35.
Williams, Richard N., and Daniel N. Robinson, eds. *Scientism: The New Orthodoxy*. London: Bloomsbury, 2015.
Williamson, Timothy. *Knowledge and Its Limits*. Oxford: Oxford University Press, 2009.
Williamson, Timothy. *Doing Philosophy: From Common Curiosity to Logical Reasoning*. Oxford: Oxford University Press, 2018.

Wilson, Catherine. *The Invisible World: Early Modern Philosophy and the Invention of the Microscope*. Princeton, NJ: Princeton University Press, 1995.
Wilson, Curtis. 'Kepler's Derivation of the Elliptical Path'. *Isis* 59, no. 1 (1968): 4–25.
Wilson, Curtis. 'Newton and Celestial Mechanics'. In *The Cambridge Companion to Newton*, edited by Robert Iliffe, 261–88. Cambridge: Cambridge University Press, 2016.
Wilson, David B. *Seeking Nature's Logic: Natural Philosophy in the Scottish Enlightenment*. University Park, PA: Pennsylvania State University Press, 2009.
Wilson, Edward O. *Sociobiology: The New Synthesis*. Cambridge, MA: Harvard University Press, 1975.
Wilson, Edward O. 'Resuming the Enlightenment Quest'. *Wilson Quarterly* 22, no. 1 (1998): 16–27.
Wilson, Edward O. *Consilience: The Unity of Knowledge*. New York: Vintage, 1999.
Wimsatt, William C. 'The Ontology of Complex Systems: Levels, Perspectives, and Causal Thickets'. *Canadian Journal of Philosophy* 20 (1994): 207–74.
Winslow, Russell. *Aristotle and Rational Discovery*. London: Continuum, 2007.
Wintroub, Michael. 'The Looking Glass of Facts: Collecting, Rhetoric and Citing the Self in the Experimental Natural Philosophy of Robert Boyle'. *History of Science* 35, no. 2 (1997): 189–217.
Wirzba, Norman. 'Christian *Theoria Physike*: On Learning to See Creation'. *Modern Theology* 32, no. 2 (2016): 211–30.
Withers, Charles W. J. *Placing the Enlightenment: Thinking Geographically about the Age of Reason*. Chicago: University of Chicago Press, 2007.
Witt, Ronald G. *In the Footsteps of the Ancients: The Origins of Humanism from Lovato to Bruni*. Leiden: Brill, 2000.
Wittgenstein, Ludwig. *Notebooks, 1914–1916*. New York: Harper, 1961.
Woelfel, James. 'Challengers of Scientism Past and Present: William James and Marilynne Robinson'. *American Journal of Theology & Philosophy* 34, no. 2 (2013): 175–87.
Wójcicki, Ryszard. 'Physics, Theoretical Knowledge and Weinberg's Grand Reductionism'. *Foundations of Science* 3, no. 1 (1998): 61–77.
Wojcik, Jan W. *Robert Boyle and the Limits of Reason*. Cambridge: Cambridge University Press, 1997.
Wojcik, Jan W. 'Pursuing Knowledge: Robert Boyle and Isaac Newton'. In *Rethinking the Scientific Revolution*, edited by Margaret J. Osler, 183–200. Cambridge: Cambridge University Press, 2000.
Wolff, Christoph. 'Bach's Music and Newtonian Science: A Composer in Search of the Foundations of His Art'. *Understanding Bach* 2 (2007): 95–106.
Woodward, James. 'Causation in Biology: Stability, Specificity, and the Choice of Levels of Explanation'. *Biology & Philosophy* 25, no. 3 (2010): 287–318.
Woodward, James. 'Scientific Explanation'. In *Physical Theory: Method and Interpretation*, edited by Lawrence Sklar, 9–39. Oxford: Oxford University Press, 2014.
Woody, Andrea. 'Re-Orienting Discussions of Scientific Explanation: A Functional Perspective'. *Studies in History and Philosophy of Science* 51, no. 1 (2015): 79–87.
Woolfolk, Robert L. 'Experimental Philosophy: A Methodological Critique'. *Metaphilosophy* 44, no. 1/2 (2013): 79–87.
Wordsworth, William. *Last Poems, 1821–1850*. Edited by Jared R. Curtis, Apryl Lea Denny-Ferris, and Jillian Heydt-Stevenson. Ithaca, NY: Cornell University Press, 1999.
Woudenberg, René van, and Joelle Rothuizen-van der Steen. 'Science and the Ethics of Belief'. *Journal for General Philosophy of Science* 47, no. 2 (2016): 349–62.

Wragge-Morley, Alexander. *Aesthetic Science: Representing Nature in the Royal Society of London, 1650–1720*. Chicago: University of Chicago Press, 2020.

Yarchin, William. 'Biblical Interpretation in the Light of the Interpretation of Nature, 1650–1900'. In *Nature and Scripture in the Abrahamic Religions: 1700–Present*, edited by Scott H. Mandelbrote and J. M. van der Meer, 39–82. Leiden: Brill, 2008.

Yeo, Richard R. *Defining Science: William Whewell, Natural Knowledge, and Public Debate in Early Victorian Britain*. Cambridge: Cambridge University Press, 1993.

Yeo, Richard R. *Notebooks, English Virtuosi, and Early Modern Science*. Chicago: University of Chicago Press, 2014.

York, Richard, and Brett Clark. 'The Science and Humanism of Stephen Jay Gould'. *Critical Sociology* 31, no. 1–2 (2005): 281–95.

Young, Spencer E. *Scholarly Community at the Early University of Paris: Theologians, Education and Society, 1215–1248*. Cambridge: Cambridge University Press, 2014.

Zachhuber, Johannes. 'World Soul and Celestial Heat: Platonic and Aristotelian Ideas in the History of Natural Philosophy'. *Archiwum Historii Filozofii i Myśli Społecznej* 57 (2012): 13–31.

Zagorin, Perez. *Francis Bacon*. Princeton, NJ: Princeton University Press, 1998.

Zagorin, Perez. 'Francis Bacon's Objectivity and the Idols of the Mind'. *British Journal for the History of Science* 34, no. 4 (2001): 379–93.

Zakai, Avihu. *Jonathan Edwards' Philosophy of Nature: The Re-Enchantment of the World in the Age of Scientific Reasoning*. London: T&T Clark, 2010.

Zambelli, Paola. *White Magic, Black Magic in the European Renaissance: From Ficino, Pico, Della Porta to Trithemius, Agrippa, Bruno*. Boston: Brill, 2007.

Zammito, John H. *The Gestation of German Biology: Philosophy and Physiology from Stahl to Schelling*. Chicago: University of Chicago Press, 2018.

Zeitz, Lisa M. 'Natural Theology, Rhetoric, and Revolution: John Ray's Wisdom of God, 1691–1704'. *Eighteenth Century Life* 18 (1994): 120–33.

Ziman, John M. *Reliable Knowledge: An Exploration of the Grounds for Belief in Science*. Cambridge: Cambridge University Press, 1978.

Ziman, John M. 'Emerging out of Nature into History: The Plurality of the Sciences'. *Philosophical Transactions: Mathematical, Physical and Engineering Sciences* 361, no. 1809 (2003): 1617–33.

Ziolkowski, Theodore. *The Sin of Knowledge: Ancient Themes and Modern Variations*. Princeton, NJ: Princeton University Press, 2000.

Index

For the benefit of digital users, indexed terms that span two pages (e.g., 52–53) may, on occasion, appear on only one of those pages.